Digital Twin Applications and Cognitive Enterprise Transformation Across Industries

Zornitsa Yordanova
https://orcid.org/0000-0002-6056-8445
University of National and World Economy, Bulgaria

Hamed Nozari
https://orcid.org/0000-0002-6500-6708
University of National and World Economy, Bulgaria

IGI Global Scientific Publishing
Publishing Tomorrow's Research Today

Vice President of Editorial	Melissa Wagner
Director of Acquisitions	Mikaela Felty
Director of Book Development	Jocelynn Hessler
Production Manager	Mike Brehm
Cover Design	Jose Rosado

Published in the United States of America by
IGI Global Scientific Publishing
701 East Chocolate Avenue
Hershey, PA, 17033, USA
Tel: 717-533-8845 | Fax: 717-533-7115
Website: https://www.igi-global.com E-mail: cust@igi-global.com

Library of Congress Cataloging-in-Publication Data

(CIP Data Pending)
ISBN13: 9798260016695
Isbn13Softcover: 9798260016701
EISBN13: 9798260016718
British Cataloguing in Publication Data
A Cataloguing in Publication record for this book is available from the British Library.

Table of Contents

Detailed Table of Contents

This research presents a conceptual and architectural framework for the evolution of low-energy smart cities based on goal-oriented agent-based intelligence. In this approach, energy goals are dynamically adjusted according to environmental conditions, and decision-making is carried out through a multi-agent negotiation mechanism and digital twin support. By combining scenario analysis, proactive forecasting, and alignment of conflicting goals, the proposed system is able to reduce energy consumption and enhance urban resilience. The results show that the transition from static optimization to intention-based guidance can provide a new foundation for sustainable energy management in future cities.

The growth of digital technologies and the expansion of the Internet of Things have made smart homes an important element in the transformation of residential environments. In this study, the conceptual foundations of resilient smart homes based on digital twins are examined. First, the concepts of smart homes and resilience in residential environments are explained, and then the digital twin paradigm in built environments is described. Next, the main components of these systems, the role of data intelligence and dynamic modeling, and mechanisms for increasing resilience in smart homes are analyzed. Finally, an integrated conceptual framework is presented that shows the interaction between physical, data, and analytical layers for the development of smart and sustainable residential systems.

This research presents a conceptual and analytical framework for the development of resilient smart buildings based on adaptive digital twins. The main focus is on

Human Digital Twins (HDTs) and Extended Reality (XR) are key emerging technologies for human-machine interaction within digital transformation of recent years. This chapter evaluates how the integration of these two technologies fosters leadership development and human-centric transformation in the context of Industry 5.0. Within this framework, the conceptual architectures of HDTs and XR are identified, and then the impact of these technologies on redefining leadership capabilities is analyzed. By synthesizing emerging discourse and technological frameworks, the findings emphasize that modern leadership values have transitioned from pure machine optimization to complex socio-technical systems. Finally, the role of the leader as a facilitator of collaborative intelligence is also redefined, establishing the foundational "flight controls" necessary for navigating contemporary industrial volatility.

Increasing climate instability and limited natural resources have created serious challenges for the sustainability of agricultural production. In such circumstances, the use of advanced digital technologies can play an important role in improving the management of agricultural systems. This research examines the role of hybrid frameworks based on digital twins and artificial intelligence in the development of climate-resilient agriculture. In this framework, smart data ecosystems, digital models, and predictive analytics tools are considered to support adaptive decision-making in farm management. The results show that the integration of digital twins with data analytics can significantly improve resource efficiency, decision-making accuracy, and sustainability of agricultural systems.

Preface

In recent decades, digital transformation has evolved from a mere technological trend to a fundamental paradigm in redefining structures, processes, and decision-making models in organizations. Among them, the concept of Digital Twin, as one of the most pioneering innovations, has played a key role in the convergence of the physical and digital worlds. By enabling simulation, real-time monitoring, and prediction of system behavior, this technology has paved the way for the formation of cognitive and intelligent organizations.

This book, titled Digital Twin Applications and Cognitive Enterprise Transformation Across Industries, is a systematic attempt to examine the applications, conceptual frameworks, and transformative implications of digital twins across a wide range of industries. This work, with an interdisciplinary approach, examines the deep connection between digital twins, artificial intelligence, agentic systems, and cognitive organizations.

The chapters cover a diverse range of innovative topics, from human-digital twin coexistence in cognitive organizations to the role of digital twins in financial risk management, smart agriculture, sustainable cities, and resilient buildings. Topics such as Industry 5.0, next-generation education, and developing entrepreneurial skills in the context of advanced technologies are also given special attention.

One of the outstanding features of this book is the focus on innovative approaches such as integrating digital twins with cognitive intelligence and advanced decision support systems. This combination not only enables more accurate data analysis, but also helps organizations make faster, smarter, and more sustainable decisions in complex and dynamic environments.

This book is designed for researchers, industrial managers, policymakers, and those interested in emerging technologies. It is hoped that this collection will serve as an inspiring reference, paving the way for future research and helping to develop innovative solutions to realize cognitive and sustainable organizations.

Finally, sincere appreciation is extended to all the authors and researchers who have enriched this work with their scientific contributions. Without their valuable efforts, this collection would not have been possible.

Zornitsa Yordanova
University of National and World Economy, Bulgaria

Hamed Nozari
University of National and World Economy, Bulgaria

Chapter 1

Human and Digital Twin Symbiosis in the Evolution of Cognitive Enterprises

Hamed Nozari
http://orcid.org/0000-0002-6500-6708
University of National and World Economy, Bulgaria

Jaanavi Bakshi
http://orcid.org/0009-0005-5096-8727
BITS Pilani, Dubai, UAE

ABSTRACT

This chapter explains the concept of human–digital twin symbiosis in the context of the evolution of cognitive organizations. As organizations move beyond mere digitalization, the need for a framework that structures the dynamic interaction between human cognition and computational intelligence has become increasingly apparent. In this paper, the layered architecture of the symbiotic organization, coevolutionary mechanisms, multi-agent dynamics, and decision intelligence logic are introduced. A staged maturity model for the transition to cognitive symbiosis is also presented. The results show that sustainable competitive advantage in the future lies not in complete automation, but in designing the evolutionary interaction between humans and digital twins.

DOI: 10.4018/979-8-2600-1669-5.ch001

1. INTRODUCTION

In recent decades, organizations have moved beyond the digitalization of processes and entered the era of operational intelligence, an era in which data, analytics, and simulation play a central role in decision-making. However, experience has shown that simply deploying digital technologies and even using advanced machine learning algorithms does not necessarily lead to the formation of a "cognitive organization." A cognitive organization is more than a digital organization; such an organization is able to dynamically learn, interpret, create scenarios, and make adaptive decisions in the face of uncertainty. In this direction, the concept of the "digital twin" has been proposed as one of the pillars of digital transformation; a concept that enables virtual, simultaneous, and dynamic representations of physical systems and organizational processes (Kinsner & Saracco, 2019).

Initially developed with the aim of monitoring and maintaining physical assets, the digital twin has gradually become a tool for forecasting, optimization, and even strategic decision-making. In new generations, digital twins not only reflect the current state of the system, but also simulate possible futures using historical data and analytical algorithms (Zheng et al., 2022). However, a fundamental challenge remains: can organizational decision-making be fully entrusted to intelligent systems? Practical experiences in the industrial, supply chain, and energy domains have shown that relying solely on computational models, without taking into account human judgment, can lead to fragile and context-insensitive decisions (Su et al., 2024).

At this point, the concept of "human-digital twin coexistence" becomes important. In this context, coexistence does not mean one-way interaction or superficial human supervision of the system, but rather a two-way, dynamic, and evolving relationship in which human cognition and computational intelligence act complementary. Humans have the ability to understand tacit knowledge, ethical considerations, contextual interpretations, and predict nonlinear outcomes; in contrast, a digital twin is able to process vast amounts of data, simulate complex scenarios in a short time, and discover patterns that are not easily observable to the human mind. The combination of these two capacities can underlie the formation of a "cognitive symbiotic organization."

A cognitive symbiotic organization is an organization in which the boundary between human decision-making and digital analysis is intelligently designed. In such an organization, the digital twin does not replace humans, but rather enhances their cognition. In other words, the twin plays the role of a cognitive infrastructure that expands the horizons of decision-makers through real-time feedback, scenario analysis, and optimization suggestions. In turn, humans guide the twin's evolution by refining models, redefining goals, and setting strategic constraints. This two-way relationship forms a kind of co-evolution in which both elements (human and digital system) continuously learn and adapt (Nozari et al., 2025a).

The transition to such a model requires a rethinking of organizational architecture. Traditional hierarchical structures based on centralized control are not sufficient to exploit the potential of cognitive coexistence. Cognitive organization requires integrated layers of data, simulation, analysis, and human interaction; layers that enable rapid feedback, decision transparency, and explainability. In addition, issues such as system trust, automation level regulation, accountability, and data governance become key elements of this transformation.

From a strategic perspective, the coexistence of humans and digital twins can create sustainable competitive advantage. Organizations that can structurally institutionalize the shared learning cycle between humans and digital models will be able to respond to environmental changes faster than their competitors, assess complex scenarios in advance, and make decisions with controlled risk. In contrast, organizations that either rely solely on human judgment or completely delegate decision-making to algorithms will face cognitive limitations or technological risks.

Accordingly, the purpose of this chapter is to explain the conceptual framework of "human-digital twin coexistence" in the context of the evolution of cognitive organizations. In what follows, the theoretical foundations of this coexistence are first described, then a proposed architecture for a symbiotic cognitive organization is presented, and finally its strategic and governance implications are discussed. The main focus is on the idea that the future of leading organizations lies not in removing humans from the decision-making cycle, but in designing a collaborative and evolutionary relationship between human cognition and digital twin infrastructure.

2. CONCEPTUAL FOUNDATIONS OF HUMAN–DIGITAL TWIN SYMBIOSIS

The coexistence of humans and digital twins has been shaped by conceptual and technological developments, during which organizations have moved from using data as a tool to creating cognitive infrastructures. The digital twin was initially conceived as a virtual representation of physical assets; a tool for observing, monitoring, and synchronizing the state of the real system with the digital model. At this stage, its main function was to create transparency and reduce the gap between operational data and management analysis. With the expansion of the Internet of Things and increased processing power, this representation has moved beyond the descriptive level and entered the realm of prediction. Analytical models have enabled the simulation of future system behavior, and decision-making has shifted from a reactive to a predictive state (Ghansah & Lu, 2025).

In the next step, digital twins have become prescriptive tools; that is, they not only model possible futures, but also suggest optimal options. At this stage, the role of the twin was upgraded from an information system to a decision-support infrastructure. However, practical experiences showed that even the most advanced analytical models are not able to fully understand the contextual nuances, ethical considerations and tacit knowledge of the organization. Decisions made solely on the basis of algorithmic output may be mathematically optimal, but may be strategically or socially challenged (Wei, 2025a). These limitations have provided the basis for rethinking the role of humans in the decision-making cycle.

On the other hand, human cognition has characteristics that cannot be easily reproduced in the form of computational models. Humans are able to tolerate ambiguity, balance between conflicting goals and interpret the long-term consequences of decisions in a value framework. In addition, a significant part of organizational knowledge is tacit in nature and is rooted in experience, culture and informal interactions. Such knowledge is not recorded in databases, but it plays a decisive role in critical moments. Therefore, removing humans from the decision-making cycle not only does not guarantee efficiency, but can also lead to a decrease in the cognitive flexibility of the organization (Wei, 2025a).

In this context, coexistence means synergy and co-evolution. The digital twin expands the cognitive capacity of the organization through real-time analysis, scenario simulation, and big data processing. In turn, humans guide the evolution of the twin by modifying model assumptions, redefining goals, and exercising strategic judgment. This two-way relationship creates a continuous feedback loop in which both elements (humans and digital systems) are exposed to learning. Models are updated based on human feedback, and humans expand their horizons through interaction with digital analytics (Modoni & Sacco, 2023).

The coexistence of humans and digital twins can also be explained in the framework of cyber-physical-social systems, where technology, physical structures and human actors are interconnected in a dynamic network. In such a system, decision-making is the result of a multilayered interaction between data, algorithms and human judgment. The importance of this view is that it considers the organization not simply as a set of technical processes, but as a cognitive ecosystem. In this ecosystem, trust, transparency and explainability play a key role. If humans cannot understand the logic of the decisions proposed by the twin, the coexistence relationship will turn into blind dependence or cognitive resistance.

Conceptually, coexistence requires defining dynamic boundaries between automation and human intervention. Not all decisions require direct human intervention, just as not all decisions should be left to algorithms. Determining the appropriate level of automation depends on factors such as the level of uncertainty, the ethical sensitivity of the decision, and its long-term consequences. In a cognitive symbiotic organization, these boundaries are not fixed, but change in accordance with environmental conditions and the level of maturity of the system. In other words, symbiosis is not a static state, but an evolutionary process that is strengthened by continuous learning.

Ultimately, the conceptual foundations of human-digital twin coexistence are based on the assumption that the future of progressive organizations does not lie in eliminating one in favor of the other, but in designing an interactive and complementary relationship between the two. The digital twin provides the computational and analytical capacity, and the human forms the semantic, value, and strategic framework. This combination can underpin the formation of organizations that are not only intelligent, but also cognitive and adaptable to the complexities of today's changing environment.

3. ARCHITECTURE OF SYMBIOTIC COGNITIVE ENTERPRISES

The architecture of the symbiotic cognitive organization is designed based on the assumption that the interaction of humans and digital twins should be organized in a layered, integrated and scalable structure. This architecture includes not only technical infrastructure, but also cognitive, interactive and governance mechanisms. The goal of this design is to create a platform in which data flow, analysis, decision and feedback continuously circulate between the physical, digital and human layers. In such an architecture, the digital twin is not just a software module, but acts as the cognitive core of the organization and is in a co-evolutionary cycle with humans.

At the lowest level is the physical layer, which includes assets, equipment, operational infrastructure, human resources and executive processes. This layer is considered the main source of operational data generation. Sensors, IoT systems, and data logging systems collect raw data and transmit it to higher layers. The quality, accuracy, and completeness of this data play a decisive role in the efficiency of the entire architecture, as any error at this level can lead to incorrect decisions at higher levels (Nozari & Yordanova, 2026a).

At the next layer, the digital twin layer is placed, which is responsible for representing, synchronizing, and simulating the state of the system. This layer consists of analytical models, predictive algorithms, and scenario-building tools. The twin at this level not only reflects the current state, but also models possible futures and evaluates the consequences of different decisions before they are implemented. In this way, the organization can examine the effects of strategic options before facing a crisis (López et al., 2026).

Above the digital twin layer is the cognitive augmentation layer, which includes decision support engines, optimization algorithms, and adaptive learning systems. At this level, the analyses generated by the twin are translated into operational and strategic recommendations. This layer can prioritize options, assess risks, and suggest optimal paths. However, the outputs of this section are not definitively implemented, but are transferred to the human interaction layer for interpretation and evaluation (He et al., 2024).

At the highest operational level is the human cognitive interaction layer. This layer includes interactive dashboards, explainability tools, and human intervention mechanisms. At this level, organizational decision makers can examine the logic of the system's recommendations, modify the model's assumptions, and make alternative decisions if necessary. The importance of this layer is that it brings the organization's value and strategic framework into the decision-making cycle and prevents the system from becoming an uncontrollable automated structure (Celik et al., 2026).

Alongside these layers, there is a governance and ethics layer that covers all levels horizontally. This layer includes data policies, automation limits, accountability mechanisms, and risk management. Without such a mechanism, coexistence may turn into over-reliance on technology or organizational resistance to it. The overall structure of this architecture is shown in Figure 1.

Figure 1. Layered architecture of a symbiotic cognitive organization

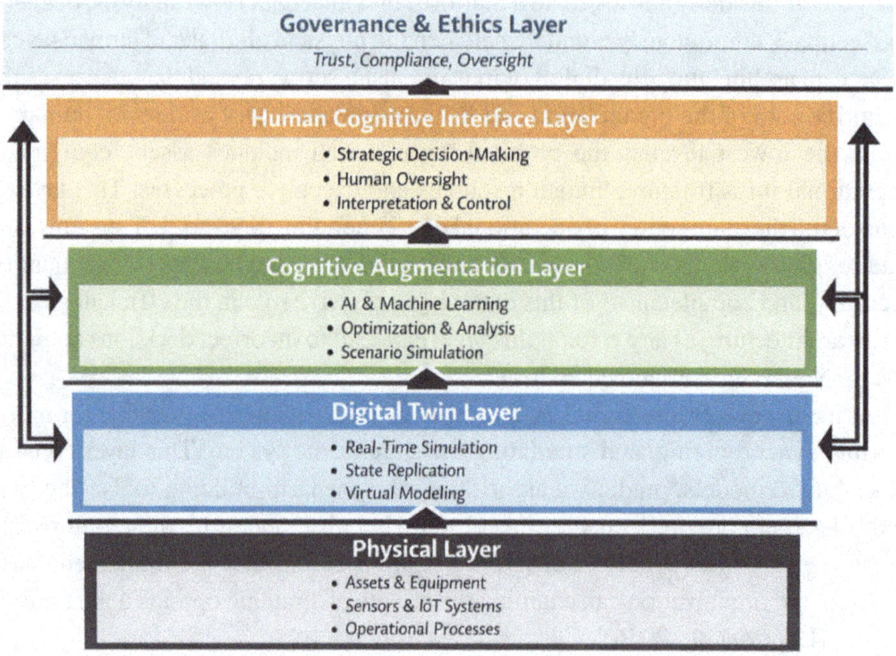

As can be seen in Figure 1, the flow of information moves from the physical layer to the digital and cognitive layers, while the flow of decisions and feedback returns from higher levels to lower ones. This bidirectional structure shows that the symbiotic cognitive organization is not a linear system, but a dynamic cycle of data, analysis, judgment, and correction. The interaction between the human and the digital twin in this architecture does not occur centrally at a specific point, but flows across all levels.

The key point in this architecture is that the boundary between automation and human intervention is designed to be flexible. Some operational decisions can be fully implemented in the digital layers, while strategic decisions require human validation and interpretation. This distinction is not static and changes as the organization matures and trust in the models increases. Therefore, the architecture of the cognitive symbiotic organization is not just a technical framework, but an evolutionary structure that allows for the gradual growth of the relationship between humans and the digital twin.

Overall, this layered architecture shows that achieving true symbiosis requires integration between data infrastructure, analytical models, human interaction mechanisms, and governance systems. Only in such a structure can we expect the digital twin to become a cognitive partner in the evolution of the organization, rather than replacing humans.

4. CO-EVOLUTION MECHANISMS BETWEEN HUMANS AND DIGITAL TWINS

The co-evolution between humans and digital twins is not simply a simple interaction between the user and the intelligent system, but a dynamic and multi-layered process that develops and deepens over time. In this process, both elements (human and digital infrastructure) simultaneously learn, adapt, and adjust their behavior based on mutual feedback. This co-evolution makes sense when the digital twin transforms from an analysis tool to a cognitive partner, and humans are promoted from a mere observer to a guiding role in the evolution of the model.

The first key mechanism in this co-evolution is two-way feedback loops. By receiving operational data, the digital twin models the state of the system and makes suggestions for improving performance. These suggestions are provided to decision makers and are modified, approved, or rejected based on human judgment. Human corrections are fed back to the model as feedback data, improving accuracy, adjusting parameters, and redefining constraints. In this way, the model learns not only from environmental data but also from human experience and interpretation. This continuous cycle forms the basis of collaborative adaptive learning (Bucci et al., 2024).

The second mechanism is learning synchronization. In many organizations, human learning and digital system learning occur separately. But in the symbiotic cognitive organization, these two processes are intertwined. As the twin simulates new scenarios, managers become familiar with the possible consequences of their decisions and gain a deeper understanding of the complexity of the system. At the same time, human reactions and choices are fed back to the model, adjusting the weights of goals, risk tolerance, and priorities. The result of this interaction is the formation of a convergent learning cycle in which the human and the model evolve synergistically (Nozari & Yordanova, 2026b).

The third mechanism is related to the dynamics of trust. Coevolution will not be sustainable without the formation of mutual trust between the human and the system. If the twin proposes decisions whose logic is not clear to the human, the likelihood of resistance or distrust increases. On the other hand, if the human relies on the model's outputs without critical evaluation, there is a risk of overdependence and reduced independent judgment. Therefore, trust must be built gradually and based on transparency, explainability, and reliable performance. This trust dynamic is one of the essential elements in stabilizing the symbiotic relationship (Hozdić & Makovec, 2023).

The fourth mechanism is the redistribution of cognitive load. In complex environments, the volume of information and the speed of change exceed the human processing capacity. The digital twin can take on a large part of the data-driven analysis, pattern recognition, and scenario evaluation, thus reducing the cognitive load. In contrast, humans can focus on strategic interpretation, valuation, and critical decisions. This dynamic division of labor allows the organization's cognitive resources to be optimally allocated and human focus to shift from repetitive to meaningful problems (He et al., 2025). The above mechanisms operate as an interactive network, as shown in Figure 2.

Figure 2. Coevolutionary mechanisms between humans and digital twins

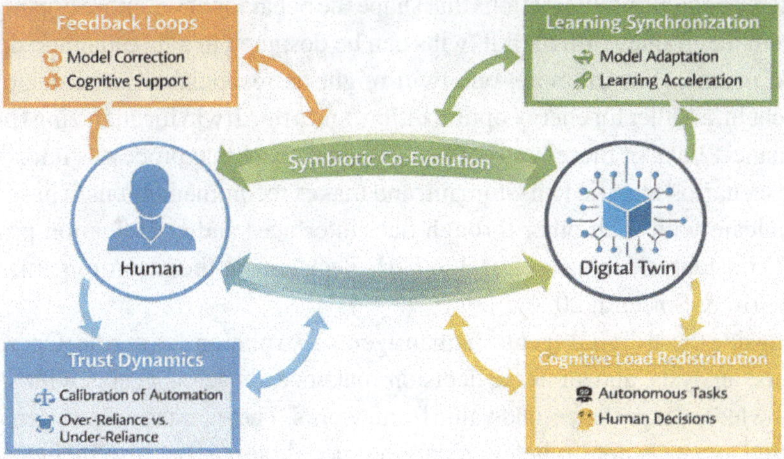

As can be seen in Figure 2, humans and digital twins are at the center of a two-way relationship that is connected through four main mechanisms (feedback, learning synchronization, trust dynamics, and cognitive load redistribution). The bidirectional arrows indicate that this interaction is not a one-way flow of data or instructions, but rather a continuous cycle of learning and adaptation.

The importance of these mechanisms is that they transform co-evolution from an abstract concept into an operational framework. An organization that can structurally design and institutionalize these cycles will be able to continuously increase its cognitive capacity. In such an organization, the digital twin is not only an analysis tool but also a learning partner, and humans are not only users of the system but also guides its evolution. The result of this interaction is the formation of a decision-making ecosystem in which human intelligence and computational intelligence grow in a common direction and lead the organization towards cognitive maturity.

5. MULTI-AGENT COGNITIVE ENTERPRISE DYNAMICS

The dynamics of a multi-agent cognitive organization are formed when digital twins, analytical systems, and human actors act not in isolation but as a collection of intelligent agents with different goals, roles, and levels of authority. In such an organization, intelligence is not centralized in a single core, but is distributed across a network of agents. Each agent(whether human or digital)has a decision domain, a learning capacity, and a level of operational autonomy. The result is a dynamic ecosystem of cognitive interactions that shape the behavior of the entire organization.

Within this framework, digital twins can be designed in a modular and domain-oriented manner. For example, one twin might be responsible for simulating the supply chain, another for energy optimization, and a third twin for analyzing financial performance. Each of these twins acts as a digital agent that processes information, models scenarios specific to its domain, and makes recommendations. These agents communicate with each other through data interfaces and coordination protocols and can consider the impact of each other's decisions at the macro-organizational level (Garro & Sorrenti, 2026).

Alongside these digital agents, human agents also play an active role. Operational managers, analysts, and strategic decision makers each act as agents with specific tacit knowledge, experience, and value frameworks. The interaction between human and digital agents is not limited to receiving model output, but also includes negotiation, goal setting, and redefining constraints. For example, if one twin proposes cost reduction as a top priority, a strategic manager may modify this proposal due to long-term considerations or credit risk. This process forms a kind of multi-agent coordination in which conflicting goals are dynamically balanced (Zhang et al., 2020).

A key feature of multi-agent dynamics is the negotiation mechanism between agents. Each agent has its own objective function, constraints, and information level. When these goals conflict, the organization needs a mechanism to resolve the conflict. This mechanism can include multi-objective optimization algorithms, dynamic priority weighting, or direct human intervention. The importance of this is that the final decision is not the result of the perspective of a single agent, but rather the result of the interaction and relative agreement among a set of agents. Such a structure increases the flexibility and resilience of the organization in the face of complexity (Nozari & Szmelter-Jarosz, 2025).

Another dynamic in multi-agent cognitive organization is the emergence of emergent behaviors. When multiple agents interact with each other, outcomes are achieved that cannot be fully predicted in advance. These emerging behaviors can take the form of performance improvements, process innovations, or even business model redefinitions. Digital twins serve as a platform for simulating and evaluating the outcomes of these interactions. On the other hand, humans guide the evolution of the system by analyzing unexpected outcomes and exercising strategic judgment (Chen & Xin, 2026).

In such an environment, temporal and informational coordination is of particular importance. If agents operate with asynchronous data or incompatible goals, the likelihood of making contradictory decisions increases. Therefore, the architecture of a multi-agent cognitive organization should include mechanisms for data synchronization, knowledge sharing, and defining levels of authority. Determining which decisions are implemented at the digital agent level and which decisions require human consensus or approval is part of the governance design of this structure.

Finally, the multi-agent dynamics of a cognitive organization can be described as a type of learning network in which each agent is both a learner and a producer of knowledge. This network matures over time as more optimal interaction patterns are formed and trust between agents is strengthened. An organization that can institutionalize such dynamics in a structural way will not only benefit from advanced analytical capacity, but will also be able to adapt to turbulent environments with greater speed and accuracy. In this framework, the coexistence of humans and digital twins goes beyond the level of two-way interaction and becomes a cognitive multi-agent ecosystem that will form the basis for the evolution of future organizations.

6. DECISION INTELLIGENCE IN SYMBIOTIC ENTERPRISES

Decision intelligence in symbiotic organizations becomes meaningful when decision-making is transformed from an ad hoc and reactive activity into a structured, continuous process based on human-digital twin interaction. In such an organization, a decision is not simply the result of data analysis or individual judgment, but rather the result of a shared cognitive cycle in which real-time data, predictive models, scenario-building, and human assessment operate in an integrated manner. This type of decision intelligence is considered the main infrastructure for the evolution of the cognitive organization.

In a symbiotic organization, the first step in the formation of decision intelligence is the integration of operational data with analytical models in the context of a digital twin. The twin represents the current state of the system and simultaneously simulates possible futures using historical data and predictive algorithms. These simulations can include optimistic, probable, and critical scenarios, revealing the impact of different decisions before they are implemented. In this way, decision makers can see and evaluate the consequences of their choices in advance, rather than reacting to events (Eirinakis et al., 2022).

The second step is to convert the simulated analyses into decision options. At this stage, optimization and multi-objective analysis engines come into play and generate a set of feasible options based on the defined constraints and objectives. These options may differ from each other in terms of cost, risk, time, or sustainability. The key point in the symbiotic organization is that these outputs are presented as "suggestions," not "definitive orders." Humans evaluate these options, taking into account contextual, strategic, and ethical considerations, and make or modify the final decision (Cao & Ming, 2025).

An important dimension of decision intelligence in this framework is explainability. If the digital twin cannot clearly present the logic behind its recommendations, cognitive engagement is impaired. Therefore, the system should be able to show which data, assumptions, and weights led to a particular outcome. This transparency not only strengthens trust, but also enables shared learning, as humans can identify and correct incorrect assumptions or potential model biases (Jagtap et al., 2026).

Another dimension of decision intelligence is the dynamism in adjusting the level of automation. In some low-risk, high-repetition operational decisions, automated execution of twin recommendations can increase efficiency. In contrast, in strategic decisions or critical situations, direct human intervention is necessary. The symbiotic organization should be able to adjust the boundary between automation and human oversight based on the level of uncertainty, ethical sensitivity, and long-term consequences. This dynamic adjustment is part of the organization's cognitive maturity (Nozari & Yordanova, 2025).

Decision intelligence in a symbiotic organization is also based on continuous learning. Each decision made generates new data that feeds back to the twin and improves the models. At the same time, the experience of decision makers is enriched and influences subsequent decisions. This feedback loop improves the quality of decisions over time, and the organization gradually achieves a higher level of predictability and consistency.

From a strategic perspective, decision intelligence in a symbiotic organization means the ability to simultaneously evaluate conflicting goals. Many organizational decisions involve trade-offs between cost and quality, speed and risk, or profitability and sustainability. The digital twin can model these trade-offs quantitatively, but the final priority determination depends on the organization's value framework and vision. The interplay between quantitative analysis and qualitative judgment reaches its peak at this point, and the true essence of symbiosis is revealed.

Ultimately, decision intelligence in symbiotic organizations can be defined as an organizational capability that arises from the synergy between computational analysis and human cognition. This capability lies neither in technology alone nor in individual experience, but in the design of an interactive structure that allows for the free flow of data, analysis, interpretation, and feedback. An organization that can institutionalize such a system will be able to make more informed, faster, and more responsible decisions in complex and changing environments, and continue its cognitive evolution path with greater stability.

7. INDUSTRIAL APPLICATION SCENARIOS

Industrial applications of human-digital twin coexistence will make sense when this conceptual framework is transferred from the theoretical level to operational contexts and tested in the face of real-world organizational complexities. Different industries face different degrees of uncertainty, risk, supply chain complexity, and technological interdependence. In such environments, cognitive symbiosis can not only increase operational efficiency, but also enhance forecasting, resilience, and strategic decision-making capacity.

In the field of smart manufacturing, digital twins can model production lines, machinery, and material flows in real time. This modeling allows for the prediction of failures, optimization of production planning, and evaluation of demand change scenarios. However, decisions related to changing product mix, adjusting inventory levels, or investing in new equipment are not made solely on the basis of model output. Manufacturing managers interpret and adjust the twin recommendations, taking into account market conditions, supplier relationships, and workforce constraints. This interaction creates a cycle of continuous improvement in which operational data, digital analytics, and human experience work together (Krupas et al., 2024).

In complex supply chains, symbiosis becomes more prominent. Demand fluctuations, logistical disruptions, and geopolitical risks make traditional planning ineffective. The digital twin can represent the supply network in a dynamic model and simulate the impact of changes in shipping routes, warehouse capacity, or ordering policies. However, determining risk tolerance, selecting a supplier diversity strategy, or prioritizing markets requires human strategic judgment. The interaction between digital scenario analysis and human decision-making in this context increases the resilience of the chain and enables faster response to crises (Schmitt, 2025).

In the energy industry and critical infrastructure, the coexistence of humans and twins takes on a dual significance. Smart energy networks, power plants and distribution systems have extensive interdependencies and social implications. A digital twin can predict network loads, consumption patterns and the effects of changing operating policies. However, deciding on energy allocation priorities in a crisis, balancing environmental sustainability with cost or managing social risks requires human and ethical considerations. In this area, coexistence means creating a decision-making framework that combines computational analysis and social responsibility (Nicoletti & Appolloni, 2025).

In the healthcare and biotechnology industries, digital twins can simulate drug manufacturing processes, the performance of medical devices or even patient-centered models. These models play a role in predicting treatment outcomes, optimizing doses or managing hospital resources. However, the final decision on the treatment pathway, patient prioritization or clinical risk management cannot be made solely on the basis of algorithms. Doctors and healthcare managers make their decisions more accurately and more informed by using digital analytics, but the responsibility and final judgment still remain in the human domain (Villegas et al., 2026).

In the field of urban management and smart infrastructure, symbiosis can also be seen as an operational framework. Urban digital twins can model traffic flows, water and energy consumption or air quality on a macro scale. These analyses make it possible to predict the impact of transport or resource management policies. However, choosing policies that strike a balance between citizen well-being, social justice and economic cost requires human participation and consideration of social values. In this context, the twin becomes a tool for supporting collective decision-making and makes the policy-making process more transparent (Nozari & Yordanova, 2025b).

What all these industrial scenarios have in common is that digital twins alone cannot create sustainable competitive advantage unless they are embedded in a cognitive symbiosis framework. Advanced analytics become operationally valuable when they are combined with human experience, judgment, and accountability. In other words, industries can leverage twins to increase the accuracy, speed, and depth of analysis, but the ultimate quality of decisions depends on the constructive interaction between computational intelligence and human cognition.

From this perspective, industrial applications of symbiosis can be understood not as isolated examples, but as manifestations of a new paradigm; one in which organizations focus on designing an interactive and evolutionary relationship between the two, rather than choosing between humans or technology. This approach can underpin the sustainable transformation of industries in an era of complexity and uncertainty.

8. TOWARD SYMBIOTIC ENTERPRISE MATURITY MODEL

The move towards a cognitive symbiotic organization is not a sudden technological leap, but a gradual and staged process in which structures, organizational culture, data infrastructure, and decision-making practices evolve simultaneously. As observed in digital transformation experiences, organizational maturity is achieved when technology, process, and human capital are aligned, and the mere deployment of advanced tools alone does not lead to sustainable transformation (Youvan, 2026). Hence, designing a maturity model for symbiotic organizations can provide a systematic framework for assessing the current state and charting the path of evolution.

At the first level of maturity, the interaction between humans and digital twins is largely instrumental and limited to receiving data-driven reports and recommendations. The twin plays the role of a decision support system, but is not integrated into the strategic structure of the organization. Data may be stored in islands and the use of analytics is more reactive. At this stage, human-technology collaboration has not yet reached the level of symbiosis and deep interdependence has not yet been established (Radanliev et al., 2022).

At the second level, operational coordination between humans and twins is strengthened. Data is synchronized and iterative optimizations are implemented at the operational level. The digital twin actively participates in production planning, resource allocation, or inventory management. However, strategic decisions are still made primarily based on human judgment, and digital analytics are used more at the tactical level.

At the third level, the organization enters the predictive decision-making stage. At this stage, scenario building, forward-looking analysis, and multi-objective assessment become part of the formal decision-making process. Managers use the digital twin to examine the possible consequences of different options, and the results of the analyses are used in strategy meetings. This level represents a deeper convergence between computational analytics and human judgment (Peruzzini et al., 2024).

The fourth level is characterized by strategic collaboration. In this stage, organizational goals, performance indicators, and macro policies are jointly defined and reconfigured at the human and digital layers. The twin not only simulates scenarios, but also plays a role in redefining business structures and innovation in operational models. Interaction between humans and systems is designed in a cyclical and institutionalized manner, and mechanisms of trust and explainability have reached maturity (Nozari, 2025).

Finally, the fifth level of maturity is "cognitive coexistence." At this stage, the organization becomes a co-learning and adaptive system in which human learning and algorithmic learning flow seamlessly. The boundary between analysis and decision is dynamically adjusted, and the organizational structure is able to respond to environmental changes with speed and coherence. The digital twin becomes a cognitive partner, and humans also play a guiding role in the evolution of the system. At this level, continuous innovation and sustainable adaptation become inherent characteristics of the organization (Nechesov et al., 2025). These maturity levels are shown in Figure 3.

Figure 3. Maturity model towards cognitive symbiosis

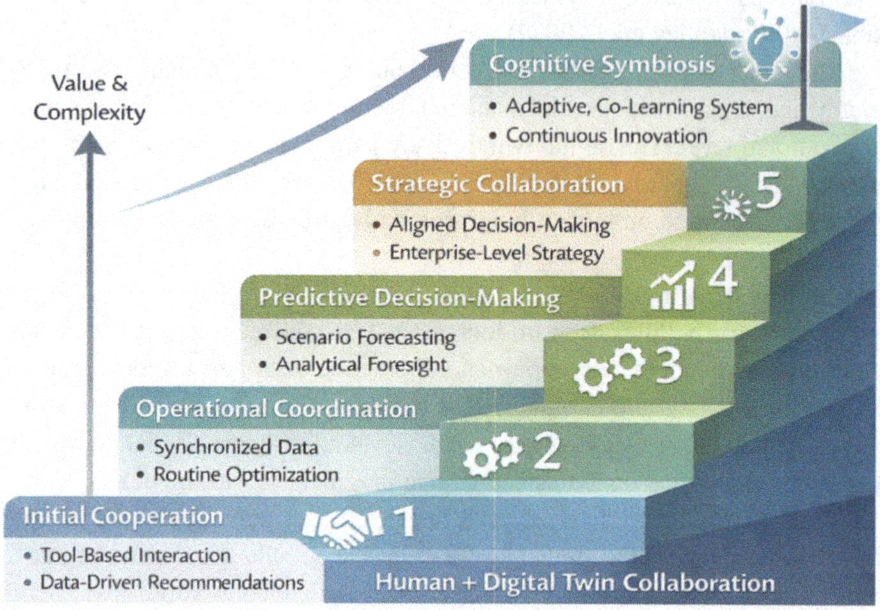

As can be seen in Figure 3, the move from early collaboration to cognitive symbiosis is accompanied by increasing value creation and complexity. Each level not only reflects technological development, but also changes in decision-making culture, governance mechanisms, and the distribution of responsibility between people and systems. This maturity model can be used as a diagnostic tool for organizations to assess their current position and identify gaps in the path to cognitive symbiosis.

At the same time, achieving higher levels of maturity requires investing in data infrastructure, developing human skills, designing trust mechanisms, and creating transparent ethical frameworks. Organizations that systematically take this path will be able to strengthen shared learning capacity and leverage the coexistence of humans and digital twins as a sustainable competitive advantage.

9. CONCEPTUAL CONTRIBUTION

The most important conceptual innovation of this chapter is the redefinition of the relationship between humans and digital twins from an instrumental interaction to a structured cognitive symbiosis. In much of the existing literature, digital twins are presented as advanced technology for simulation, prediction, or optimization, and the human role is largely defined as that of a user or observer. In contrast, this chapter argues that the future of leading organizations lies not in increasing the level of mere automation, but in designing a co-evolutionary mechanism between human cognition and digital infrastructure.

The first conceptual contribution of this research is to present the framework of the "symbiotic cognitive organization." In this framework, the digital twin is defined as a cognitive partner that participates alongside humans in the production, interpretation, and reorganization of organizational knowledge. This view elevates the twin beyond the level of an analytical tool and elevates it to part of the cognitive architecture of the organization. At the same time, humans are not only the ultimate decision-maker, but also the driver of model evolution and the setter of the boundaries of automation. This redefinition creates a paradigm shift in the understanding of human-technology interaction.

The second conceptual innovation is the explanation of coevolutionary mechanisms as a dynamic and multilayered system. Instead of focusing on a specific point of interaction, this chapter shows that coexistence is formed in the form of feedback loops, learning synchronization, trust dynamics, and cognitive load redistribution. These four mechanisms create an interactive network in which human decisions, models, and experiences continuously redefine each other. Such a framework allows for a more detailed analysis of organizational maturity and the design of institutional mechanisms to enhance coexistence.

The third contribution is the development of a staged maturity model for the transition to a cognitive symbiotic organization. This model shows that coexistence is not a zero-one situation, but rather encompasses a range of levels of cooperation, from initial instrumental interaction to adaptive co-learning at the strategic level. The presentation of this maturity model allows for a structural assessment of the state of organizations and the identification of gaps in the path of cognitive evolution. In this way, coexistence is transformed from an abstract concept into an operational roadmap.

The fourth innovation is the introduction of the logic of "symbiotic decision intelligence". In this logic, the decision is not the result of a single agent, but the result of multi-agent interaction between digital twins, analytical systems and human actors. This approach shows that the quality of decisions in complex organizations depends on the design of mechanisms of coordination, explainability and dynamic adjustment between quantitative analysis and qualitative judgment. In this way, decision-making is redefined as a networked and interactive process.

Finally, this chapter establishes the nexus between technology, cognition, and governance by presenting an integrated layered architecture. Unlike approaches that define digital architecture solely at the data infrastructure or algorithm level, this framework demonstrates that achieving coexistence requires the simultaneous design of physical, digital, cognitive, and governance layers. This conceptual integration provides a foundation for developing organizations that are able to operate adaptively and responsibly in complex and changing environments.

10. CONCLUSION

The transformation of organizations in the age of complexity is no longer limited to the deployment of digital technologies. What distinguishes leading organizations from others is their ability to design a structured, dynamic, and evolutionary relationship between humans and intelligent infrastructures. Focusing on the concept of human-digital twin coexistence, this chapter showed that the future of cognitive organizations lies neither in removing humans from the decision-making cycle nor in resisting automation, but in the synergy of the two.

First, the conceptual foundations of coexistence were explained and it became clear that the digital twin has evolved from a simple monitoring tool to an advanced cognitive infrastructure. However, even the most advanced analytical models lack human contextual understanding, value judgment, and ethical sensitivity. Hence, coexistence was proposed as a framework in which human cognition and computational intelligence act complementary. This perspective turns the twin into a cognitive partner and the human into a driver of system evolution.

Next, the architecture of the cognitive symbiotic organization was introduced; a layered architecture that starts with the physical infrastructure and reaches the digital layers, cognitive augmentation, human interaction and governance. This architecture showed that achieving symbiosis requires the integration of data, analysis, decision and feedback. An organization that designs these layers separately will not be able to fully exploit the potential of symbiosis. In contrast, an organization that institutionalizes interaction between these levels can create a sustainable cycle of shared learning.

Coevolutionary mechanisms were also introduced as the backbone of this relationship. Two-way feedback loops, learning synchronization, trust dynamics and cognitive load redistribution showed that symbiosis is not a static state, but a gradual process dependent on continuous interaction. These mechanisms allow the organization to improve the quality of its decisions over time and benefit from human experience and digital analytics simultaneously.

In the section on multi-agent dynamics, it was emphasized that organizational intelligence in future structures will operate in a distributed manner. Domain-centric digital twins, analytical systems, and human actors interact with each other in the form of a network of agents, and the final decision is formed from the outcome of these interactions. Such a structure increases the flexibility, resilience, and innovation capacity of the organization. In this framework, decision-making is no longer an individual or linear act, but a networked and dynamic process.

The presented industrial applications showed that coexistence has operational capabilities in the areas of manufacturing, supply chain, energy, health, and urban management. In all these areas, digital analytics becomes a sustainable value when it is integrated with human judgment and the organization's value framework. These findings show that coexistence is not an abstract idea, but can be implemented as an operational paradigm in various industries.

Finally, the cognitive symbiotic organization maturity model showed that the transition to this level of evolution requires a series of distinct stages. Organizations must move beyond initial instrumental interaction to strategic collaboration and, ultimately, adaptive co-learning. This path requires investing in data infrastructure, developing human skills, building trust mechanisms, and designing transparent governance frameworks. Symbiosis is not a mere technological goal, but rather a cultural and cognitive transformation at the organizational level.

Overall, this chapter argues that the organizations of the future will be those that can balance the processing capacity of digital twins with the ability to interpret and judge humans. In such organizations, technology does not replace humans, but rather expands their cognitive horizons. The result of this synergy is the formation of organizations that are not only intelligent, but also adaptive, responsible, and innovative. Human-digital twin symbiosis is the foundation for the evolution of cognitive organizations in an era of uncertainty and continuous change.

REFERENCES

Bucci, I., Fani, V., & Bandinelli, R. (2024). Towards human-centric manufacturing: Exploring the role of human digital twins in industry 5.0. *Sustainability (Basel), 17*(1), 129. DOI: 10.3390/su17010129

Cao, Y., & Ming, X. (2025). Digital Twin-Driven Low-Carbon Service Design and Modularization in Central Air Conditioning Ecosystems: A Multi-Criteria and Co-Intelligence Approach. *Sustainability (Basel), 17*(21), 9877. DOI: 10.3390/su17219877

Celik, S. E., Chen, Z., Yildizbasi, A., & Sarkis, J. (2026). Generative artificial intelligence for adopting circular economy in supply chains: Opportunities, challenges, and future pathways. *International Journal of Generative Artificial Intelligence in Business, 1*(1-2), 8–35. DOI: 10.1504/IJGAIB.2026.151818

Chen, Y., & Xin, J. (2026). Human-Digital Symbiosis and Green Intelligent Manufacturing: Organizational Digital Acceptance and Proficiency. *Journal of Global Information Management, 34*(1), 1–43. DOI: 10.4018/JGIM.401375

Eirinakis, P., Lounis, S., Plitsos, S., Arampatzis, G., Kalaboukas, K., Kenda, K., Lu, J., Rožanec, J. M., & Stojanovic, N. (2022). Cognitive digital twins for resilience in production: A conceptual framework. *Information (Basel), 13*(1), 33. DOI: 10.3390/info13010033

Garro, A., & Sorrenti, A. (2026). Designing an ISO 23247-compliant Hybrid Digital Twin Architecture for industry. *Information and Software Technology, 193*, 108039. DOI: 10.1016/j.infsof.2026.108039

Ghansah, F. A., & Lu, W. (2025). Cyber-physical systems and digital twins for "cognitive building" in the construction industry. *Construction Innovation, 25*(3), 787–818. DOI: 10.1108/CI-07-2022-0164

He, Q., Li, L., Li, D., Peng, T., Zhang, X., Cai, Y., Zhang, X., & Tang, R. (2024). From digital human modeling to human digital twin: Framework and perspectives in human factors. *Chinese Journal of Mechanical Engineering, 37*(1), 9. DOI: 10.1186/s10033-024-00998-7

He, Q., Wang, P., Peng, T., Wang, B., & Tang, R. (2025). Human digital twin in human-centric smart manufacturing. In *Human-Centric Smart Manufacturing Towards Industry 5.0* (pp. 59–76). Springer Nature Switzerland. DOI: 10.1007/978-3-031-82170-7_3

Hozdić, E., & Makovec, I. (2023). Evolution of the human role in manufacturing systems: On the route from digitalization and cybernation to cognitization. *Applied System Innovation*, *6*(2), 49. DOI: 10.3390/asi6020049

Jagtap, N., Singh, T., & Sonawane, E. (2026). The pivotal role of artificial intelligence in digital twins: A case study. In Transforming Industries, Empowering Societies (pp. 187-198). Elsevier. DOI: 10.1016/B978-0-443-32878-7.00014-6

Kinsner, W., & Saracco, R. (2019, July). Towards evolving symbiotic cognitive education based on digital twins. In 2019 IEEE 18th International Conference on Cognitive Informatics & Cognitive Computing (ICCI* CC) (pp. 13-21). IEEE. DOI: 10.1109/ICCICC46617.2019.9146095

Krupas, M., Kajati, E., Liu, C., & Zolotova, I. (2024). Towards a human-centric digital twin for human–machine collaboration: A review on enabling technologies and methods. *Sensors (Basel)*, *24*(7), 2232. DOI: 10.3390/s24072232 PMID: 38610442

López, E. J., Ruiz, J. E. P., Chávez, O. L., Muñoz, F., Velásquez, L. A. G., & Lugo, J. G. C. (2026). The Evolution of Mechatronics Engineering and Its Relationship with Industry 3.0, 4.0, and 5.0. *Technologies*, *14*(2), 81. DOI: 10.3390/technologies14020081

Modoni, G. E., & Sacco, M. (2023). A human digital-twin-based framework driving human centricity towards industry 5.0. *Sensors (Basel)*, *23*(13), 6054. DOI: 10.3390/s23136054 PMID: 37447903

Nechesov, A., Dorokhov, I., & Ruponen, J. (2025). Virtual cities: From digital twins to autonomous AI societies. *IEEE Access : Practical Innovations, Open Solutions*, *13*, 13866–13903. DOI: 10.1109/ACCESS.2025.3531222

Nicoletti, B., & Appolloni, A. (2025). A digital twin framework for enhancing human–agentic AI–machine collaboration. *Journal of Intelligent Manufacturing*, •••, 1–17.

Nozari, H. (2025). Integrating Neuromorphic Intelligence with Fuzzy Multi-Objective Models in Resilient Green Supply Chain Design. Process Integration and Optimization for Sustainability, 1-14.

Nozari, H., & Szmelter-Jarosz, A. (2025). A Predictive Maintenance Approach for Composting Plants Based on ERP and Digital Twin Integration. *Machines (Basel)*, *13*(12), 1123. DOI: 10.3390/machines13121123

Nozari, H., Szmelter-Jarosz, A., & Weiland, D. (2025). A Fuzzy Multi-Objective Sustainable and Agile Supply Chain Model Based on Digital Twin and Internet of Things with Adaptive Learning Under Environmental Uncertainty. *Applied Sciences (Basel, Switzerland)*, *15*(19), 10399. DOI: 10.3390/app151910399

Nozari, H., & Yordanova, Z. (2025). A fuzzy–digital twin optimization framework for simultaneous management of waste and energy consumption in sustainable manufacturing. An International Journal of Optimization and Control: Theories & Applications, 25360152.

Nozari, H., & Yordanova, Z. (2025). An intelligent digital twin approach for optimizing multi-channel supply chains in uncertain environments. Green Technologies and Sustainability, 100309.

Nozari, H., & Yordanova, Z. (2026). Digital twin–driven decision support for human capital management under uncertainty. Green Technologies and Sustainability, 100353.

Nozari, H., & Yordanova, Z. (2026). An Adaptive Fuzzy Multi-Objective Digital Twin Framework for Multi-Depot Cold-Chain Vehicle Routing in Agri-Biotech Supply Networks. *Logistics (Basel)*, *10*(2), 27. DOI: 10.3390/logistics10020027

Peruzzini, M., Prati, E., & Pellicciari, M. (2024). A framework to design smart manufacturing systems for Industry 5.0 based on the human-automation symbiosis. *International Journal of Computer Integrated Manufacturing*, *37*(10-11), 1426–1443. DOI: 10.1080/0951192X.2023.2257634

Radanliev, P., De Roure, D., Nicolescu, R., Huth, M., & Santos, O. (2022). Digital twins: Artificial intelligence and the IoT cyber-physical systems in Industry 4.0. *International Journal of Intelligent Robotics and Applications*, *6*(1), 171–185. DOI: 10.1007/s41315-021-00180-5

Schmitt, U. (2025, October). Platform Powering Cognitive and Memetic Digital Twins: To Boldly Go Where Memes May Live Long and Prosper. In International Conference on Cooperative Design, Visualization and Engineering (pp. 112-122). Cham: Springer Nature Switzerland.

Su, C., Tang, X., Han, Y., Wang, T., & Jiang, D. (2024). Cognitive digital twin in manufacturing process: Integrating the knowledge graph for enhanced human-centric Industry 5.0. *International Journal of Production Research*, 1–22. DOI: 10.1080/00207543.2024.2435583

Villegas, L. F., Romero, D., Macchi, M., & Skoogh, A. (2026). Towards a Digital Twin Lifecycle Management Framework. *Enterprise Information Systems*, •••, 2616681. DOI: 10.1080/17517575.2026.2616681

Wei, Y. (2025). A digital twin model for grain enterprise financial shared service centers based on distributed deep learning and neural symbolic reasoning. *Scientific Reports*, *15*(1), 40558. DOI: 10.1038/s41598-025-24350-8 PMID: 41254019

Wei, Y. (2025). A digital twin model for grain enterprise financial shared service centers based on distributed deep learning and neural symbolic reasoning. *Scientific Reports*, *15*(1), 40558. DOI: 10.1038/s41598-025-24350-8 PMID: 41254019

Youvan, D. C. (2026). From Ancient Mnemonics to Neural Symbiosis: The Historical and Futuristic Evolution of Cognitive Tools Shaping Human Thought.

Zhang, N., Bahsoon, R., & Theodoropoulos, G. (2020, October). Towards engineering cognitive digital twins with self-awareness. In 2020 IEEE international conference on systems, man, and cybernetics (SMC) (pp. 3891-3891). IEEE.

Zheng, X., Lu, J., & Kiritsis, D. (2022). The emergence of cognitive digital twin: Vision, challenges and opportunities. *International Journal of Production Research*, *60*(24), 7610–7632. DOI: 10.1080/00207543.2021.2014591

Chapter 2
Perceived AI-Driven Digital Twin Work Environments and Workforce Adaptation:
The Mediating Role of Employee Preparedness

S. Vijayarani
http://orcid.org/0000-0001-7027-3109
SRM Institute of Science and Technology, India

Anto Pravin Singh D.
http://orcid.org/0000-0002-4221-8142
SRM Institute of Science and Technology, India

M. Sowmiya
SRM Institute of Science and Technology, India

ABSTRACT

This chapter examines the relationship between perceived AI-driven digital twin work environments, employee preparedness, and workforce adaptation. Drawing on Socio-technical Systems Theory and Human Capital Theory, a quantitative cross-sectional design was adopted using data from 150 employees across technology-driven industries. Structural Equation Modeling (SEM) results indicate that digital twin environments significantly influence workforce adaptation both directly and indirectly through employee preparedness. The findings highlight that technological effectiveness depends on employees' readiness and capability to engage with digital

DOI: 10.4018/979-8-2600-1669-5.ch002

systems. Employee preparedness emerges as a key mediating mechanism linking technology and adaptation. From a managerial perspective, the study emphasizes the need to align digital investments with skill development, training, and user-centric system design. Overall, the study demonstrates that successful digital transformation requires integration of technological systems with human capabilities.

1. INTRODUCTION

1.1 Background

The rapid advancement of digital technologies has fundamentally transformed the way organizations operate, compete, and manage their workforce. Among these innovations, Artificial Intelligence (AI) and digital twin technologies have emerged as powerful enablers of organizational transformation. Digital twins, defined as virtual representations of physical systems, processes, or environments, enable real-time data integration and simulation to improve decision-making and operational efficiency (Tao et al., 2019). When integrated with AI, these systems evolve into intelligent, adaptive, and predictive ecosystems capable of real-time analytics and optimization (Rojek et al., 2025).

AI-driven digital twin work environments are reshaping traditional workplace structures by enabling continuous monitoring, simulation, and optimization of operations. These technologies allow organizations to anticipate disruptions, enhance decision accuracy, and improve operational performance through predictive analytics (Kritzinger et al., 2018). As digital twins become more intelligent and adaptive, they are increasingly central to Industry 4.0 and Industry 5.0 transformations, facilitating data-driven and human-centric organizational models (Attaran, 2023).

This technological shift has significantly altered workforce expectations. Employees are no longer confined to routine tasks but are required to engage with complex digital systems, interpret data, and make informed decisions in dynamic environments. The integration of AI into workplace processes has led to a transformation in job roles, emphasizing higher-order cognitive skills such as problem-solving, collaboration, and critical thinking (Sunkar, 2025). While organizations report productivity gains through AI adoption, these advancements simultaneously introduce challenges related to workforce adaptation and skill readiness.

In this context, workforce adaptation has emerged as a critical determinant of organizational success. Employees must continuously update their competencies, interact with advanced technologies, and adjust to evolving work processes. However, the effectiveness of AI-driven digital twin environments is not solely dependent on technological infrastructure but also on how employees perceive and engage with these systems. The concept of perceived AI-driven digital twin work environments highlights employees' subjective evaluation of these technologies, including their usefulness, usability, and relevance to job performance.

Furthermore, employee preparedness plays a crucial role in bridging the gap between technological implementation and workforce outcomes. Preparedness encompasses employees' digital skills, knowledge, confidence, and cognitive readiness required to function effectively in AI-enabled environments. As highlighted by recent studies, successful digital twin adoption requires not only technological integration but also workforce upskilling and capability development (Verhoef et al., 2021).

1.2 Research Problem

Despite the growing adoption of AI-driven digital twin technologies, existing research has predominantly focused on their technical and engineering applications, such as system optimization, predictive maintenance, and operational efficiency. Studies by Tao et al. (2019) and Kritzinger et al. (2018) emphasize the technological benefits of digital twins but provide limited insights into their implications for employee behavior and workforce outcomes.

Similarly, research on digital transformation has largely concentrated on technology adoption, digital readiness, and performance outcomes, often overlooking the human capability mechanisms through which technological environments influence employee adaptability (Vial, 2019; Verhoef et al., 2021). While digital technologies are recognized as key drivers of organizational change, there remains a lack of integrated frameworks that explain how these technologies translate into workforce adaptation.

Another critical gap lies in the limited understanding of the mechanism linking technological environments to adaptive outcomes. Although prior research highlights the importance of employee skills and competencies, the mediating role of employee preparedness in connecting perceived technological environments with workforce adaptation remains underexplored.

Furthermore, most existing studies are concentrated in developed economies, with limited empirical evidence from emerging contexts such as India. Variations in digital infrastructure, organizational maturity, and workforce capabilities may significantly influence how employees perceive and adapt to AI-driven digital twin environments.

To address these gaps, the present study adopts a human-centric perspective by examining the relationship between perceived AI-driven digital twin work environments, employee preparedness, and workforce adaptation, while explicitly investigating the mediating role of preparedness.

1.3 Objectives of the Study

- To examine the impact of perceived AI-driven digital twin work environments on workforce adaptation
- To analyze the influence of perceived AI-driven digital twin work environments on employee preparedness
- To assess the effect of employee preparedness on workforce adaptation
- To investigate the mediating role of employee preparedness in the relationship between perceived AI-driven digital twin work environments and workforce adaptation
- To provide practical insights for organizations to enhance employee readiness and facilitate the effective adoption of digital twin technologies

1.4 Research Questions

- How do perceived AI-driven digital twin work environments influence workforce adaptation?
- What is the impact of perceived AI-driven digital twin work environments on employee preparedness?
- How does employee preparedness affect workforce adaptation?
- Does employee preparedness mediate the relationship between perceived AI-driven digital twin work environments and workforce adaptation?

1.5 Significance of the Chapter

This study contributes to both theory and practice. From a theoretical perspective, it extends digital transformation literature by integrating technological perception with workforce outcomes. By incorporating employee preparedness as a mediating variable, the study provides a comprehensive understanding of how digital twin environments influence workforce adaptation.

From a practical perspective, the study offers actionable insights for organizations implementing AI-driven digital twin technologies. It emphasizes workforce upskilling, training interventions, and perception management to ensure successful technology adoption and workforce adaptability.

1.6 Structure of the Chapter

The chapter begins with an introduction outlining the background, research problem, objectives, and significance of the study. This is followed by a literature review, methodology, data analysis, discussion, implications, limitations, and conclusion.

2. LITERATURE REVIEW

2.1 Conceptual Background

2.1.1 Digital Twin Work Environment

Digital Twin (DT) technology refers to real-time digital representations of physical systems, enabled through continuous data exchange, synchronization, and predictive analytics (Tao et al., 2019). Unlike conventional digital systems, DT environments involve bidirectional interaction between physical and virtual entities, allowing dynamic monitoring, simulation, and optimization.

Early research primarily emphasizes technical and operational applications, including predictive maintenance, process optimization, and system efficiency (Kritzinger et al., 2018; Ivanov & Dolgui, 2021). These studies establish DT as a core component of Industry 4.0 but largely adopt a technology-centric perspective, focusing on system performance rather than human interaction.

Subsequent advancements integrate Artificial Intelligence (AI), enabling DT systems to support real-time analytics, predictive modeling, and autonomous decision-making (Negri et al., 2021; Qi & Tao, 2018). These developments transform DT environments into intelligent, adaptive systems capable of managing complex operational processes.

Recent contributions further extend DT applications to decision support, workforce modeling, and optimization under uncertainty (Nozari & Yordanova, 2026; Nozari, 2026). For instance, AI-driven DT frameworks facilitate simulation-based decision-making and enhance organizational responsiveness. Despite these advancements, existing research continues to emphasize system-level efficiency, with limited attention to how employees interact with and respond to such environments.

Moreover, DT environments differ significantly from general digital systems due to their continuous feedback loops, predictive capabilities, and data-intensive nature, which impose higher cognitive and behavioral demands on employees. However, this distinction is often insufficiently addressed in the literature, leading to conceptual ambiguity.

Thus, while DT research has advanced technologically, it lacks a human-centric perspective, particularly in explaining how AI-driven Digital Twin environments influence employee preparedness and workforce adaptation.

2.1.2 Workforce Adaptation

Workforce adaptation refers to employees' ability to adjust skills, behaviors, and performance in response to technological and organizational changes. In digital transformation contexts, adaptation is critical as employees must engage with evolving technologies and complex work environments.

Existing research suggests that adaptation is influenced by factors such as perceived usefulness, organizational support, and digital readiness (Verhoef et al., 2021; Tarafdar et al., 2019). These studies highlight that employees are more likely to adapt when technologies are perceived as beneficial and when organizations provide adequate support mechanisms.

However, much of this research is grounded in general digital environments, such as enterprise systems or basic automation, and does not fully capture the complexity of AI-driven DT systems. DT environments require employees to interpret real-time data, predictive insights, and simulation outputs, which significantly increases cognitive and analytical demands.

Additionally, prior studies identify barriers such as resistance to change, skill gaps, and technostress as key challenges to adaptation (Tarafdar et al., 2019). While these insights are valuable, they often treat adaptation as a direct consequence of technology adoption, overlooking the processes through which employees develop the capability to adapt.

Emerging research indicates that DT environments can facilitate adaptation through simulation-based learning and data-driven decision-making, enabling employees to experiment with scenarios and improve performance. However, empirical evidence explaining how these environments translate into adaptive behavior remains limited. Therefore, there is a need to examine workforce adaptation in AI-driven DT contexts through underlying mechanisms rather than direct relationships.

2.1.3 Employee Preparedness

Employee preparedness represents a multidimensional construct, encompassing digital skills, learning agility, cognitive readiness, and psychological willingness to engage with technology. It reflects employees' ability to effectively operate in technology-driven environments.

While preparedness is widely recognized in digital transformation research, it is often conceptualized as an independent factor, lacking integration with specific technological contexts (Sousa & Rocha, 2019; Van Laar et al., 2017). This limits understanding of how preparedness evolves in response to advanced technologies.

Recent studies suggest that DT systems actively contribute to preparedness through simulation, real-time feedback, and continuous learning environments (Dolgui et al., 2020; Nozari & Yordanova, 2026). These systems enable employees to develop competencies in a controlled, data-driven setting, improving decision-making and performance.

Additionally, research highlights the importance of continuous learning and capability development in digital environments, emphasizing that employees must continuously upgrade their skills to keep pace with technological advancements (Brynjolfsson & McAfee, 2014).

However, despite these insights, empirical research linking preparedness to workforce adaptation in DT environments remains limited. Furthermore, conceptual overlap with constructs such as digital readiness and technology acceptance creates ambiguity, reducing theoretical clarity.Thus, the role of employee preparedness as a mediating mechanism linking DT environments to workforce adaptation remains insufficiently explored.

2.2 Theoretical Framework: Socio-Technical Systems Theory and Human Capital Theory

The rapid advancement of digital technologies such as artificial intelligence (AI) and digital platforms has significantly transformed organizational processes and workforce dynamics. These technologies not only improve efficiency but also redefine employee roles, skills, and adaptability. To understand workforce adaptation in such environments, this study integrates Socio-technical Systems Theory (STS) and Human Capital Theory (HCT), offering a comprehensive explanation of how technological systems and human capabilities interact to influence organizational outcomes.

2.2.1 Socio-Technical Systems Theory (STS Theory)

Socio-technical Systems Theory, developed by Trist and Bamforth (1951), proposes that organizations consist of two interdependent subsystems: the technical subsystem (tools, technologies, processes) and the social subsystem (people, relationships, structures). Organizational effectiveness depends on the alignment or joint optimization of these subsystems.

In digital transformation contexts, STS is particularly relevant as new technologies reshape work processes and decision-making structures. However, successful implementation depends not only on technological capabilities but also on how well these systems align with human needs. Misalignment can lead to resistance, inefficiencies, and reduced performance.

Scholars such as Orlikowski (1992) emphasize that technology and human action are mutually shaping, while Mumford (2006) highlights the importance of human-centered system design. Pasmore et al. (1982) further show that organizations aligning social and technical systems achieve higher adaptability and performance. Contemporary studies reinforce this view, indicating that digital transformation requires both technological upgrades and organizational change, including employee readiness and cultural adaptation.

Thus, STS explains workforce adaptation by emphasizing the need for balance between technological systems and human elements in organizations.

2.2.2 Human Capital Theory (HCT)

Human Capital Theory, developed by Becker (1964), focuses on the role of employee competencies in enhancing productivity and organizational performance. It views employees as valuable assets whose knowledge, skills, and abilities drive success.

In digital environments, this theory becomes increasingly important as employees must continuously upgrade their skills to keep pace with technological advancements. Early contributions by Schultz (1961) and Mincer (1974) highlight that investments in education and training improve productivity. More recent studies emphasize that technology alone does not lead to performance gains unless supported by skilled employees.

Scholars such as Brynjolfsson and McAfee (2014) and Westerman et al. (2011) argue that organizations must build digital capabilities among employees to fully leverage technology. Similarly, Sousa and Rocha (2019) and Van Laar et al. (2017) stress the importance of continuous learning, adaptability, and digital skills.

Within this framework, employee preparedness is a key construct, referring to employees' readiness, skills, and ability to use digital technologies effectively. Prepared employees are more likely to adopt new systems and adapt to changing work environments.

2.2.3 Integration of STS and HCT

The integration of STS and HCT provides a unified perspective on workforce adaptation. STS explains the importance of aligning technological and social systems, while HCT explains how this alignment is achieved through employee skill development.

Together, these theories suggest that technological systems and human capabilities are mutually reinforcing. While digital technologies enhance efficiency, their effectiveness depends on employees' preparedness to use them. Thus, employee preparedness acts as a mediating mechanism linking technology to workforce adaptation.

This integrated framework highlights that organizational success in digital environments depends not only on technology adoption but also on continuous investment in human capabilities, providing strong theoretical support for examining workforce adaptation.

2.3 Review of Empirical Studies

2.3.1 Digital Twins and Workforce Outcomes

Empirical research on Digital Twin (DT) technology has largely emphasized technical and operational outcomes, particularly in predictive maintenance, system optimization, and process efficiency. For example, Tao et al. (2019) conceptualize DT as enabling real-time monitoring and simulation, while Kritzinger et al. (2018) focus on its classification within Industry 4.0 applications. Although these studies establish the technological capabilities of DT systems, they predominantly adopt a system-centric perspective, offering limited insight into employee-level implications.

Similarly, Attaran (2023) highlights improvements in operational visibility and decision accuracy through DT integration. However, such studies reinforce a consistent pattern in the literature—prioritizing performance outcomes over human interaction with technology. In contrast, more recent work by Rojek et al. (2025) begins to shift toward a behavioral perspective, suggesting that AI-enabled DT systems enhance employees' decision-making capabilities through predictive analytics and real-time insights. This indicates a gradual transition from purely technical analyses to more human-oriented considerations.

Despite this shift, existing studies remain fragmented. While some acknowledge that DT systems support simulation-based learning and problem-solving, they do not sufficiently examine how employees adapt their behaviors and competencies in response to these environments. Furthermore, research suggests that the effectiveness of DT systems is contingent upon employees' ability to interpret and utilize system outputs (Vial, 2019), yet this dependency is often treated implicitly rather than being empirically investigated.

Additionally, studies grounded in digital transformation highlight that technology adoption is influenced by perceived usefulness and ease of use (Verhoef et al., 2021). However, these frameworks are typically applied to general digital technologies and do not account for the complex, real-time, and data-intensive nature of Digital Twin environments.

Thus, while existing research demonstrates the operational benefits of Digital Twin technologies, it lacks a comprehensive understanding of how such environments influence workforce adaptation at the employee level, particularly in AI-driven contexts.

2.3.2 Employee Preparedness and Workforce Adaptation

The relationship between employee preparedness and workforce adaptation has been widely examined in digital transformation literature, with preparedness consistently identified as a key determinant of adaptation. Research rooted in Human Capital Theory (Becker, 1964) emphasizes that employees with higher levels of knowledge and skills are better equipped to respond to technological change. Supporting this view, Verhoef et al. (2021) argue that digital transformation necessitates continuous upskilling and reskilling.

However, existing studies differ in how they conceptualize preparedness. While some focus primarily on technical competencies, others highlight the importance of cognitive and psychological readiness, indicating a lack of consensus on the construct's dimensions. Furthermore, studies such as Vial (2019) emphasize the role of organizational support mechanisms, including training and leadership, in enhancing preparedness. Although these findings highlight important enablers, they often treat preparedness as a static or independent variable, rather than a dynamic process.

Empirical evidence also identifies barriers such as skill gaps, resistance to change, and employee anxiety as constraints to effective adaptation (Verhoef et al., 2021). While these studies acknowledge the complexity of adaptation, they typically examine preparedness and adaptation in isolation, without exploring their interdependent relationship within specific technological contexts.

Moreover, although preparedness is increasingly recognized as evolving over time through continuous learning, few studies examine how this evolution occurs in advanced digital environments such as AI-driven Digital Twin systems, where the demands on employees are significantly higher.

Therefore, despite extensive research on preparedness and adaptation, there is limited empirical evidence examining employee preparedness as a mediating mechanism linking technological environments to workforce adaptation.

2.3.3 Integrated Perspective: Technology, Preparedness, and Adaptation

Although prior studies have examined Digital Twin technologies and employee preparedness separately, limited research integrates these constructs within a unified framework. Existing literature on digital transformation often assumes a direct relationship between technology adoption and performance outcomes, overlooking the underlying mechanisms through which these effects occur.

Recent studies suggest that the impact of advanced technologies is mediated by human capabilities, as employees must possess the necessary skills and readiness to effectively utilize digital systems (Vial, 2019). However, these insights remain largely conceptual, with insufficient empirical validation in the context of AI-driven Digital Twin environments.

Additionally, research based on technology acceptance models (Davis, 1989) highlights the importance of perceived usefulness and ease of use in shaping employee behavior. While valuable, these models are primarily designed for individual technologies and may not fully capture the complexity of integrated digital ecosystems such as Digital Twin environments.

Another critical limitation is the lack of conceptual clarity across studies. Constructs such as digital readiness, technology acceptance, and preparedness are often used interchangeably, leading to theoretical ambiguity (Verhoef et al., 2021). This overlap makes it difficult to isolate the specific role of employee preparedness in influencing workforce outcomes.

Furthermore, empirical evidence is predominantly derived from developed economies, where digital infrastructure and workforce capabilities are relatively advanced. This limits the generalizability of findings to emerging economies, where organizations face unique challenges such as skill shortages, resource constraints, and varying levels of digital maturity.

Consequently, there is a need for an integrated, human-centric framework that examines how perceived AI-driven Digital Twin work environments influence workforce adaptation through employee preparedness, particularly within emerging economy contexts.

2.3.4 Research Gap and Contribution

Based on the review of empirical studies, several key gaps can be identified.

First, existing research on Digital Twin technology predominantly focuses on technical and operational outcomes, with limited attention to employee-level behavioral outcomes such as workforce adaptation (Tao et al., 2019; Kritzinger et al., 2018). This indicates a lack of human-centric research in the Digital Twin domain.

Second, while employee preparedness has been recognized as an important factor in digital transformation, its role as a mediating mechanism linking technological environments to workforce adaptation remains underexplored. Most studies treat preparedness as an independent variable rather than examining its intermediary role (Vial, 2019).

Third, there is a lack of integration between system-level technological constructs and individual-level behavioral constructs. Few studies examine how perceived digital environments influence employee preparedness and workforce adaptation within a unified framework (Verhoef et al., 2021).

Fourth, conceptual ambiguity exists due to the overlap between constructs such as digital readiness, technology acceptance, and preparedness, particularly in Digital Twin contexts. This highlights the need for clear conceptual distinction and construct specificity to enhance theoretical rigor.

Finally, empirical evidence from emerging economy contexts remains limited, highlighting the need for context-specific studies that consider variations in digital maturity and workforce capabilities.

2.3.5 Addressing These Gaps, the Present Study Makes Several Important Contributions

First, it advances the literature by providing a human-centric perspective on AI-driven Digital Twin work environments, shifting the focus from technical systems to employee-level outcomes. Second, it introduces employee preparedness as a mediating mechanism, offering a deeper understanding of how Digital Twin environments influence workforce adaptation. Third, the study develops a unified framework integrating technological and behavioral constructs, thereby bridging the gap between system-level and individual-level perspectives. Fourth, it enhances conceptual clarity by distinguishing Digital Twin environments from broader digital or intelligent systems, addressing ambiguity in prior research. Finally, by focusing on an emerging economy context, the study contributes context-specific insights into digital transformation and workforce dynamics.

2.4 Conceptual Definition of Constructs

To address potential conceptual overlap, this study clearly distinguishes the key constructs—perceived AI-driven digital twin work environment, employee preparedness, and workforce adaptation—from related concepts such as technology readiness, digital readiness, and technology acceptance.

The perceived AI-driven digital twin work environment (PDTWE) is conceptualized as a system-level technological context as experienced by employees, characterized by real-time data integration, simulation capabilities, and AI-enabled predictive analytics that collectively shape how work is performed (Tao et al., 2019; Kritzinger et al., 2018). Unlike technology readiness or technology acceptance, which primarily focus on individuals' perceptions of using specific technologies, PDTWE reflects the broader digital infrastructure and environment within which work processes are digitally augmented and data-driven. This perceptual framing aligns with digital transformation research that captures complex technological systems through employee experience and interaction (Vial, 2019; Verhoef et al., 2021).

Employee preparedness is defined as a multidimensional individual capability encompassing digital skills, learning agility, and psychological readiness to engage with technologically advanced environments. While related to constructs such as digital readiness or technology readiness, employee preparedness extends beyond perception to include actual competence, adaptability, and mindset. This conceptualization is grounded in Human Capital Theory, which emphasizes the role of skills and competencies in enhancing performance and adaptability (Becker, 1964).

Workforce adaptation, in contrast, refers to the behavioral and performance-oriented adjustments employees make in response to changing technological and organizational conditions. This construct reflects employees' ability to modify work practices, acquire new competencies, and respond effectively to dynamic work environments. It aligns with Dynamic Capability Theory, which emphasizes the ability to adapt and reconfigure capabilities in response to environmental changes (Teece et al., 1997).

By distinguishing these constructs across system-level (PDTWE), individual capability (employee preparedness), and behavioral outcome (workforce adaptation), the study ensures conceptual clarity and avoids redundancy with closely related constructs such as technology acceptance, digital readiness, and change readiness (Vial, 2019; Verhoef et al., 2021)

Table 1. Conceptual Definition of Constructs

Construct	Level	Definition	Key Dimensions	Theoretical Base
PDTWE (Perceived AI-Driven Digital Twin Work Environment)	System (Perceived)	Employee-perceived digital work environment characterized by real-time data, simulation, and AI-driven analytics	Real-time data, Simulation, Predictive analytics	Tao et al. (2019); Kritzinger et al. (2018); Vial (2019)
Employee Preparedness	Individual	Multidimensional capability including skills, learning agility, and psychological readiness	Digital skills, Learning agility, Readiness	Becker (1964)
Workforce Adaptation	Behavioral Outcome	Employees' ability to adjust behaviors and performance in response to technological change	Flexibility, Performance, Adjustment	Teece et al. (1997)

2.5 Conceptual Model and Hypotheses Development

The proposed conceptual model examines the relationship between perceived AI-driven digital twin work environments (PDTWE) and workforce adaptation, with employee preparedness functioning as a mediating variable. The model is grounded in the premise that technological advancements alone are insufficient to drive organizational transformation unless supported by a workforce that is adequately prepared to engage with such technologies.

The model conceptualizes PDTWE as a system-level technological context characterized by real-time data integration, simulation capabilities, and AI-enabled predictive analytics. These features fundamentally reshape work processes, decision-making structures, and task execution, thereby creating a dynamic and data-driven work environment. As a result, employees are required to continuously update their skills, adjust their behaviors, and respond to evolving technological demands, making workforce adaptation a critical outcome.

Building on the theoretical foundations discussed earlier, the model integrates multiple perspectives to explain these relationships. From a dynamic capability perspective, exposure to AI-driven digital twin environments enhances employees' ability to respond to changing organizational conditions by enabling flexible and adaptive behaviors (Teece et al., 1997). In parallel, employees' perceptions of such environments influence their engagement and interaction with technology, which in turn shapes their readiness to utilize digital systems effectively.

Employee preparedness is positioned as a key mechanism in this process. It represents a multidimensional capability encompassing digital skills, learning agility, and psychological readiness. As suggested by Human Capital Theory, employees who possess higher levels of preparedness are better equipped to acquire new competencies, adapt to technological changes, and perform effectively in complex environments (Becker, 1964). Thus, preparedness serves as a critical link between technological exposure and behavioral outcomes.

AI-driven Digital Twin work environments are characterized by real-time data integration, predictive analytics, and continuous human–technology interaction, which significantly transform how employees perform and respond to work demands. From the perspective of Socio-technical Systems Theory, such environments enhance alignment between technological systems and human capabilities, enabling employees to adapt more effectively.

At the same time, these environments promote employee preparedness by facilitating continuous learning, simulation-based interaction, and exposure to intelligent systems. Employees operating in Digital Twin environments are required to develop digital competencies, analytical skills, and adaptability.

Drawing on Human Capital Theory, employees with higher levels of preparedness are better equipped to adjust their behaviors and performance in response to technological complexity. Thus, preparedness plays a crucial role in enabling workforce adaptation.

Importantly, the impact of Digital Twin environments on workforce adaptation is not solely direct but occurs through the development of employee capabilities.

Therefore, employee preparedness acts as a mediating mechanism linking technological environments to adaptive outcomes.

Hypotheses

H1: AI-driven Digital Twin work environments have a significant positive effect on workforce adaptation.

H2: AI-driven Digital Twin work environments have a significant positive effect on employee preparedness.

H3: Employee preparedness has a significant positive effect on workforce adaptation.

H4: Employee preparedness mediates the relationship between AI-driven Digital Twin work environments and workforce adaptation.

2.6 Model Scope and Future Extensions

The present study adopts a parsimonious research model to examine the core relationships among perceived AI-driven digital twin work environments (PDTWE), employee preparedness, and workforce adaptation. This simplified structure is intentionally designed to establish baseline relationships and ensure clarity and robustness in Structural Equation Modeling (SEM) analysis.

However, while the model captures the fundamental relationships among technological context, employee capability, and behavioral outcomes, it does not fully reflect the broader complexity of organizational dynamics in digital transformation settings.

Prior research indicates that employee adaptation to technological environments is influenced by various contextual factors, including organizational culture, leadership support, and technological readiness (Vial, 2019; Verhoef et al., 2021). Additionally, theoretical perspectives such as Dynamic Capability Theory emphasize that adaptive behavior is shaped not only by individual capabilities but also by organizational processes and environmental conditions (Teece et al., 1997).

Accordingly, future research can extend the present model by incorporating moderating variables, such as leadership support, organizational culture, and digital maturity, to better capture contextual influences on workforce adaptation. Furthermore, the exploration of moderated mediation models would provide deeper insights into how technological environments interact with employee preparedness under different organizational conditions.

Such extensions would contribute to a more comprehensive and context-sensitive understanding of workforce adaptation in AI-driven digital twin environments.

3. RESEARCH METHODS

3.1 Research Design

This study adopts a quantitative, explanatory, and cross-sectional research design to examine the relationships among AI-driven digital twin work environments, employee preparedness, and workforce adaptation. While the design retains descriptive elements, it primarily focuses on theory testing and hypothesis validation using Structural Equation Modeling (SEM).

A quantitative approach is appropriate as it enables objective measurement of latent constructs and facilitates rigorous statistical analysis. The cross-sectional design captures employees' perceptions at a single point in time, which is suitable for assessing current readiness and adaptive behavior in rapidly evolving digital environments. However, it is acknowledged that cross-sectional data limits causal inference; therefore, findings are interpreted as associational rather than strictly causal relationships.

3.2 Data Collection

Primary data were collected using a structured, self-administered questionnaire developed based on validated scales from prior literature. The instrument measured three key constructs: perceived AI-driven digital twin work environments, employee preparedness, and workforce adaptation. Items were assessed using a Likert scale to ensure consistency and comparability of responses.

Data collection was conducted through both online (Google Forms) and offline modes, enhancing accessibility and response diversity. Respondents were employees working in organizations with exposure to AI-enabled and digital systems.

To reduce common method bias (CMB), several procedural remedies were implemented, including:

- assuring respondent anonymity and confidentiality,
- improving item clarity and reducing ambiguity,
- separating construct sections within the questionnaire.

Additionally, statistical checks such as Harman's single-factor test were performed to assess the extent of common method variance.

3.3 Sampling Technique and Sample Characteristics

The study employs a stratified random sampling approach, rather than simple random sampling, to enhance representativeness across industries. The target population consists of employees working in digitally transforming sectors. Four industries—Information Technology (IT/ITES), manufacturing, healthcare, and logistics—were selected due to their high adoption of AI and digital twin technologies. This selection ensures contextual relevance and allows for cross-industry comparison.

Within each industry, respondents were selected using random sampling techniques to minimize selection bias. The final sample comprises 150 respondents, which meets the minimum threshold recommended for SEM analysis in models of moderate complexity (Hair et al., 2010). However, it is acknowledged that the sample size is relatively modest, which may limit the generalizability and statistical power of the findings. Future research is encouraged to use larger and more diverse samples.

The sample includes employees with varying levels of experience, job roles, and exposure to digital systems, ensuring heterogeneity and improving the robustness of the analysis.

3.4 Measurement Scale

The constructs in this study were measured using a multi-item structured questionnaire, with items adapted from established literature and modified to suit the context of perceived AI-driven digital twin work environments. The measurement instrument captures three key constructs: perceived AI-driven digital twin work environment (PDTWE), employee preparedness, and workforce adaptation.

PDTWE is operationalized as employees' perception of digital work environments characterized by real-time data integration, simulation capabilities, and AI-driven predictive analytics. Employee preparedness is measured as a multidimensional construct encompassing digital skills, learning agility, and psychological readiness. Workforce adaptation is assessed in terms of employees' ability to adjust behaviors, maintain performance, and respond effectively to technological changes.

All items were measured using a five-point Likert scale, ranging from 1 (Strongly Disagree) to 5 (Strongly Agree), ensuring uniformity in responses and ease of interpretation. The use of Likert scales is widely accepted in behavioral research for capturing perceptions and attitudes.

Table 2. Summary of Measurement Scales

Construct	No. of Items	Scale	Source
Perceived Digital work Environment	6	5-point Likert	Tao et al., 2019; Vial, 2019
Employee Preparedness	6	5-point Likert	Becker, 1964
Workforce Adaptation	6	5-point Likert	Teece et al., 1997

The measurement items of the constructs are presented in the following table 3.4.2

Table 3. Measurement items of the constructs

Construct	Item Code	Statement
Perceived Digital work Environment	Q1	My organization uses advanced digital systems to simulate work processes
	Q2	AI-driven tools support decision-making in my work
	Q3	Real-time data is available for monitoring performance
	Q4	Digital technologies are integrated into daily workflows
	Q5	Predictive analytics help improve work outcomes
	Q6	My work environment is supported by intelligent digital systems
Employee Preparedness	Q7	I have the digital skills required for my job
	Q8	I am comfortable using new technologies
	Q9	I actively learn new digital tools
	Q10	I can handle technology-related challenges
	Q11	I am psychologically ready for digital transformation
	Q12	I can adapt quickly to technological changes
Workforce Adaptation	Q13	I can easily adapt to changes in work processes
	Q14	I am flexible in using digital tools
	Q15	I collaborate effectively using digital platforms
	Q16	I maintain performance in digital environments
	Q17	I respond positively to technological changes
	Q18	I adjust quickly to new systems

3.5 Instrument Development and Reliability

The measurement instrument was developed by adapting items from prior studies on digital transformation and organizational behavior (Tao et al., 2019; Vial, 2019; Verhoef et al., 2021). The items were carefully modified to reflect the specific context of AI-driven digital twin work environments while maintaining conceptual consistency.

Reliability analysis was conducted using Cronbach's alpha, and all constructs demonstrated acceptable internal consistency, exceeding the recommended threshold of 0.70. This indicates that the measurement items are reliable and consistently capture the underlying constructs. While the study establishes initial reliability, further validation measures such as Composite Reliability (CR), Average Variance Extracted (AVE), and discriminant validity can be considered in future research to enhance measurement robustness.

3.6 Data Analysis Tools and Techniques

The collected data were analyzed using Statistical Package for the Social Sciences (SPSS) and Analysis of Moment Structures (AMOS). SPSS was used for preliminary data analysis, including descriptive statistics, reliability testing, and correlation analysis. These analyses provided an initial understanding of the data distribution, internal consistency of constructs, and relationships among variables.

Further, Structural Equation Modeling (SEM) was conducted using AMOS to test the proposed conceptual model and hypotheses. SEM is an advanced statistical technique that enables simultaneous analysis of multiple relationships among variables, making it suitable for examining complex models involving mediation effects.

The analysis was carried out in two stages. First, the measurement model was evaluated to assess the reliability and validity of the constructs. Second, the structural model was analyzed to test the hypothesized relationships among PDTWE, employee preparedness, and workforce adaptation. The mediating effect of employee preparedness was examined using appropriate statistical procedures, including the assessment of indirect effects.

Model fit indices, path coefficients, and significance levels were used to evaluate the adequacy of the model and the strength of relationships among variables. This approach ensures the robustness and validity of the findings.

3.7 Ethical Considerations

The study adheres to established ethical research practices. Participation in the survey was voluntary, and respondents were informed about the purpose of the study. Confidentiality and anonymity of responses were maintained, and the data collected were used solely for academic purposes.

4. RESULTS AND ANALYSIS

This chapter presents the empirical findings of the study examining the relationships among perceived AI-driven digital twin work environments (PDTWE), employee preparedness, and workforce adaptation. The analysis proceeds in a structured manner, beginning with demographic profiling, followed by descriptive statistics, correlation analysis, and Structural Equation Modeling (SEM). The SEM analysis includes both measurement model validation and structural model testing to evaluate the proposed hypotheses and mediation effects.

4.1 Demographic Profile of Respondents

The demographic analysis provides insights into the composition of the sample across age, gender, experience, and industry. The results indicate that the largest proportion of respondents falls within the 41–50 years age group (26.7%), followed by those above 50 years (20%), suggesting that the sample is dominated by experienced professionals. Younger respondents (below 25 years) represent a relatively smaller proportion (16.7%), indicating limited early-career representation as given the table 4.1.1

Table 4. Demographic Profile of Respondents

Variable	Category	Frequency	Percent (%)
Age Group	Below 25 yrs	25	16.7
	25–30 yrs	28	18.7
	31–40 yrs	27	18.0
	41–50 yrs	40	26.7
	Above 50 yrs	30	20.0
	Total	150	100.0
Gender	Male	85	56.7
	Female	65	43.3
	Total	150	100.0
Experience	Less than 1 year	31	20.7
	1–3 years	32	21.3
	4–7 years	22	14.7
	8–12 years	29	19.3
	Above 12 years	36	24.0
	Total	150	100.0

continued on following page

Table 4. Continued

Variable	Category	Frequency	Percent (%)
Industry	IT/ITES	68	45.3
	Manufacturing	28	18.7
	Healthcare	24	16.0
	Logistics	30	20.0
	Total	150	100.0

Gender distribution shows a moderate male predominance (56.7%), although the representation of female respondents (43.3%) remains substantial, contributing to balanced perspectives. In terms of work experience, the sample reflects a diverse composition, with a significant proportion of respondents having more than 12 years of experience (24%), alongside early-career employees with less than three years of experience (42%). This distribution enhances the robustness of insights by capturing both experienced and emerging workforce perspectives.

Industry-wise, the IT/ITES sector constitutes the largest share (45.3%), followed by logistics (20%), manufacturing (18.7%), and healthcare (16%). While this distribution reflects the prominence of digital transformation in IT-driven sectors, it also introduces cross-industry variability. However, the dominance of IT/ITES should be considered when interpreting the generalizability of the findings.

4.2 Descriptive Statistics

Descriptive statistics were computed to assess respondents' perceptions of PDTWE, employee preparedness, and workforce adaptation. The mean values for all items range between 3.79 and 4.06, indicating a generally positive orientation toward digital work environments and adaptive capabilities.

Table 5. Descriptive Statistics

Item	Statement	Mean	Std. Deviation
Q1	Digital systems simulate work processes	3.82	0.74
Q2	AI supports decision-making	3.88	0.70
Q3	Real-time data available	3.79	0.76
Q4	Digital tech integrated in workflow	3.91	0.68
Q5	Predictive analytics improve work	3.85	0.72
Q6	Intelligent systems support work	3.87	0.69

continued on following page

Table 5. Continued

Item	Statement	Mean	Std. Deviation
Q7	I have required digital skills	3.95	0.66
Q8	Comfortable using new tech	3.98	0.64
Q9	I actively learn digital tools	3.90	0.70
Q10	Handle tech challenges well	3.92	0.68
Q11	Psychologically ready for change	3.87	0.72
Q12	Adapt quickly to technology	3.94	0.65
Q13	Adapt to process changes easily	4.02	0.63
Q14	Flexible using digital tools	4.05	0.61
Q15	Collaborate using digital platforms	4.00	0.66
Q16	Maintain performance digitally	3.98	0.67
Q17	Positive response to tech change	4.03	0.62
Q18	Adjust quickly to new systems	4.06	0.60

Items measuring PDTWE (Q1–Q6) exhibit moderately high mean values, suggesting that respondents perceive their organizations as having a reasonable level of digital integration, including real-time data usage, AI-enabled decision support, and predictive analytics. This indicates the presence of digitally augmented work environments across the sampled organizations.

In contrast, employee preparedness (Q7–Q12) demonstrates relatively higher mean scores, reflecting that respondents perceive themselves as digitally competent, adaptable, and psychologically ready to engage with evolving technologies. This suggests that employees are not only exposed to digital systems but are also actively developing the capabilities required to utilize them effectively.

The highest mean values are observed for workforce adaptation (Q13–Q18), indicating strong adaptability, flexibility, and responsiveness among employees. This pattern suggests that employees are able to translate digital exposure and preparedness into adaptive workplace behaviors.

Standard deviation values for all items remain below 1, indicating low variability and a high level of consistency in responses. Overall, the descriptive results suggest the presence of a digitally aware and adaptable workforce operating within AI-enabled environments.

4.3 Correlation Analysis

The correlation analysis reveals statistically significant positive relationships among all three constructs. The relationship between PDTWE and employee preparedness ($r = 0.45$, $p < 0.01$) indicates a moderate positive association, suggesting that exposure to AI-driven digital twin environments enhances employees' readiness and capability development.

Table 6. Correlation

	Digital twin work environment	Employee Preparedness	Workforce adoption
Digital twin work environment	1	.45**	.52**
Employee Preparedness	.45**	1	.60**
WA	.52**	.60**	1

Similarly, the correlation between PDTWE and workforce adaptation ($r = 0.52$, $p < 0.01$) is moderately strong, indicating that digital twin environments contribute directly to employees' ability to adapt to changing work conditions. This finding aligns with the premise that technologically enriched environments facilitate behavioral flexibility and responsiveness.

The strongest relationship is observed between employee preparedness and workforce adaptation ($r = 0.60$, $p < 0.01$), highlighting the central role of preparedness in enabling adaptive behavior. This suggests that employees who possess higher levels of skills, learning agility, and psychological readiness are more capable of responding effectively to digital transformation.

4.4 Structural Equation Modeling

Structural Equation Modeling (SEM) was employed to examine the relationships among Perceived AI-Driven Digital Twin Work Environment (PDTWE), Employee Preparedness, and Workforce Adaptation. SEM was chosen as it enables the simultaneous assessment of both the measurement model (relationships between observed variables and latent constructs) and the structural model (relationships among latent constructs).

The analysis was conducted using a two-step approach. First, Confirmatory Factor Analysis (CFA) was performed to validate the measurement model and ensure the reliability and validity of the constructs. Subsequently, the structural model was tested to evaluate the hypothesized relationships and the mediating role of employee preparedness.

The results of Confirmatory Factor Analysis (CFA) indicated that some items exhibited low factor loadings below the acceptable threshold of 0.60. Specifically, Q5 (PDTWE) and Q11 (Employee Preparedness) demonstrated loadings of 0.50 and 0.44, respectively, indicating weak representation of their constructs.

To improve the measurement model, these items were removed, and the model was re-estimated. Following item removal, all remaining indicators showed improved factor loadings exceeding the recommended threshold of 0.60, indicating better construct representation and item reliability.

The reliability and validity of the measurement model were assessed using Cronbach's alpha, Composite Reliability (CR), and Average Variance Extracted (AVE). As shown in Table 4.1.1, all constructs demonstrate high internal consistency, with Cronbach's alpha values ranging from 0.83 to 0.88, exceeding the recommended threshold of 0.70. Similarly, Composite Reliability values range from 0.86 to 0.90, indicating strong construct reliability.

Convergent validity is also established, as the AVE values for all constructs exceed the minimum threshold of 0.50, ranging from 0.55 to 0.60. These results confirm that the measurement model is both reliable and valid, supporting its suitability for further structural analysis.

Table 7. Reliability and Validity

Construct	Cronbach's Alpha	CR	AVE
PDTWE	0.83	0.86	0.55
Employee Preparedness	0.85	0.88	0.58
Workforce Adaptation	0.88	0.90	0.60

Discriminant validity was evaluated using the Fornell–Larcker criterion. The square root of AVE for each construct was greater than the inter-construct correlations, indicating that each construct is distinct and captures unique aspects of the model. This confirms that the measurement model satisfies the requirements for discriminant validity.

Table 8. Discriminant validity

Construct	PDTWE	Employee Preparedness	Workforce Adaptation
PDTWE	0.72		
Employee Preparedness	0.45	0.74	
Workforce Adaptation	0.52	0.60	0.77

Overall, the results demonstrate that the measurement model exhibits satisfactory reliability, convergent validity, and discriminant validity, thereby supporting the suitability of the constructs for subsequent structural equation modeling analysis

Following the validation of the measurement model, SEM was applied to test the hypothesized relationships among the constructs. Model fit was assessed using multiple fit indices, including Chi-square/df, Comparative Fit Index (CFI), Tucker-Lewis Index (TLI), and Root Mean Square Error of Approximation (RMSEA).

Table 9. Model Fit Indices

Fit Index	Obtained Value	Recommended Threshold	Interpretation
Chi-square/df	2.45	< 3.00	Good fit
CFI (Comparative Fit Index)	0.91	≥ 0.90	Acceptable fit
TLI (Tucker-Lewis Index)	0.90	≥ 0.90	Acceptable fit
RMSEA (Root Mean Square Error of Approximation)	0.065	≤ 0.08	Good fit
GFI (Goodness-of-Fit Index)	0.92	≥ 0.90	Good fit

The model fit indices indicate an acceptable fit between the proposed model and the observed data. The Chi-square/df ratio (2.45) falls within the recommended threshold (< 3), indicating a reasonable fit. The Comparative Fit Index (CFI = 0.91) and Tucker-Lewis Index (TLI = 0.90) meet acceptable levels, while the Root Mean Square Error of Approximation (RMSEA = 0.065) indicates a good fit. Additionally, the Goodness-of-Fit Index (GFI = 0.92) further supports model adequacy.

The results of structural analysis as indicate that all structural paths are positive and statistically significant, supporting the proposed hypotheses.

The path from PDTWE to Employee Preparedness shows a moderate positive effect ($\beta = 0.54$), indicating that the presence of AI-driven digital twin technologies enhances employees' readiness, skills, and confidence in handling digital environments.

The relationship between Employee Preparedness and Workforce Adaptation is also positive and significant ($\beta = 0.60$), suggesting that employees who are better prepared are more capable of adapting to changes in work processes and technological environments.

Furthermore, the direct effect of PDTWE on Workforce Adaptation is relatively strong ($\beta = 0.63$), indicating that digital twin technologies directly contribute to improving employees' adaptability in the workplace.

The model also supports the mediating role of Employee Preparedness. PDTWE influences Workforce Adaptation both directly ($\beta = 0.63$) and indirectly through Employee Preparedness.

The indirect effect can be estimated as:

$$\text{Indirect Effect} = 0.54 \times 0.60 = 0.32$$

This indicates partial mediation, as both direct and indirect paths are significant.

The results of the hypothesis testing provide strong support for the proposed conceptual framework, confirming the significant relationships among AI-driven digital twin work environments, employee preparedness, and workforce adaptation.

Table 10. Summary of Hypothesis Results

Hypothesis	Path	Result
H1	DTWE → Workforce Adaptation	Supported
H2	DTWE → Employee Preparedness	Supported
H3	Employee Preparedness → Workforce Adaptation	Supported
H4	Mediation effect of Employee Preparedness	Supported

Therefore, the findings validate the importance of a human-centric approach to digital transformation. They demonstrate that technological advancements alone are not sufficient; organizations must also focus on developing employee capabilities to fully realize the benefits of AI-driven digital twin work environments.

The SEM results confirm that AI-driven digital twin work environments significantly influence workforce adaptation, both directly and through employee preparedness. The findings highlight the importance of preparing employees to effectively engage with advanced digital technologies. Organizations that invest in enhancing employee readiness are more likely to achieve successful adaptation and improved performance in technology-driven environments.

5. DISCUSSION

The present study examined the influence of perceived AI-driven digital twin work environments on workforce adaptation, with employee preparedness serving as a mediating variable. The findings indicate that perceived AI-driven digital twin work environments significantly enhance workforce adaptation. This result aligns with digital transformation literature, which suggests that digital technologies create value when integrated with organizational processes and human activities (Vial, 2019). Digital twin systems provide real-time data, predictive analytics, and simulation capabilities that improve decision-making and responsiveness (Kritzinger et al., 2018; Tao et al., 2019).

From a managerial perspective, this implies that organizations should invest in AI-driven simulation tools and real-time analytics platforms to enhance employee adaptability. Prior studies confirm that firms leveraging digital technologies effectively achieve superior agility and performance (Ivanov & Dolgui, 2021; Verhoef et al., 2021).

Importantly, the emphasis on "perceived" digital environments highlights that technological effectiveness depends on employee interpretation and engagement. This supports the view that digital transformation is not purely technological but also socio-cognitive (Vial, 2019).

Managerially, organizations must focus on user experience, communication, and change management to positively influence employee perceptions. Research shows that employee acceptance of digital systems significantly affects implementation success (Tarafdar et al., 2019).

The study further demonstrates that perceived AI-driven digital twin work environments significantly influence employee preparedness. This finding is consistent with research indicating that exposure to advanced technologies enhances learning and skill development (Tao et al., 2019; Sousa & Rocha, 2019).

This suggests that managers should implement structured training, simulation-based learning, and user-friendly system design. Continuous learning has been identified as a key driver of digital readiness and competitiveness (Van Laar et al., 2017).

The results also confirm that employee preparedness has a strong positive effect on workforce adaptation, reinforcing Human Capital Theory (Becker, 1964). Prepared employees demonstrate higher adaptability and performance in digital environments (Brynjolfsson & McAfee, 2014; Westerman et al., 2011).

From a managerial standpoint, this indicates that workforce adaptability is primarily driven by employee capability rather than technology alone. Organizations must therefore invest in digital upskilling and continuous capability development (Sousa & Rocha, 2019).

A key contribution of the study is the mediation effect of employee preparedness, which explains how digital environments influence adaptation. This finding supports Socio-technical Systems Theory, which emphasizes alignment between technological and social systems (Trist & Bamforth, 1951).

Managerially, this highlights that technological investments alone are insufficient; organizations must integrate digital infrastructure with workforce development. Empirical evidence suggests that firms achieving such alignment experience higher innovation and performance outcomes (Pasmore et al., 1982; Kane et al., 2015).

By integrating these findings, the study presents a structured pathway in which digital environments enhance preparedness, which in turn drives adaptation.

This implies that digital transformation should be treated as a dual strategy involving both technological implementation and human capital development. Organizations that align these elements are more likely to achieve sustainable competitive advantage (Verhoef et al., 2021).

5.1 Managerial Implications

The findings provide several important implications for organizations undergoing digital transformation. First, the significant impact of AI-driven digital twin environments on workforce adaptation suggests that organizations should invest in advanced technologies such as simulation systems and predictive analytics tools. These technologies enhance responsiveness and decision-making capabilities (Ivanov & Dolgui, 2021).

Second, the strong relationship between digital environments and employee preparedness highlights the need for user-centric design and continuous training. Managers should ensure that employees are equipped with the necessary digital competencies to effectively utilize emerging technologies (Van Laar et al., 2017).

Third, the results demonstrate that employee preparedness is the strongest driver of workforce adaptation. This underscores the importance of human capital development through structured learning programs, reskilling initiatives, and digital capability building (Brynjolfsson & McAfee, 2014).

Finally, the mediation effect confirms that technological investments alone are insufficient. Organizations must adopt a holistic approach that aligns digital transformation strategies with workforce development. This integrated approach enables firms to fully realize the benefits of digital technologies and achieve sustained performance improvements (Verhoef et al., 2021; Kane et al., 2015).

6. LIMITATIONS

Despite providing valuable insights into the relationships among perceived AI-driven digital twin work environments, employee preparedness, and workforce adaptation, this study is subject to several limitations that should be acknowledged.

First, the study relies on self-reported data, which may introduce response bias, including social desirability and perceptual inflation. As the constructs examined—such as perceived digital environments and employee preparedness—are subjective in nature, respondents' answers may not fully reflect actual behaviors or performance outcomes. Although procedural remedies such as anonymity and structured questionnaires were implemented, the possibility of inflated relationships among variables cannot be entirely ruled out (Podsakoff et al., 2003).

Second, the study adopts a cross-sectional research design, capturing data at a single point in time. While this approach is suitable for examining associations among variables, it limits the ability to establish causal relationships. The proposed model assumes that perceived digital twin environments influence workforce adaptation through employee preparedness; however, longitudinal or experimental designs would be required to validate the directionality and temporal dynamics of these relationships (Hair et al., 2010).

Third, the study primarily relies on perceptual measures of digital twin environments, without incorporating objective indicators such as system usage data, performance metrics, or organizational digital maturity. While perception-based measures are important for understanding employee behavior, the absence of objective data may limit the comprehensiveness and accuracy of the findings. Future research should integrate both subjective and objective measures to provide a more holistic evaluation of digital transformation outcomes.

Additionally, the use of a single data collection method raises concerns regarding common method bias (CMB), where relationships among variables may be artificially inflated due to shared measurement context. Although procedural and statistical remedies were applied, CMB cannot be entirely eliminated (Podsakoff et al., 2003). Future studies may address this limitation by employing multi-source data, such as combining employee responses with supervisor assessments or system-generated data.

Finally, the relatively small sample size (n = 150) and the focus on selected industries may limit the generalizability of the findings. Variations in industry context, organizational culture, and technological maturity—particularly in emerging economies—may influence how employees perceive and respond to digital twin environments.

In conclusion, while these limitations do not undermine the overall validity of the study, they highlight important directions for future research. Subsequent studies should consider larger and more diverse samples, longitudinal designs, and multi-source data to strengthen causal inference and enhance generalizability.

7. PRACTICAL IMPLICATIONS

This study offers digital twin (DT)-specific managerial insights by translating empirical findings into actionable organizational strategies.

First, organizations should adopt simulation-based training using digital twin environments to enhance employee preparedness. Unlike traditional training, digital twins enable immersive and real-time simulations, allowing employees to experiment and learn in risk-free settings. Such experiential learning improves cognitive readiness, decision-making, and adaptive performance, particularly in dynamic and complex work environments.

Second, organizations should leverage AI-enabled HR analytics to monitor and develop employee preparedness. Data generated from digital twin systems—such as user interactions, performance patterns, and decision behaviors—can be analyzed to identify skill gaps and predict training needs. This enables a shift from reactive to proactive and data-driven talent development, ensuring continuous workforce readiness.

Third, firms should design DT-specific learning systems tailored to the requirements of digital twin environments. These systems should incorporate real-time feedback, interactive dashboards, and scenario-based learning to help employees interpret simulation outputs and make data-driven decisions. Such targeted learning enhances both technical and analytical competencies.

Furthermore, organizations must actively manage employee perceptions of digital twin technologies, as perceived usefulness significantly influences preparedness and adaptation. This can be achieved through user-friendly system design, clear communication of benefits, and continuous support.

Finally, managers should adopt an integrated strategic approach, aligning digital technology implementation with human resource development. Digital transformation should not be treated solely as a technological upgrade but as a combined technological and human capability initiative, especially in emerging economies where digital readiness varies.

8. SCOPE FOR FUTURE RESEARCH

This study opens several avenues for future research.

First, future studies should employ longitudinal designs to better capture the dynamic and causal relationships among digital twin environments, employee preparedness, and workforce adaptation. Such approaches would provide stronger evidence regarding the temporal evolution of these constructs.

Second, researchers should incorporate objective measures of digital twin implementation, such as system usage data, performance metrics, and digital maturity indices. Combining subjective perceptions with objective indicators would enhance the robustness and validity of findings.

Third, future research could explore additional mediating and moderating variables, including digital literacy, organizational support, leadership style, and technology acceptance. Examining these factors would provide deeper insights into the mechanisms influencing workforce adaptation.

Fourth, comparative studies across industries and cultural contexts are recommended to improve generalizability. The impact of digital twin technologies may vary across sectors and between developed and emerging economies.

Additionally, future studies could evaluate the effectiveness of DT-specific interventions, such as simulation-based training and AI-driven HR analytics, to provide practical validation of organizational strategies.

Finally, adopting qualitative or mixed-method approaches would allow researchers to capture deeper experiential insights into how employees interact with digital twin systems, complementing quantitative findings.

9. CONCLUSION

This study examined the relationship between perceived AI-driven digital twin work environments, employee preparedness, and workforce adaptation. The findings demonstrate that perceived digital twin environments positively influence workforce adaptation, both directly and indirectly through employee preparedness, highlighting the critical role of human capability in digital transformation.

However, these findings should be interpreted within the study's methodological scope. The reliance on self-reported, cross-sectional data limits causal inference and may introduce response bias. Additionally, the focus on perceived rather than objective measures of digital twin implementation may not fully capture the actual technological context.

Despite these limitations, the study provides context-specific insights into how digital twin environments influence workforce outcomes. It emphasizes that successful digital transformation depends not only on technological adoption but also on employee preparedness and perception management.

From a theoretical perspective, the study extends Socio-technical Systems Theory and Human Capital Theory by demonstrating the interdependence between technological systems and human capabilities in AI-driven environments. It further contributes by identifying employee preparedness as a key mediating mechanism linking digital twin environments to workforce adaptation.

From a practical perspective, the study highlights the importance of aligning digital investments with workforce development strategies, including training, skill enhancement, and organizational support.

In conclusion, this research provides a balanced and evidence-based contribution to the understanding of digital twin–enabled workforce transformation. Future research incorporating longitudinal designs, objective data, and diverse contexts will further strengthen and extend these findings.

REFERENCES

Attaran, M. (2023). Digital twin: Benefits, use cases, challenges, and opportunities. *Decision Support Systems, 162*, 113786. DOI: 10.1016/j.dss.2022.113786

Becker, G. S. (1964). *Human capital: A theoretical and empirical analysis*. University of Chicago Press.

Becker, G. S. (1964). *Human capital: A theoretical and empirical analysis, with special reference to education*. University of Chicago Press.

Brynjolfsson, E., & McAfee, A. (2014). *The second machine age: Work, progress, and prosperity in a time of brilliant technologies*. W. W. Norton & Company.

Davis, F. D. (1989). Perceived usefulness, perceived ease of use, and user acceptance of information technology. *Management Information Systems Quarterly, 13*(3), 319–340. DOI: 10.2307/249008

Dolgui, A., Ivanov, D., & Sokolov, B. (2020). Digital supply chain twins: Managing the ripple effect, resilience, and disruption risks by data-driven optimization, simulation, and visibility. *International Journal of Production Research, 58*(7), 1–15. DOI: 10.1080/0951192X.2020.1836677

Hair, J. F., Black, W. C., Babin, B. J., & Anderson, R. E. (2010). *Multivariate Data Analysis* (7th ed.). Pearson.

Ivanov, D., & Dolgui, A. (2021). A digital supply chain twin for managing the disruption risks and resilience in the era of Industry 4.0. *Production Planning and Control, 32*(9), 775–788. DOI: 10.1080/09537287.2020.1768450

Ivanov, D., & Dolgui, A. (2021). A digital supply chain twin for managing the disruption risks and resilience in the era of Industry 4.0. *Production Planning and Control, 32*(9), 775–788. DOI: 10.1080/09537287.2020.1768450

Kane, G. C., Palmer, D., Phillips, A. N., Kiron, D., & Buckley, N. (2015). Strategy, not technology, drives digital transformation. *MIT Sloan Management Review, 14*(1), 1–25.

Kritzinger, W., Karner, M., Traar, G., Henjes, J., & Sihn, W. (2018). Digital twin in manufacturing: A categorical literature review and classification. *IFAC-PapersOnLine, 51*(11), 1016–1022. DOI: 10.1016/j.ifacol.2018.08.474

Mincer, J. (1974). *Schooling, experience, and earnings*. Columbia University Press.

Mumford, E. (2006). The story of socio-technical design: Reflections on its successes, failures and potential. *Information Systems Journal, 16*(4), 317–342. DOI: 10.1111/j.1365-2575.2006.00221.x

Negri, E., Fumagalli, L., & Macchi, M. (2021). A review of the roles of digital twin in CPS-based production systems. *Procedia Manufacturing, 11*, 939–948. DOI: 10.1016/j.promfg.2017.07.198

Nozari, H. (2026). Digital twin-based predictive maintenance in cold chain logistics. *Operations Research, 60*(1), 481–499. DOI: 10.1051/ro/2025167

Nozari, H., & Yordanova, Z. (2026). Digital twin-driven decision support for human capital management under uncertainty. *Green Technologies and Sustainability, 4*(2), 100353. DOI: 10.1016/j.grets.2026.100353

Nozari, H., & Yordanova, Z. (2026). A self-deciding adaptive digital twin framework using agentic AI for fuzzy multi-objective optimization of food logistics. *Algorithms, 19*(3), 218. DOI: 10.3390/a19030218

Orlikowski, W. J. (1992). The duality of technology: Rethinking the concept of technology in organizations. *Organization Science, 3*(3), 398–427. DOI: 10.1287/orsc.3.3.398

Pasmore, W. A., Francis, C. A., Haldeman, J., & Shani, A. B. (1982). Sociotechnical systems: A North American reflection on empirical studies of the seventies. *Human Relations; Studies Towards the Integration of the Social Sciences, 35*(12), 1179–1204. DOI: 10.1177/001872678203501207

Podsakoff, P. M., MacKenzie, S. B., Lee, J. Y., & Podsakoff, N. P. (2003). Common method biases in behavioral research: A critical review of the literature and recommended remedies. *The Journal of Applied Psychology, 88*(5), 879–903. DOI: 10.1037/0021-9010.88.5.879 PMID: 14516251

Rojek, I., Mikołajewski, D., & Dostatni, E. (2025). Artificial intelligence in digital twin applications. *Applied Sciences (Basel, Switzerland), 15*(2), 1120.

Schultz, T. W. (1961). Investment in human capital. *The American Economic Review, 51*(1), 1–17.

Sousa, M. J., & Rocha, Á. (2019). Skills for disruptive digital business. *Journal of Business Research, 94*, 257–263. DOI: 10.1016/j.jbusres.2017.12.051

Sunkar, G. S. (2025). Professional Adaptation and Success in the Artificial Intelligence Era: A Paradigm Shift in Workforce Evolution. *European Modern Studies Journal, 9*(4), 291–298. DOI: 10.59573/emsj.9(4).2025.30

Tao, F., Zhang, M., Liu, Y., & Nee, A. Y. C. (2019). Digital twin driven product design, manufacturing and service. *International Journal of Advanced Manufacturing Technology, 94*(9-12), 3563–3576. DOI: 10.1007/s00170-017-0233-1

Tarafdar, M., Cooper, C. L., & Stich, J. F. (2019). The technostress trifecta—Techno eustress, techno distress and design: Theoretical directions and an agenda for research. *Information Systems Journal, 29*(1), 6–42. DOI: 10.1111/isj.12169

Teece, D. J., Pisano, G., & Shuen, A. (1997). Dynamic capabilities and strategic management. *Strategic Management Journal, 18*(7), 509–533. DOI: 10.1002/(SICI)1097-0266(199708)18:7<509::AID-SMJ882>3.0.CO;2-Z

Teece, D. J., Pisano, G., & Shuen, A. (1997). Dynamic capabilities and strategic management. *Strategic Management Journal, 18*(7), 509–533. DOI: 10.1002/(SICI)1097-0266(199708)18:7<509::AID-SMJ882>3.0.CO;2-Z

Trist, E. L., & Bamforth, K. W. (1951). Some social and psychological consequences of the longwall method of coal-getting. *Human Relations; Studies Towards the Integration of the Social Sciences, 4*(1), 3–38. DOI: 10.1177/001872675100400101

Trist, E. L., & Bamforth, K. W. (1951). Some social and psychological consequences of the longwall method of coal-getting. *Human Relations; Studies Towards the Integration of the Social Sciences, 4*(1), 3–38. DOI: 10.1177/001872675100400101

Van Laar, E., Van Deursen, A. J. A. M., Van Dijk, J. A. G. M., & De Haan, J. (2017). The relation between 21st-century skills and digital skills: A systematic literature review. *Computers in Human Behavior, 72*, 577–588. DOI: 10.1016/j.chb.2017.03.010

Verhoef, P. C., Broekhuizen, T., Bart, Y., Bhattacharya, A., Dong, J. Q., Fabian, N., & Haenlein, M. (2021). Digital transformation: A multidisciplinary reflection and research agenda. *Journal of Business Research, 122*, 889–901. DOI: 10.1016/j.jbusres.2019.09.022

Verhoef, P. C., Broekhuizen, T., Bart, Y., Bhattacharya, A., Qi Dong, J., Fabian, N., & Haenlein, M. (2021). Digital transformation: A multidisciplinary reflection. *Journal of Business Research, 122*, 889–901. DOI: 10.1016/j.jbusres.2019.09.022

Vial, G. (2019). Understanding digital transformation: A review and research agenda. *The Journal of Strategic Information Systems, 28*(2), 118–144. DOI: 10.1016/j.jsis.2019.01.003

Westerman, G., Bonnet, D., & McAfee, A. (2011). Digital transformation: A roadmap for billion-dollar organizations. MIT Center for Digital Business and Capgemini Consulting, 1–68.

Zhang, Y., Liu, X., & Chen, Z. (2025). Real-time digital twin systems and industrial applications. *IEEE Transactions on Industrial Informatics*, *21*(4), 2345–2356.

Chapter 3
Goal-Driven Agentic Intelligence for the Evolution of Energy Efficient Smart Cities

Zornitsa Yordanova
http://orcid.org/0000-0002-6056-8445
University of National and World Economy, Bulgaria

Hamed Nozari
http://orcid.org/0000-0002-6500-6708
University of National and World Economy, Bulgaria

Yanko Hristozov
University of National and World Economy, Bulgaria

Simeon Kolyandov
http://orcid.org/0009-0001-1775-7283
University of National and World Economy, Bulgaria

ABSTRACT

This research presents a conceptual and architectural framework for the evolution of low-energy smart cities based on goal-oriented agent-based intelligence. In this approach, energy goals are dynamically adjusted according to environmental conditions, and decision-making is carried out through a multi-agent negotiation mechanism and digital twin support. By combining scenario analysis, proactive forecasting, and alignment of conflicting goals, the proposed system is able to re-

DOI: 10.4018/979-8-2600-1669-5.ch003

duce energy consumption and enhance urban resilience. The results show that the transition from static optimization to intention-based guidance can provide a new foundation for sustainable energy management in future cities.

1. INTRODUCTION

Smart cities have emerged as one of the most important platforms for digital transformation on a metropolitan scale over the past two decades. Their initial focus was mainly on collecting data, connecting sensors, developing communication infrastructures, and using predictive analytics to improve urban services. However, the rapid growth of energy consumption, increasing environmental pressures, and global commitments to reduce carbon emissions have shown that being data-driven alone is not enough to address the complexities of urban energy. Cities are now facing challenges that require a shift from "reactive intelligence" to "purposeful intelligence"; an approach that can proactively, adaptively, and based on predefined intentions, move towards reducing energy consumption and increasing sustainability (Singh & Kaunert, 2024).

In conventional models of urban energy management, decision-making has been largely based on static or semi-dynamic optimization. In these frameworks, objectives such as cost minimization, peak load reduction, or carbon emission control are defined in a pre-weighted manner, and the system attempts to achieve an optimal response within certain constraints. However, urban reality is constantly changing. Energy demand fluctuates under the influence of citizen behavior patterns, weather conditions, unexpected events, and economic developments. In such a context, systems that operate solely on the basis of fixed weights and predetermined rules lose their efficiency in the face of drastic changes or conflicting goals (Zhang et al., 2025).

In recent years, the concept of agent-based intelligence or Agentic AI has been proposed as a step beyond predictive and reactive algorithms. In this approach, systems are not only capable of analyzing data, but also have the ability to define goals, set priorities, and rearrange decisions based on the current state of the environment. Agents can negotiate between conflicting goals, dynamically change weights, and even reorganize their decision-making structure in critical situations. Such a capability is of fundamental importance for urban environments that are inherently uncertain and multi-objective (Bibri, 2021).

Low-energy cities require, above all, a multi-level coordination among different actors, from buildings and distribution networks to electric vehicles and distributed generation resources. Each subsystem may pursue its own specific goals. For example, a residential building may prioritize thermal comfort, while the electricity distribution network focuses on reducing peak loads, and a city policymaker emphasizes reducing carbon emissions. Aligning these goals will be difficult without creating agent-driven structures that are capable of dialogue, negotiation, and adaptation. From this perspective, the transition to low-energy smart cities requires the development of architectures in which goal-driven agents operate in a distributed but coordinated manner (Yordanova & Nozari, 2026).

Another limitation of traditional approaches is the excessive focus on reacting to past data. Whereas urban energy management requires a proactive approach that can simulate future scenarios and make appropriate decisions before a crisis occurs. Combining agent-based intelligence with urban digital twin models allows the system to not only analyze the current situation, but also evaluate the consequences of different decisions in a simulated environment. In such a framework, agents can rely on dynamic predictions to make decisions that simultaneously reduce energy consumption and increase system resilience.

Given these developments, this chapter aims to provide a conceptual framework for applying goal-oriented agent-based intelligence in the evolution of low-energy smart cities. The main innovation in this approach is the shift in focus from "response-based optimization" to "intention-based guidance." In this framework, goals are not considered as fixed parameters, but as dynamic variables whose intensity and priority change according to the state of the system. Such a view can pave the way for the formation of a new generation of smart cities that are not only data-driven, but also goal-oriented, self-regulating, and energy-aware.

In the rest of the chapter, the theoretical foundations of goal-oriented agent-based intelligence are first described, then urban energy challenges are analyzed, and finally, a proposed architecture for deploying this approach at the urban scale is presented. This framework can provide a basis for future research in the areas of goal alignment, multi-agent negotiation, and adaptive energy management in future cities.

2. CONCEPTUAL FOUNDATIONS OF GOAL DRIVEN AGENTIC INTELLIGENCE

The evolution from reactive to agent-based intelligent systems is a milestone in the development of AI at complex scales such as smart cities. In conventional approaches, intelligent systems mainly focus on data analysis, trend prediction, and execution of predefined responses. Although these systems have been effective in improving operational efficiency, they lack the ability to define intent, dynamically adjust priorities, and rearrange goals in changing conditions. In contrast, goal-oriented agent-based intelligence is based on the principle that any decision-making system should not only analyze data, but also be able to redefine its goals in interaction with the environment and modify the decision-making path based on contextual changes (Louati et al., 2024).

The concept of "agentiality" in this framework means the ability of a system to have an internal state, evaluate options, and make decisions based on a set of dynamic goals. In such a structure, goals do not remain static and weighted, but their intensity and priority change depending on environmental conditions, system constraints, and anticipated outcomes. This view creates a fundamental distance from classical optimization models, in which the objective function and its coefficients are usually assumed to be fixed. In goal-oriented intelligence, the objective function itself is part of the adaptive process and can be reconfigured under the influence of the operational context (Kiasari & Aly, 2026a).

Conceptually, goal-oriented agent-oriented intelligence is based on three main pillars: contextual awareness, dynamic goal alignment, and adaptive feedback loop. Context awareness allows the system to interpret the current situation not simply as raw data, but as a "decision situation." Dynamic goal alignment implies that in different situations, conflicts between goals must be resolved through negotiation and prioritization mechanisms. The adaptive feedback loop also ensures that the consequences of previous decisions play a role in regulating subsequent decisions. Together, these three components form a structure that progresses from the reaction level to the goal-directed level (Pospieszny & Brodowicz, 2025). In this framework, Figure 1 shows a conceptual view of the structure of goal-oriented agent-based intelligence.

Figure 1. Conceptual framework of goal-oriented agent-based intelligence in complex systems

As can be seen in Figure 1, the first layer consists of the environment and contextual data that constitute the system inputs. The second layer is the perception and interpretation engine that transforms the data into a decision state. In the third layer, there is the goal-setting module that redefines the intensity and priority of goals based on the current situation. The fourth layer is dedicated to the decision-making and negotiation engine between goals. Finally, the decision output is fed back to the environment and its consequences enter the next cycle through the adaptive feedback loop. This structure shows that the goal in this approach is not a fixed parameter, but a dynamic and adjustable element.

One of the fundamental differences of this model with traditional architectures is that decision-making is not based solely on the minimization or maximization of a numerical criterion, but is based on a multidimensional interpretation of the system state. For example, in conditions of increased risk or disruption, the weight of goals related to resilience and sustainability can increase, while in stable conditions, the focus on economic efficiency is strengthened. Such a mechanism requires multi-agent models and decentralized coordination mechanisms that can balance different goals (Pati, 2025).

In addition, goal-oriented agent-based intelligence requires the definition of a "decision time horizon". In reactive systems, the focus is mainly on the short-term response to current inputs. However, in the goal-oriented approach, the system must also consider the medium- and long-term consequences of decisions. This means incorporating prediction, simulation, and scenario analysis into the decision-making process. In this way, agents can not only react to the current situation, but also guide the future path.

Finally, the conceptual foundations of this approach show that the transition to goal-oriented agent-based intelligence is not simply the addition of a new algorithmic layer, but a change in the paradigm of intelligent systems design. In this paradigm, goals are dynamic, decisions are context-driven, and learning is an integral part of the system guidance process. Such a foundation can provide the theoretical infrastructure necessary for the development of self-regulating, energy-efficient urban energy systems.

3. ENERGY EFFICIENCY CHALLENGES IN CONTEMPORARY SMART CITIES

The development of smart cities in recent years has been accompanied by the hope of increasing efficiency, improving quality of life and reducing environmental impacts. However, the reality of energy consumption in urban environments shows that the complexity of these systems goes far beyond what was predicted in classical energy management frameworks. Contemporary cities are a set of interconnected subsystems that include buildings, electricity distribution networks, electric transportation, digital infrastructure and data centers. Each of these components has its own consumption patterns and the interaction between them often leads to nonlinear and unpredictable behaviors. In such a context, achieving energy efficiency is not a simple technical issue, but a multi-layered systemic challenge (Nozari, 2026a).

The first major challenge is the continuous growth of energy demand in cities. The increase in urban population, the spread of digital technologies, the development of advanced heating and cooling systems and the spread of electric vehicles have all led to an increase in the load on the grid. Although smart technologies can optimize consumption, at the same time, new digital infrastructures are also energy consumers themselves. Data centers, communication networks, and processing systems contribute significantly to the total urban energy consumption. Therefore, the smart city faces the paradox of increasing technological burden while trying to reduce consumption (Poornima & Nasurudeen Ahamed, 2025).

The second challenge is the inherent multi-objective nature of urban energy management. Urban policymakers seek to reduce carbon emissions, grid operators focus on stability and preventing voltage fluctuations, citizens demand thermal comfort and low energy costs, and investors emphasize economic returns. These objectives often conflict with each other. For example, reducing consumption through load shedding can lead to reduced comfort, or increasing renewable generation may create short-term instability in the grid. Managing these conflicts requires mechanisms beyond single-objective optimization.

The third challenge is the high uncertainty and volatility of the urban environment. Climate change, heat waves, grid disruptions, energy price fluctuations, and unpredictable consumer behavior all affect the urban energy balance. In such situations, models based on historical data may be ineffective in the face of unusual events. Energy efficiency in smart cities requires systems that can adapt to sudden shocks and reorganize their decision-making structures (Zhao et al., 2026).

Another important challenge is the centralization of energy control structures. Many urban energy management systems rely on centralized models in which a control center makes key decisions. Although this structure is efficient in steady-state conditions, it faces processing limitations and decision-making delays at large scales and in the face of large and variable data. Furthermore, centralization can reduce the flexibility and resilience of the system, since any disruption in the central core may affect the entire network.

Another challenge relates to the integration of distributed generation and renewable energy sources. Solar and wind power introduce uncertainty into the grid due to their intermittent nature. Energy storage, load management, and accurate generation forecasting play a key role in reducing this uncertainty, but coordination between these components is complex. Without adaptive and distributed mechanisms, achieving maximum efficiency in such a grid will be difficult (Bandi et al., 2025).

In addition to technical issues, institutional and behavioral challenges are also important. Urban consumer behavior strongly influences energy patterns. Even the most advanced control systems cannot achieve the desired results without social participation and acceptance. Also, regulatory frameworks sometimes do not keep pace with the pace of technological developments, which can hinder innovation in energy management.

In summary, energy efficiency in contemporary smart cities faces a set of intertwined challenges, including growing demand, multi-objectives, uncertainty, centralization, volatility of renewable resources, and behavioral factors. These challenges demonstrate that static and purely optimization-oriented approaches do not address complex urban needs. The transition to goal-oriented and agent-oriented architectures can provide a platform for adaptive management, dynamic coordination, and alignment of conflicting goals, a topic that will be structurally explored in the following sections.

4. ARCHITECTURE OF A GOAL DRIVEN AGENTIC URBAN ENERGY SYSTEM

Designing an urban energy system that can operate in a goal-oriented and agent-oriented manner requires moving from linear and centralized architectures to layered, distributed and adaptive structures. In this approach, urban energy is considered not as a mere physical flow, but as a dynamic decision-making ecosystem in which data, goals and actions are in continuous interaction with each other. The proposed architecture should be able to simultaneously guarantee three features: accurate contextual perception, dynamic adjustment of goals and coordinated execution of decisions at the level of urban subsystems (Tran et al., 2026).

At the first level of this architecture is the perception and data layer. This layer includes energy sensors, climate data, load demand information, network status, renewable resource generation and user behavioral data. The fundamental difference between this layer and traditional systems is in the way data is interpreted. Data is not considered simply as numerical input, but is transformed into a "decision situation"; That is, a combination of key indicators that represent the current state of the system. This transformation provides the basis for goal-oriented decision-making in higher layers.

The second layer is the city digital twin layer. In this section, a dynamic virtual model of the city's energy system is created that is able to simulate the behavior of the network, buildings, and generation resources. The digital twin allows for scenario analysis, short-term forecasting, and evaluation of the consequences of different decisions. Thus, the system examines the effect of each action in the simulated environment before implementing it in the real world. This approach reduces the risk of costly or unstable decisions and increases the proactive capacity of the system (Nozari, 2026b).

In the third layer, there is the goal-setting engine. In this section, goals such as reducing energy consumption, minimizing carbon emissions, maintaining thermal comfort, and network stability are dynamically set. This engine redefines the intensity and priority of each objective based on the decision state extracted from the first layer and the analyses of the second layer. For example, in conditions of increasing peak load, the objective of load reduction is prioritized; while in stable conditions, the focus may be on economic optimization. This mechanism forms the heart of the goal-oriented architecture.

The fourth layer includes the multi-agent decision-making and negotiation engine. In this section, different agents representing urban subsystems interact with each other. The building agent, the network agent, the renewable generation agent, and the policy agent each have their own preferences and constraints. Through negotiation mechanisms, these agents reach a dynamic agreement in which a balance is established between the different objectives. This process allows for distributed and flexible coordination rather than imposing a centralized decision (Mishra & Singh, 2023).

Finally, there is the execution and feedback layer. The decisions made in the negotiation layer are sent to the physical infrastructure; Including building load regulation, energy storage management, renewable generation control and smart tariff regulation. The outcome of these actions is fed back to the perception layer through an adaptive feedback loop, completing the decision-making cycle. This feedback loop ensures that the system continuously learns from its results and modifies the decision-making parameters (Yigitcanlar et al., 2020). The proposed architecture is shown in Figure 2.

Figure 2. Layered architecture of a city energy system based on agent-based goal-oriented intelligence

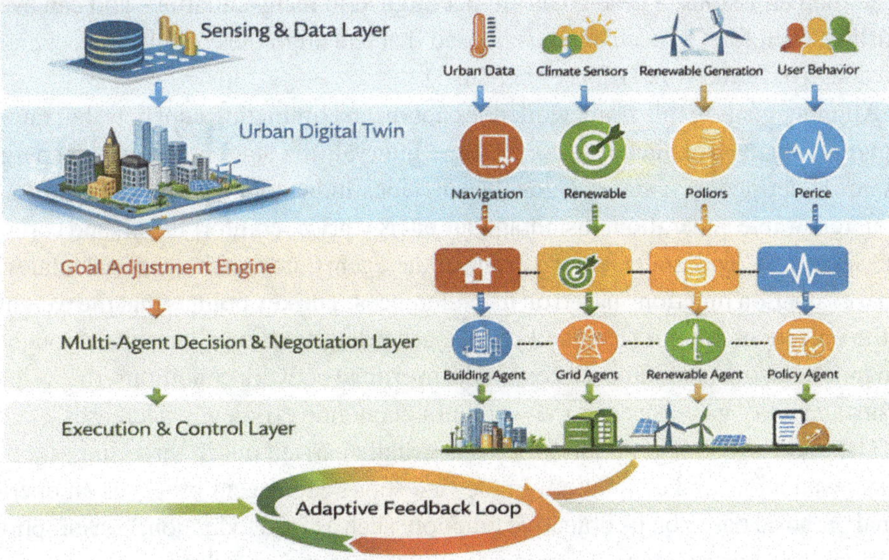

As can be seen in Figure 2, the architecture is designed horizontally and in layers. The information flow from the data layer is transmitted to the digital twin and then to the goal-setting engine. After that, a multi-agent negotiation process forms the final decision and finally, execution is carried out at the physical infrastructure level. The adaptive feedback loop establishes the connection between execution and perception and ensures that the system is adaptable in the face of environmental changes.

This architecture not only enables energy savings, but also provides a framework for aligning conflicting goals and increasing urban resilience. Its layered structure is scalable and can be expanded to regional or metropolitan levels. Also, due to the distributed nature of decision-making, the risk of centralization and single points of failure is reduced.

5. MULTI AGENT GOAL ALIGNMENT AND NEGOTIATION MECHANISMS

In urban energy systems designed based on goal-oriented agent-based intelligence, the main challenge is not just decision-making, but also the dynamic alignment of goals among different agents. Each agent represents an urban subsystem with its own preferences, constraints, and performance indicators. For example, the building agent seeks to maintain comfort and reduce costs, the network agent focuses on sustainability and avoiding overload, the renewable energy agent emphasizes maximizing the use of clean resources, and the policy agent may prioritize reducing carbon emissions. These goals are not aligned in many situations and can even conflict. Therefore, a mechanism is needed that can align these goals in a coherent and adaptive framework (Bibri et al., 2023).

Aligning goals in this framework does not mean eliminating conflicts, but rather managing them dynamically. At each time interval, the state of the urban energy system is defined as a state vector that includes indicators such as the grid load level, renewable generation, instantaneous energy price, comfort index, and carbon emission level. This state vector is sent to the agents, and each agent calculates a "goal evaluation function" based on its preferences. The key point is that the weights of the objectives are not fixed and change according to the system state through a dynamic adjustment module. For example, in critical network conditions, the weight of sustainability increases and the weight of economic efficiency decreases.

The multi-agent negotiation process is usually carried out in three stages: proposal, evaluation, and adaptation. In the first stage, each agent proposes an operational proposal based on its objective function; such as load reduction, consumption shifting, increasing storage, or energy injection into the grid. In the second stage, these proposals are evaluated in a common space to assess their compatibility with the metropolitan goals. In the third stage, through a consensus mechanism, the proposals are modified or combined to lead to a collective decision. This process can be repeated until a convergence criterion is met (Fang et al., 2023).

One effective approach in this regard is the use of mechanisms based on the dynamic collective welfare function. In this method, instead of one agent dominating the others, a composite function is defined that calculates the weighted sum of the agents' objectives, taking into account the current priorities. This function is updated in each negotiation iteration, and the agents modify their proposals with the aim of maximizing it. As a result, the negotiation process leads to a gradual convergence that establishes a balance between conflicting objectives (Nozari & Yordanova, 2026).

An important feature of this mechanism is that the final decision is not necessarily the absolute optimal one from the point of view of a specific objective, but rather an adaptive equilibrium in which the greatest compatibility with the system conditions is achieved. This approach moves away from centralization and allows the agents to act within the framework of metropolitan objectives while maintaining relative independence. Furthermore, the negotiation mechanism can be improved by reinforcement learning or by adjusting the weighting parameters over time, so that the system gradually behaves more efficiently (Abdi & Nozari, 2026).

To clarify the operational logic of this process, Pseudocode 1 shows the general framework of goal alignment and multi-agent negotiation.

Pseudocode 1. Goal alignment and multi-agent negotiation algorithm in urban energy system

```
Initialize system_state S
Initialize agents A = {Building, Grid, Renewable, Policy}
Initialize dynamic_goal_weights W(t)

While system is active:

Update S from sensing layer
Update W(t) based on S and global priorities

For each agent ai in A:
Compute local_objective_i = f_i(S, W(t))
Generate proposal_i

Collect all proposals P = {proposal_1, ..., proposal_n}

Compute collective_score = F(P, W(t))

While convergence not achieved:
For each agent ai:
Adjust proposal_i to improve collective_score
Recalculate collective_score

Select final_decision D*
Execute D* in control layer

Observe system response
Update learning parameters

End While
```

As can be seen in Pseudocode 1, first the system state is updated and the weights of the goals are adjusted based on the current conditions. Then each agent presents its proposal. These proposals are examined in the form of a collective evaluation function and modified until convergence is achieved. The final decision, after execution, enters the learning loop to improve the performance in subsequent iterations.

6. DIGITAL TWIN INTEGRATION FOR PROACTIVE ENERGY MINIMIZATION

In the architecture of goal-oriented urban energy systems, the digital twin plays a pivotal role in the transition from reactive to proactive management. The digital twin can be considered a dynamic and real-time representation of the physical infrastructure of the city that simulates the behavior of the network, buildings, generation resources, and consumption patterns in a computational environment. This representation is not simply a static model, but is continuously updated with sensor data and is able to evaluate the consequences of different decisions before they are implemented in the real environment (Neffati et al., 2021).

In conventional energy management approaches, decision-making is often based on the analysis of past data and responding to the current situation. In other words, the system intervenes when signs of load increase, voltage drop, or generation fluctuation are observed. In contrast, the integration of the digital twin into the goal-oriented agent-based intelligence framework allows for the prediction of short- and medium-term trends. In this way, the system can evaluate different scenarios and select the most appropriate decision before a peak load or sudden increase in consumption occurs. This proactive capability is the basis for energy minimization at the city level.

The digital twin in this architecture has three main functions. First, it synchronizes real-world data with the virtual model. Real-time data related to energy demand, grid status, renewable generation, and climate conditions are fed into the model and represent the current state of the system. Second, it performs scenario analysis. In this step, agents can test several decision-making options in the virtual environment; for example, load shifting, tariff change, storage activation, or generation adjustment. Third, the evaluation of consequences is based on dynamic goals. Each scenario is scored based on indicators such as total energy consumption, carbon emissions, comfort index, and grid stability, and the result is fed back to the goal-setting engine (Kiasari & Aly, 2026a).

A key feature of this integration is the two-way interaction between the digital twin and the multi-agent negotiation engine. The agents do not simply execute their proposals in the real environment, but first simulate them in the twin. This interaction makes the negotiation process based on the prediction of consequences, not just the theoretical preferences of the agents. For example, if the building agent proposes to reduce the load, the twin can estimate the impact of this reduction on the comfort and stability index of the network and feed the result back to the negotiation process. As a result, the final decision is made based on a combination of real data and scenario prediction.

In the context of energy minimization, this mechanism is of particular importance. Many consumption reduction measures may incur costs or reduce comfort in the short term, but lead to reduced emissions and increased efficiency in the long term. The digital twin allows for the assessment of these long-term consequences by providing a wider time horizon. Hence, decision-making is promoted from a reactive level to a strategic guidance level (Nozari & Yordanova, 2025).

In addition, the digital twin can act as a platform for continuous learning. The actual results of decision implementation are compared with the model predictions, and the differences between them are used to improve the model parameters. This process makes the twin more accurate and consistent with urban reality over time. Such a mechanism is doubly important in complex and changing urban environments where consumption patterns change rapidly.

Digital twin integration also allows for simultaneous management of different scales. From the level of a building to the level of a region or the entire city, multi-scale models can operate in harmony. This multi-scale structure increases the flexibility and scalability of the system and allows for local optimization within the framework of macro-goals (Maldonado-Romo et al., 2026).

In short, the digital twin in the framework of agent-based intelligence is not a peripheral tool, but the core of the prediction and evaluation of the urban energy system. By relying on this infrastructure, the smart city will be able to make corrective decisions and actively guide the consumption path before instability or energy waste occurs. This approach is a fundamental step in realizing low-energy, resilient and sustainable cities.

7. SCENARIO ANALYSIS AND URBAN APPLICATION CASES

One of the fundamental advantages of the goal-oriented architecture based on agent-based intelligence is the ability to analyze scenarios in complex urban environments. As a dynamic system, the city is always exposed to sudden changes in energy demand, climate conditions, consumer behavior, and economic policies. Therefore, evaluating the system performance only in a fixed state cannot provide a realistic picture of its efficiency. Scenario analysis allows the system behavior to be examined in diverse and sometimes critical conditions and its adaptive capacity to be measured.

In the proposed framework, scenario analysis is performed based on the interaction between the digital twin and the multi-agent negotiation mechanism. Each scenario is defined as a hypothetical situation including a set of input variables; such as a sudden increase in temperature, a decrease in renewable generation, a rapid growth in electric transportation demand, or a change in energy tariffs. These situations are implemented in a simulation environment and the response of the agents is evaluated. The evaluation criteria include total energy consumption, carbon emission intensity, comfort index, network stability, and operating cost.

Scenario 1: Residential Area With Peak Load Due to Heat Wave

In a dense residential area, a sharp increase in temperature can lead to a sudden increase in cooling system consumption. In the traditional approach, the network response is usually done in the form of increasing production or imposing consumption restrictions. However, in the goal-driven architecture, agents intervene before the critical point is reached. The digital twin predicts the temperature increase and the consumption growth trend, and the goal-setting engine increases the grid stability weight. The building agent participates by shifting the load, intelligently adjusting the indoor temperature and activating energy storage. The grid agent also adjusts consumption through voltage management and dynamic tariffs. The result of this interaction is to reduce the intensity of peaks and prevent overloads without a significant loss of comfort.

Scenario 2: Commercial Area With Sharp Energy Price Fluctuations

In a commercial area where energy consumption is directly dependent on economic activity, energy price fluctuations can affect the decisions of operators. In this scenario, the electricity price increase occurs during peak demand hours. The policy and economic agents increase the cost weight in the objective function, while the network agent continues to focus on sustainability. Through multi-agent negotiation, a decision is made to shift some of the high-demand activities to off-peak hours. This action not only reduces energy costs, but also helps to smooth the network load.

Scenario 3: Industrial Area With the Goal of Reducing Carbon Emissions

In an industrial area, the city policy emphasizes reducing carbon emissions over a certain period of time. In this situation, the target setting engine increases the weight of the carbon index. The renewable agent tries to increase the share of clean energy, and the energy storage agent steps in to compensate for fluctuations. In the event of a shortage of clean generation, the industrial agent reduces fossil energy consumption by optimizing the production process or scheduling operations. The result is a reduction in emission intensity within the set targets without serious disruption to production.

Scenario 4: Sudden Disruption in the Distribution Network

In this scenario, a part of the distribution network goes out of service due to a technical failure. The agent-based system prioritizes resilience by quickly identifying the situation. The agents maintain balance by reconfiguring the load, activating local resources, and using energy storage. The digital twin simulates alternative energy supply paths and suggests the best option. This coordinated response reduces the system recovery time and prevents the spread of the disruption.

Analysis of these scenarios shows that the agent-based goal-oriented architecture exhibits consistent and adaptive performance not only in stable conditions, but also in critical situations. The main difference between this approach and traditional systems is that decision-making is done dynamically and based on changing goal priorities, rather than simply implementing fixed rules. In addition, the use of the digital twin allows decisions to be evaluated before implementation and reduces the risk of error.

Finally, the scenario analysis shows that urban energy efficiency is not limited to numerical reductions in consumption, but also depends on the system's ability to align diverse goals, manage conflicts, and maintain stability in changing conditions. Such capabilities will be the foundation for developing energy-efficient and resilient cities in the future.

8. GOVERNANCE, ETHICS AND SYSTEM RESILIENCE

The development of urban energy systems based on goal-oriented agent-based intelligence is not just a technical development, but a fundamental change in the urban governance paradigm. When decision-making is elevated from the level of simple algorithmic responses to the level of dynamic negotiation between intelligent agents, questions of accountability, transparency, justice, and security arise. Therefore, the success of such an architecture depends on the simultaneous design of governance mechanisms and ethical frameworks commensurate with its complexity.

The first dimension of governance in this framework is the transparency of decision-making. In goal-oriented systems, the priority of objectives can change dynamically; for example, in crisis situations, the weight of sustainability prevails over the weight of cost. These changes, if made without explanation or documentation, can lead to distrust among citizens or operators. Therefore, there needs to be a mechanism to record and explain the logic of changing priorities. The explain ability of decisions is not only a technical requirement, but also a governance imperative. The system should be able to show why a particular decision was made at a particular point in time and how it was aligned with the metropolitan goals.

The second dimension is energy equity. In smart cities, load distribution, dynamic tariffs and consumption management can have different impacts on different groups in society. If agent-based mechanisms focus solely on economic efficiency, low-income households or marginalized areas may be more affected than others. Therefore, the goal-setting engine should also consider social equity indicators. In other words, equity should be defined as a dynamic objective alongside sustainability and efficiency. This approach ensures that energy optimization does not come at the cost of increasing social inequality.

The third dimension is cybersecurity and data protection. Urban energy systems are based on widespread sensors, real-time communications and behavioral data analytics. Any intrusion or manipulation of data can lead to incorrect decisions and even widespread disruption in the network. Since the proposed architecture is based on multi-agent interaction and continuous information exchange, it is essential to design multi-level security layers. These layers should include data encryption, agent authentication, anomaly monitoring, and rapid recovery mechanisms.

System resilience is the fourth dimension of this section. Resilience means the system's ability to absorb shocks, maintain basic performance, and recover quickly after a disruption. In a goal-oriented architecture, resilience is not an added feature, but rather part of the decision-making process. When instability indicators increase, the weight of the resilience goal in the goal-setting engine increases, and the negotiation process between agents is directed toward maintaining stability. This dynamic mechanism causes the system to react before reaching a breaking point and prevent the spread of disruption.

Furthermore, the distributed decision-making structure naturally enhances resilience. In centralized systems, the failure of a decision-making core can disrupt the entire network. In a multi-agent architecture, however, decision-making is distributed among several independent agents. If one component fails, other agents can temporarily take over or reconfigure the decision structure. This feature makes the system more resilient to technical errors and external disturbances.

Another issue is regulatory compliance. Many existing legal frameworks for energy management are based on traditional, centralized structures. The deployment of agent-driven goal-oriented systems requires a review of regulations regarding data ownership, liability for automated decisions, and how human intervention should occur in emergencies. In other words, smart governance must evolve in step with technological change.

Finally, public trust is a determining factor in the acceptance of such systems. Citizens will engage with smart energy management when they feel that decisions are fair, explainable, and in the public interest. Transparency, public participation, and access to information can strengthen this trust.

In conclusion, this section suggests that purpose-driven, agent-centered architecture can only lead to sustainable energy reductions and increased urban efficiency if it is accompanied by a coherent framework of governance, ethics, and resilience. Combining technology with social responsibility and structural security is a necessary condition for the formation of energy-efficient, equitable, and sustainable cities of the future.

9. CONCLUSION

The evolution of smart cities in recent decades has shown that relying solely on data, sensor connectivity, and predictive analytics is not enough to address the complex challenges of urban energy. What is observed today at the urban scale is a network of interconnected subsystems, conflicting goals, and changing environmental conditions, in which static or rule-based decision-making cannot meet long-term needs. By introducing the concept of goal-oriented agent-based intelligence, this chapter attempts to provide a theoretical and architectural framework for the transition from reactive management to goal-oriented guidance in urban energy systems.

In this approach, the goal is not considered a fixed parameter, but a dynamic variable whose intensity and priority change in interaction with the state of the system. This paradigm shift creates a significant gap with classical optimization models. In traditional architectures, the objective function is predefined and the system seeks to find the optimal point within the framework of specific constraints. But in the agent-driven goal-oriented architecture, the goal structure itself is part of the decision-making process and can be reconfigured based on climatic, social, economic or technical conditions. Such a view allows for managing the conflict between economic efficiency, network sustainability, emission reduction and energy justice.

One of the important achievements of this framework was the integration of the digital twin with a multi-agent negotiation mechanism. The digital twin enables scenario analysis and outcome prediction, and elevates the negotiation process between agents from the level of theoretical preferences to the level of practical evaluation. Before implementing each decision, the agents examine its effect in the virtual environment and modify their proposals based on the results. This interaction paves the way for proactive decision-making and reduces the risk of error in the real environment. In this way, the urban energy system will be able to guide the consumption path and prevent instability before a crisis occurs.

The analysis of applied scenarios also showed that this architecture is adaptable and flexible in various conditions, from heat waves and peak loads to energy price fluctuations or network disruptions. The main difference between this system and traditional approaches is that responses are not formed in a command-and-control manner, but through dynamic interaction between independent but coordinated agents. Such a structure, in addition to increasing efficiency, also strengthens the resilience of the system; because distributed decision-making reduces the risk of single points of failure.

Along with the technical aspects, this chapter also emphasized the importance of governance, ethics and energy justice. Although goal-oriented systems have the ability to dynamically adjust goals, they may face social challenges without transparent, explainable and fair frameworks. Fairness in load and cost distribution, protection of user data and ensuring cybersecurity are among the basic requirements for the practical deployment of such systems. Technology can lead to a sustainable reduction in energy consumption when it is accompanied by social responsibility and public trust.

Conceptually, this research suggests that the evolution of low-power smart cities requires a shift in the way we think about "intelligence." Intelligence is not simply about faster data analysis, but also about the ability to define intentions, set priorities, and dynamically coordinate in a multi-objective environment. In such a framework, the city becomes a self-regulating system in which decisions are made not only based on the past, but also with the future and long-term consequences in mind.

However, the way forward requires deeper research in several key areas. Developing rigorous mathematical models for goal alignment, designing adaptive learning algorithms to improve the negotiation process, and creating common standards for data exchange between agents are among the topics that can strengthen this framework. It is also important to examine the social and institutional implications of deploying such systems on a metropolitan scale.

Ultimately, the transition to smart, energy-efficient cities is not achieved by simply adding new technologies, but requires a redefinition of the urban decision-making architecture. Goal-oriented agent-based intelligence is an effective step in this direction by providing a platform for dynamically adjusting goals, negotiating between subsystems, and predicting consequences. This approach can pave the way for a new generation of cities in which energy consumption is managed not reactively, but in a directed and purposeful manner; cities that are simultaneously efficient, sustainable, equitable, and resilient.

REFERENCES

Abdi, H., & Nozari, H. (2026). Energy challenges in transformative technologies-based super-smart city implementation. In *Energy-Efficient Transformative Technologies for Data-Driven Smart Cities* (pp. 71–90). Elsevier. DOI: 10.1016/B978-0-443-27618-7.00002-9

Bandi, A., Kongari, B., Naguru, R., Pasnoor, S., & Vilipala, S. V. (2025). The rise of agentic ai: A review of definitions, frameworks, architectures, applications, evaluation metrics, and challenges. *Future Internet*, *17*(9), 404. DOI: 10.3390/fi17090404

Bibri, S. E. (2021). A novel model for data-driven smart sustainable cities of the future: The institutional transformations required for balancing and advancing the three goals of sustainability. *Energy Informatics*, *4*(1), 4. DOI: 10.1186/s42162-021-00138-8

Bibri, S. E., Alexandre, A., Sharifi, A., & Krogstie, J. (2023). Environmentally sustainable smart cities and their converging AI, IoT, and big data technologies and solutions: An integrated approach to an extensive literature review. *Energy Informatics*, *6*(1), 9. DOI: 10.1186/s42162-023-00259-2 PMID: 37032812

Fang, B., Yu, J., Chen, Z., Osman, A. I., Farghali, M., Ihara, I., Hamza, E. H., Rooney, D. W., & Yap, P. S. (2023). Artificial intelligence for waste management in smart cities: A review. *Environmental Chemistry Letters*, *21*(4), 1959–1989. DOI: 10.1007/s10311-023-01604-3 PMID: 37362015

Kiasari, M., & Aly, H. (2026). Agentic Artificial Intelligence for Smart Grids: A Comprehensive Review of Autonomous, Safe, and Explainable Control Frameworks. *Energies*, *19*(3), 617. DOI: 10.3390/en19030617

Kiasari, M., & Aly, H. (2026). Agentic Artificial Intelligence for Smart Grids: A Comprehensive Review of Autonomous, Safe, and Explainable Control Frameworks. *Energies*, *19*(3), 617. DOI: 10.3390/en19030617

Louati, A., Louati, H., Kariri, E., Neifar, W., Hassan, M. K., Khairi, M. H., Farahat, M. A., & El-Hoseny, H. M. (2024). Sustainable smart cities through multi-agent reinforcement learning-based cooperative autonomous vehicles. *Sustainability (Basel)*, *16*(5), 1779. DOI: 10.3390/su16051779

Maldonado-Romo, J., Bradley, R., Anthony, B., Juárez, L., Yoon, J., Amador, A., Guerrero, E., Bergmann, N., Montesinos, L., & Ponce, P. (2026). A collaborative and sustainable engineering design framework enabled by agentic AI. *International Journal of Sustainable Engineering*, *19*(1), 2624166. DOI: 10.1080/19397038.2026.2624166

Mishra, P., & Singh, G. (2023). Energy management systems in sustainable smart cities based on the internet of energy: A technical review. *Energies*, *16*(19), 6903. DOI: 10.3390/en16196903

Neffati, O. S., Sengan, S., Thangavelu, K. D., Kumar, S. D., Setiawan, R., Elangovan, M., & Velayutham, P. (2021). Migrating from traditional grid to smart grid in smart cities promoted in developing country. *Sustainable Energy Technologies and Assessments*, *45*, 101125. DOI: 10.1016/j.seta.2021.101125

Nozari, H. (2026). Neuromorphic-driven energy optimization in smart urban logistics. In *Energy-Efficient Transformative Technologies for Data-Driven Smart Cities* (pp. 211–228). Elsevier. DOI: 10.1016/B978-0-443-27618-7.00015-7

Nozari, H. (2026). Transformative AI for energy-efficient lean cities. In *Energy-Efficient Transformative Technologies for Data-Driven Smart Cities* (pp. 229–252). Elsevier. DOI: 10.1016/B978-0-443-27618-7.00016-9

Nozari, H., & Yordanova, Z. (2025). A fuzzy–digital twin optimization framework for simultaneous management of waste and energy consumption in sustainable manufacturing. An International Journal of Optimization and Control: Theories & Applications, 25360152.

Nozari, H., & Yordanova, Z. (2026). An Adaptive Fuzzy Multi-Objective Digital Twin Framework for Multi-Depot Cold-Chain Vehicle Routing in Agri-Biotech Supply Networks. *Logistics (Basel)*, *10*(2), 27. DOI: 10.3390/logistics10020027

Pati, A. K. (2025). Agentic AI: A comprehensive survey of technologies, applications, and societal implications. *IEEE Access : Practical Innovations, Open Solutions*, *13*, 151824–151837. DOI: 10.1109/ACCESS.2025.3585609

Poornima, G., & Nasurudeen Ahamed, N. (2025). Reimagining Autonomy: Agentic AI in the Age of Human-Centric Innovation. In *The Power of Agentic AI: Redefining Human Life and Decision-Making: In Industry 6.0* (pp. 291–307). Springer Nature Switzerland. DOI: 10.1007/978-3-031-89424-4_19

Pospieszny, P., & Brodowicz, D. P. (2025). Toward Agentic Environments: GenAI and the Convergence of AI, Sustainability, and Human-Centric Spaces. *Sustainable Development (Bradford)*, sd.70544. DOI: 10.1002/sd.70544

Singh, B., & Kaunert, C. (2024). Dynamic landscape of artificial general intelligence (AGI) for advancing renewable energy in urban environments: synergies with SDG 11—Sustainable cities and communities lensing policy and governance. In *Artificial general intelligence (AGI) security: Smart applications and sustainable technologies* (pp. 247–270). Springer Nature Singapore.

Tran, G. T., Le Dinh, T., & Le, T. D. (2026). Agentic AI Across Technologies, Applications, and Development Domains: A Systematic Literature Review. In *International Conference on Data Analytics & Management* (pp. 587-601). Springer, Cham. DOI: 10.1007/978-3-032-03769-5_44

Yigitcanlar, T., Desouza, K. C., Butler, L., & Roozkhosh, F. (2020). Contributions and risks of artificial intelligence (AI) in building smarter cities: Insights from a systematic review of the literature. *Energies*, *13*(6), 1473. DOI: 10.3390/en13061473

Yordanova, Z., & Nozari, H. (2026). Cross-layer optimization for energy efficiency in super-smart cities of the future. In *Energy-Efficient Transformative Technologies for Data-Driven Smart Cities* (pp. 125–144). Elsevier. DOI: 10.1016/B978-0-443-27618-7.00013-3

Zhang, H., Lin, Y., & Nguyen, G. (2025). AI-driven Sustainable Urban Intelligence: Integrating Smart Technologies for Efficient, Resilient, and Inclusive City Management. Journal of Science. *Innovation & Social Impact*, *1*(1), 201–208.

Zhao, C., Liu, G., Zhang, R., Liu, Y., Wang, J., Kang, J., Niyato, D., Li, Z., Shen, X., Han, Z., Sun, S., Yuen, C., & Kim, D. I. (2026). Edge General Intelligence Through World Models, Large Language Models, and Agentic AI: Fundamentals, Solutions, and Challenges. *IEEE Transactions on Cognitive Communications and Networking*, *12*, 5649–5675. DOI: 10.1109/TCCN.2026.3658762

Chapter 4
Conceptual Foundations of Digital Twin–Enabled Smart and Resilient Homes

Manuel Villasalero
http://orcid.org/0000-0001-8436-4714
University of Castilla-La Mancha, Spain

Kalin Dimtchev
University of National and World Economy, Bulgaria

ABSTRACT

The growth of digital technologies and the expansion of the Internet of Things have made smart homes an important element in the transformation of residential environments. In this study, the conceptual foundations of resilient smart homes based on digital twins are examined. First, the concepts of smart homes and resilience in residential environments are explained, and then the digital twin paradigm in built environments is described. Next, the main components of these systems, the role of data intelligence and dynamic modeling, and mechanisms for increasing resilience in smart homes are analyzed. Finally, an integrated conceptual framework is presented that shows the interaction between physical, data, and analytical layers for the development of smart and sustainable residential systems.

DOI: 10.4018/979-8-2600-1669-5.ch004

1. INTRODUCTION

The rapid growth of digital technologies in recent decades has led to fundamental changes in the way residential environments are designed, operated, and managed. In this regard, the concept of smart homes has emerged as one of the most important manifestations of digital transformation in everyday life. Smart homes, using the Internet of Things, smart sensors, communication systems, and data analysis platforms, are able to continuously monitor environmental conditions and the performance of various equipment and make control decisions based on the received data. Such capabilities enable optimal management of energy consumption, increase safety levels, improve occupant comfort, and improve the efficiency of home systems. However, the increasing dependence of residential systems on digital and energy infrastructures has raised the need for approaches that can ensure the resilience and sustainability of these systems in addition to making them smart (Adibi et al., 2024).

Resiliency in residential environments refers to the ability of home systems to maintain stable performance in the face of various disturbances. These disruptions can be caused by a variety of factors, such as power grid fluctuations, extreme weather conditions, equipment failures, or even cyber threats. In such situations, a home equipped with only simple automation capabilities may not perform well in the face of crises. Hence, the concept of resilient smart homes has gradually gained attention; homes that, in addition to smart capabilities, have the ability to predict, adapt, and recover their performance in critical situations. Achieving such a level of intelligence requires the use of technologies that can analyze environmental data and dynamically model system behavior (Yordanova & Nozari, 2025).

In the meantime, digital twin technology has been proposed as one of the most important tools for the transformation of cyber-physical systems. A digital twin refers to a virtual representation of a physical system that is continuously updated with real-world data and allows for the analysis and simulation of system behavior. This technology was initially used in advanced industries and complex engineering systems, but its application has gradually expanded to built environments and urban systems. In the field of residential environments, a digital twin can create a dynamic picture of the state of the house, in which sensor data, equipment performance, and environmental conditions are analyzed in an integrated manner. Such a capability allows for predicting the state of the system, identifying anomalies, and making optimal decisions for home management (Afolabi et al., 2025).

One of the key features of a digital twin is the creation of a feedback loop between the physical space and the digital space. In this structure, data collected from environmental sensors is transferred to a digital model and, after analysis, the results are returned to the physical system in the form of control recommendations or management decisions. This process allows the house to not only monitor its current state, but also to predict the future behavior of the system and, if necessary, make the necessary adjustments. From an energy management perspective, such a capability can help reduce energy consumption, improve equipment efficiency, and increase the stability of home systems. In addition, digital twins allow for the simulation of different scenarios, which can help to assess the effects of environmental changes or potential disturbances before they occur (Hossain et al., 2026a).

Despite significant advances in smart homes and digital twin technology, there is still a need to develop coherent conceptual frameworks to combine the two. Many existing studies focus on the technical or functional aspects of smart systems, while less attention has been paid to the integrated study of the interaction between smartization, data analytics, and resilience in residential environments. Understanding this interaction is of great importance for designing a new generation of smart homes that can manage environmental and technological complexities. For this reason, explaining the conceptual foundations of resilient smart homes based on digital twins can be an important step towards developing analytical frameworks and designing smart residential systems of the future.

Accordingly, the aim of this chapter is to examine the conceptual foundations and main structures of resilient smart homes based on digital twins. In this regard, first, the concepts related to smart homes and resilience in residential environments are explained, then the digital twin paradigm in built environments is examined, and then the key components and analytical mechanisms of such systems are described. This approach allows for the presentation of an integrated conceptual framework that can be a basis for the development of smart residential systems in future generations.

2. SMART HOMES AND RESIDENTIAL RESILIENCE: CONCEPTUAL BACKGROUND

Technological developments in the field of Internet of Things, smart communications and data analytics have made the concept of smart homes one of the main axes of digital transformation in residential environments. A smart home refers to an environment in which a set of digital equipment and systems operate in an integrated manner and enable automatic monitoring and control of different parts of the home. These systems usually include environmental sensors, network-connected devices, data management platforms and control systems that can monitor the internal conditions of the home in real time and make the necessary adjustments based on the received data. In such an environment, heating and cooling systems, lighting, security, energy management and even home appliances operate intelligently and in harmony, and their main goal is to increase efficiency, reduce resource consumption and improve the quality of life of residents (Nozari et al., 2025a).

In recent years, the spread of smart homes has not been limited to increasing the level of automation, and researchers have also drawn attention to the concept of resilience in residential environments. Resilience refers to the ability of a system to maintain stable performance under conditions of disturbance or crisis. In residential environments, this concept means the ability of a house to cope with various disturbances such as energy fluctuations, equipment failure, extreme climatic conditions, or disruption of communication infrastructure. In such situations, a resilient house should be able to detect the state of the system, respond appropriately, and restore its performance in the event of a disturbance. For this reason, combining smart technologies with resilience approaches has been proposed as one of the important paths for the development of future residential systems (Al-Ali et al., 2020).

From a conceptual perspective, resilient smart houses can be considered as systems that, in addition to control and automation capabilities, also have the ability to analyze data and predict the state of the system. In such systems, data collected from various sensors can help identify environmental changes or functional anomalies. This information is then made available to analytical systems so that they can make appropriate management decisions. For example, in the event of fluctuations in the energy grid, a smart home can automatically adjust energy consumption or use alternative sources. Such capabilities make the residential system not only intelligent, but also adaptable to changing environmental conditions (Piras et al., 2024).

In this context, the relationship between smart home capabilities and different dimensions of resilience becomes very important. Some smart home capabilities, such as environmental data monitoring, automatic equipment control, and energy consumption analysis, can play a direct role in increasing the resilience of the system. For example, energy management systems can adjust consumption patterns when energy demand increases or supply decreases. Also, equipment condition monitoring systems can prevent widespread disruptions by detecting failures early. This relationship between smart capabilities and resilience goals is conceptually presented in Table 1.

Table 1. Relationship between smart home capabilities and residential resilience dimensions

Smart Home Capability	Functional Description	Contribution to Residential Resilience
Environmental sensing	Continuous monitoring of temperature, humidity, air quality, and other environmental parameters	Early detection of abnormal conditions and environmental risks
Energy management systems	Optimization and control of electricity consumption and energy distribution	Improved stability during energy fluctuations or supply disruptions
Automated control systems	Intelligent control of lighting, heating, cooling, and appliances	Adaptive response to environmental or operational changes
Equipment monitoring	Real-time tracking of device performance and operational status	Early identification of faults and prevention of system failures
Data analytics platforms	Processing and analysis of operational data from multiple sensors	Predictive insights for maintaining stable system performance

As shown in Table 1, many of the key capabilities of smart homes are directly related to the goals of resilience in residential environments. Environmental sensors can detect abnormal conditions, such as rising temperatures or deteriorating air quality, while energy management systems can adjust energy consumption based on network conditions. Equipment monitoring systems can also detect potential failures by analyzing performance data. Together, these capabilities enable the home to respond appropriately to environmental changes or technical disruptions and maintain stable performance.

However, achieving this level of coordination between data, devices, and control decisions requires more advanced frameworks for analyzing and managing information. Without an integrated analytics framework, smart home systems may be limited to a set of automation tools and unable to use the collected data to predict or manage crises. Therefore, many studies have emphasized the necessity of using analytical models and cyber-physical systems to enhance the capabilities of smart homes (Nozari et al., 2025b).

In such circumstances, technologies that can combine environmental data with analytical models and digital simulations become very important. One emerging approach in this field is the use of digital twins, which allow for the creation of a dynamic representation of the physical system. Such an approach can strengthen the connection between smart homes and resilience goals and pave the way for the development of a new generation of smart residential environments.

3. DIGITAL TWIN PARADIGM IN BUILT ENVIRONMENTS

In recent years, the concept of digital twin has become one of the important approaches in the management and analysis of complex systems. Digital twin generally refers to a virtual representation of a physical system that is continuously updated using real-time data and allows for the analysis of the system's behavior in real time. In this framework, a dynamic connection is established between the physical space and the digital model, and data collected from sensors are transferred to the analytical model to assess the state of the system. Such a structure allows for accurate monitoring of system performance, identification of anomalies, and prediction of future changes (Hossain et al., 2026a).

The application of digital twins was initially proposed in areas such as aerospace engineering, advanced manufacturing, and industrial infrastructure, but gradually the use of this technology expanded to built environments and urban systems. In built environments, a digital twin can act as a dynamic model of a building or physical space that stores information about the state of the structure, environmental conditions, and equipment performance. This virtual model can help system managers better understand the current state of the building and make appropriate management decisions. For this reason, the digital twin has been considered as one of the key tools in the intelligent management of buildings and urban infrastructure (Nozari, 2026a).

One of the important features of the digital twin in built environments is the ability to integrate diverse data into a single analytical framework. Modern buildings are usually equipped with a set of sensors and control systems that record information about energy consumption, environmental conditions, equipment performance, and the state of the structure. The digital twin can collect this data and use it in the form of a unified analytical model. As a result, system administrators can use this virtual model to simulate different building performance scenarios and evaluate the impact of different decisions before implementation (Bedi & Wani, 2026).

In residential environments and smart buildings, the digital twin can play the role of an advanced analytics layer that combines data obtained from home sensors with simulation models and analytical algorithms. In such a structure, information related to temperature, energy consumption, air quality, equipment status, and even occupant usage patterns of different spaces in the house are recorded in the digital model. This information is then used to analyze system performance and provide management recommendations. This process allows the residential system to continuously assess its status and make necessary adjustments if necessary (Nozari, 2025).

From a system architecture perspective, the digital twin in built environments usually consists of several main components. First, the physical layer, which includes the building, equipment, and operational infrastructure. Second, the layer of sensors and data acquisition systems that record information about the state of the system.

Third, the layer of data processing and modeling that analyzes the collected data and creates a digital model of the system. Finally, there is the decision-making and feedback layer that returns the results of the analysis to the physical system in the form of control actions or management recommendations. This connection between the different layers allows the system to dynamically adapt to environmental changes. The conceptual structure of this interaction between the physical system, sensor data, and the digital model is shown in Figure 1.

Figure 1. Conceptual cycle of digital twin interaction in built environments

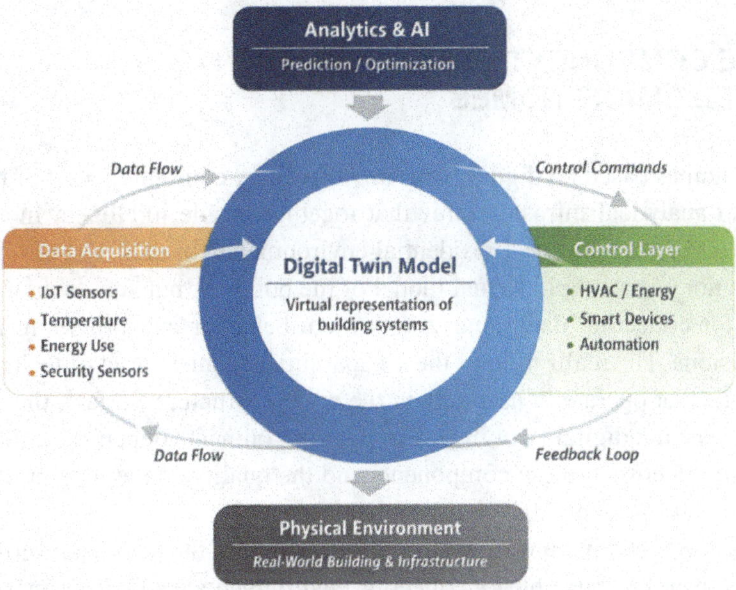

In this conceptual framework, the digital twin performance cycle begins with the physical system, where data is collected through sensors and monitoring equipment. This data is then transferred to data analysis platforms and processed in the digital model. Using data analysis and simulation methods, the digital model can assess the state of the system and identify its behavioral patterns. The results of these analyses are fed back to the physical system in the form of control decisions or management recommendations, thus forming a feedback loop between the physical and digital spaces.

The existence of such a loop allows the building or residential system to continuously monitor its status and respond appropriately to environmental changes. For example, if sensor data shows that energy consumption in a part of the building has increased, the digital model can analyze the cause of this change and suggest appropriate adjustments. Similarly, in the event of an abnormality in the performance of equipment, the system will be able to predict the probability of failure and take preventive measures using data analysis.

Overall, the digital twin paradigm in built environments enables the creation of a comprehensive and data-driven view of building performance. This approach not only helps improve building management, but also paves the way for the development of smarter and more resilient systems. Especially in residential environments, the use of digital twins can play an important role in integrating smart home capabilities with advanced analytics, paving the way for a new generation of data-driven smart homes.

4. CORE COMPONENTS OF DIGITAL TWIN–ENABLED SMART HOMES

Smart homes based on digital twins are formed by combining a set of physical, digital and analytical infrastructures that together enable intelligent monitoring, modeling and management of residential environments. In such systems, the digital twin is not just a simple virtual model of the building, but acts as an integrated platform for collecting data, analyzing system behavior and providing management decisions. The realization of these capabilities requires a set of technological components that operate in harmony in the system structure. In fact, the effective functioning of the digital twin in residential environments depends on the degree of integration between these components and the quality of the data produced by them (Widanage & Kim, 2026).

One of the first key components in this structure is the sensor and IoT device layer. This layer is responsible for collecting environmental and operational data and acts as a connection point between the physical space of the house and the digital space. Various sensors can record information such as temperature, humidity, air quality, energy consumption, lighting status, and equipment performance. The data from these sensors is continuously processed in the system and forms the basis for subsequent analyses. The rapid development of Internet of Things technology has made it possible to install a large number of sensors at a relatively low cost, and as a result, more accurate data on the state of residential environments is available to analytical systems (Okimi et al., 2026).

Another important component in the architecture of digital twin-based homes is the communication and data transmission infrastructure. The data generated by the sensors must be transmitted to data processing platforms via reliable communication networks. This infrastructure can include wireless networks, local communication systems, and cloud platforms that enable data exchange between different devices. The presence of a stable communication infrastructure ensures that system data is provided to the digital model without delay and with minimal errors. This is especially important in systems that rely on real-time analytics (Nozari & Yordanova, 2025b).

In addition to these two components, data processing and modeling platforms play an important role in the performance of the digital twin. These platforms process the collected data and evaluate the system behavior using analytical models and digital simulations. In many cases, these models can analyze energy consumption patterns, equipment performance, or changes in environmental conditions, providing valuable information for decision-making. In addition, the use of data analysis methods and predictive algorithms can help identify long-term trends in system performance and allow for better planning for resource management (Rizo-Maestre et al., 2026).

To better understand the main components of such systems, Table 2 shows the most important components of the digital twin-based smart home architecture and the role each plays in the system performance.

Table 2. Core components of digital twin–enabled smart homes

Component	Description	Role in Digital Twin System
IoT sensing layer	Network of environmental and operational sensors installed within the home	Collects real-time data from physical environment
Communication infrastructure	Wired and wireless networks connecting sensors, devices, and data platforms	Enables reliable data transmission across the system
Data processing platforms	Computational platforms that aggregate and manage large volumes of sensor data	Organizes and prepares data for analysis and modeling
Digital twin modeling environment	Virtual model representing the physical structure and systems of the home	Simulates and analyzes the behavior of the physical system
Intelligent decision layer	Analytical tools and algorithms supporting prediction and optimization	Generates recommendations or control actions for system management

As can be seen in Table 2, the successful operation of a digital twin–enabled smart home depends on the coordinated interaction of several technological layers. The sensor layer provides the required data, while the communication infrastructure ensures the transmission of this data. Data processing platforms transform raw information into analyzable data, and the digital modeling environment allows the simulation of system behavior. Finally, the intelligent decision-making layer can analyze the modeling results to provide suggestions for improving system performance or activate the necessary control measures.

The interaction between these components allows the home to continuously monitor its status and make decisions based on real-time data. For example, if sensor data indicates that energy consumption in a part of the home has increased, the system can analyze the available information to identify the cause of this change and make the necessary adjustments to the relevant equipment. Similarly, if a device shows signs of declining performance, the system will be able to use data analytics to predict the likelihood of failure and take preventive measures.

Overall, the digital twin-based smart home architecture can be considered as a multi-layered system in which data, models, and decisions interact in a continuous cycle. Such a structure provides the necessary basis for the development of smarter and more sustainable residential environments, enabling more efficient management of resources and equipment in future residential environments.

5. DATA INTELLIGENCE AND DYNAMIC MODELING IN RESIDENTIAL DIGITAL TWINS

One of the most important features of digital twins in residential environments is their ability to analyze data and dynamically model the behavior of physical systems. Today's smart homes are equipped with a vast network of sensors, connected devices, and control systems that continuously generate data about environmental conditions, energy consumption, and equipment performance. Although this vast volume of data has high information value, it can only help improve system performance if it is analyzed effectively. This is where the concept of "data intelligence" comes into play; it refers to the use of analytical methods to extract meaningful patterns from environmental data. In the digital twin framework, these analyses lead to the creation of models that are able to simulate and predict the behavior of the residential system under different conditions (Nozari & Yordanova, 2025c).

Data intelligence in digital twin-based homes depends on several main steps. First, raw data is collected by environmental sensors and connected devices. This data may include information about temperature, humidity, air quality, energy consumption, lighting status, or the performance of heating and cooling systems. Next, the data is organized and cleaned in data processing platforms to prepare it for analysis. This step is important because sensor data often contains noise or incomplete values that can affect the accuracy of analytical models. After data preparation, analytical models and predictive algorithms are used to identify patterns in system behavior (Patel et al., 2026).

In this process, dynamic modeling plays a key role. Dynamic models are able to take into account changes in the system over time and predict its future behavior. For example, by analyzing energy consumption data throughout the day, a digital model can identify the consumption patterns of residents and predict when energy consumption will increase. Based on these predictions, the system can make the necessary adjustments to optimize energy consumption. Such capabilities make the digital twin not only a monitoring tool, but also a decision-support system for managing residential environments (Nozari & Yordanova, 2025d).

In many digital twin systems, analytical models are continuously updated with new data. This feature allows the digital model to react to changes in the behavior of residents or environmental conditions. For example, if the pattern of use of household appliances changes, the digital model can take this change into account in its analyses and provide more accurate predictions. This type of adaptive modeling is considered one of the important elements in the development of new-generation smart homes, since the system will be able to adapt to changing environmental conditions (Ristvej et al., 2026). The general structure of the data analysis process and dynamic modeling in the digital twin of residential homes is shown in Figure 2.

Figure 2. Data intelligence and dynamic modeling workflow in residential digital twins

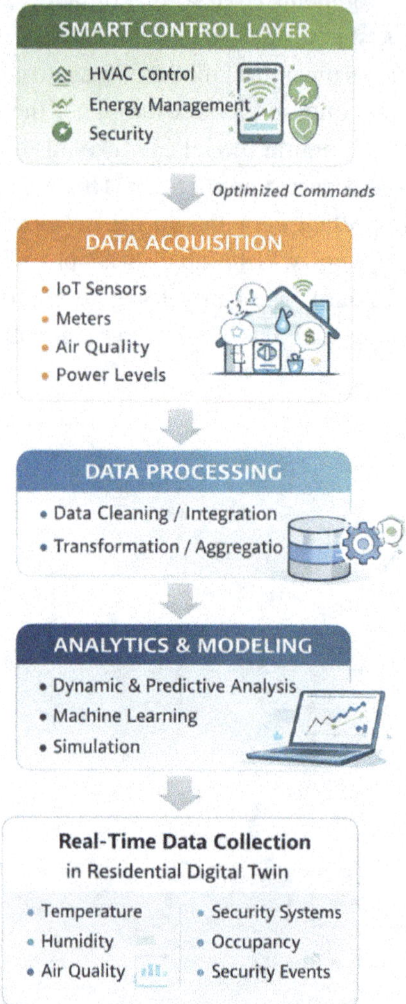

In this conceptual framework, data collected from sensors first enters the data processing layer, where data cleaning, integration, and preparation operations are performed. After this stage, the data is transferred to the analysis and modeling layer, and analytical models are used to identify patterns in the system's performance. The results of these analyses are provided to the digital model in the form of predictions and performance indicators. Ultimately, this information can be used to optimize equipment performance, manage energy consumption, or improve the environmental conditions of the home.

Such a cycle of data analysis and dynamic modeling allows the digital twin to continuously assess the status of the residential system and provide more accurate management decisions. For example, if sensor data shows that energy consumption in a part of the home has increased unusually, the analytical model can identify the cause of the change and suggest solutions to reduce consumption. Similarly, if patterns are observed that indicate the possibility of equipment failure, the system can suggest preventive measures.

Overall, the combination of data intelligence and dynamic modeling makes the digital twin a powerful tool for analyzing and managing residential environments. These capabilities not only enable accurate monitoring of home performance, but also provide the necessary basis for intelligent and optimal decision-making in the management of resources and equipment. As a result, digital twins can play an important role in the development of sustainable and resilient smart homes in the future.

6. RESILIENCE MECHANISMS IN DIGITAL TWIN SMART HOMES

In recent years, the concept of resilience has become a key component in the design and management of residential systems. The increasing complexity of urban infrastructure, the increasing dependence of homes on energy and communication networks, as well as the occurrence of unexpected environmental events have made traditional building management approaches no longer responsive to new needs. In such circumstances, smart homes can only be considered as stable and reliable environments when, in addition to control and automation capabilities, they also have mechanisms to maintain performance in disturbed conditions. Digital twin technology in this regard provides a platform in which environmental data, analytical models, and decision-making systems interact with each other in a dynamic cycle, enabling the strengthening of the resilience of residential systems (Nozari & Szmelter-Jarosz, 2025).

One of the most important resilience mechanisms in digital twin-based homes is continuous monitoring of the system status. In this structure, a set of sensors continuously record information related to environmental conditions, energy consumption, and equipment performance. This data is transferred to the digital model and analyzed there. Continuous monitoring of data allows for the rapid identification of unusual changes in system performance. For example, a sudden increase in energy consumption in one part of the house can be a sign of equipment malfunction. In such situations, the digital model is able to detect this change and suggest corrective actions. This type of proactive monitoring is one of the main foundations for increasing resilience in smart home systems (Divya et al., 2026).

Another important mechanism in enhancing the resilience of smart homes is the ability to predict and analyze possible scenarios. Using historical data and analytical models, the digital twin can simulate the behavior of the system under different conditions. This capability allows for the examination of the possible consequences of various disturbances. For example, scenarios such as sudden increases in energy consumption, power grid power reduction, or extreme changes in ambient temperature can be simulated in a digital model and their effects on the performance of the residential system can be examined. The results of these analyses can help system managers design appropriate solutions to manage the situation before a crisis occurs. Such an approach allows the residential system to not only react to disturbances, but also to predict them and be prepared to deal with them (Yordanova & Nozari, 2026a).

Another important element in resilience mechanisms is the system's adaptability. Digital twin-based homes can dynamically change system settings by analyzing environmental data and occupant behavior. For example, if the ambient temperature increases, the system can automatically adjust the performance of cooling equipment or reduce the consumption of some unnecessary equipment when the load on the energy grid increases. This type of adaptability allows the system to maintain stable performance in changing environmental conditions. Additionally, the use of analytical algorithms and learning models can help to gradually improve system performance, making the home more adaptable to environmental conditions and occupant behavior patterns over time.

From an operational perspective, resilience in digital twin-based homes is not limited to predicting and responding to disturbances, but also includes the ability to quickly recover the system. In the event of a failure in one part of the system, the digital model can provide the information necessary to identify the cause of the problem and accelerate the process of returning the system to a stable state. For example, if a key equipment in the heating or cooling system malfunctions, the data recorded in the digital model can help to quickly identify the location of the failure and plan for its repair. This feature reduces system downtime and increases the stability of the home's performance (Li et al., 2026).

In addition, digital twins can help improve coordination between different parts of the residential system. In many smart homes, different devices operate independently, with limited interaction between them. But in a digital twin-based architecture, data generated by different parts of the system is collected into a unified model. This data integration allows for comprehensive analysis of system performance and enables management decisions to be made based on a holistic view of the home's condition. Such an approach can prevent conflicts between different system settings and ensure coordinated operation of devices.

Finally, one of the most important advantages of using digital twins in smart homes is the ability to create a continuous learning cycle in the system. Data collected from system performance over time can be used to improve analytical models. This process causes the digital model to gradually become more accurate and more capable of predicting system behavior. As a result, the residential system can continuously improve its performance and achieve a higher level of resilience to environmental changes and potential disturbances.

In summary, the resilience mechanisms in smart homes based on digital twins can be summarized in four main axes: continuous data monitoring, prediction and simulation of disturbance scenarios, dynamic adaptation to environmental changes, and rapid recovery of system performance. The combination of these capabilities enables smart residential environments to not only have high efficiency under normal conditions, but also to maintain stable performance when faced with unexpected conditions. Such features play an important role in the development of a new generation of smart homes that can provide a safer, more stable, and more reliable environment for residents by using data and digital models.

7. INTEGRATED CONCEPTUAL FRAMEWORK FOR DIGITAL TWIN–ENABLED RESILIENT HOMES

Given the increasing complexity of residential systems and their increasing dependence on digital infrastructure, the development of a unified conceptual framework for smart homes based on digital twins has become of great importance. Such a framework should be able to organize the interaction between the physical, data, and analytical layers in a way that not only increases the system's efficiency, but also strengthens its resilience. In fact, smart homes can effectively respond to environmental and technical disturbances when information about the system's state is collected and processed in a coherent analytical structure. Conceptual frameworks based on digital twins play an important role in this regard, as they enable a dynamic link between the physical environment and the digital model, creating the necessary context for data-driven analysis and decision-making (Yordanova & Nozari, 2026b).

In such a structure, the physical layer of the home includes all the equipment, sensors, and operational infrastructure that are located in the real environment. This layer is considered the starting point of the system information cycle, as the initial data related to the environmental state and equipment performance are generated in this section. Environmental sensors can collect information about temperature, humidity, air quality, energy consumption, and the status of various equipment. This data is then transmitted to higher layers of the system through communication infrastructure. The quality and accuracy of the data collected at this stage plays a decisive role in the performance of the digital model, as subsequent analyses are based on this information (Trivedi & Juyal, 2026).

After the data transfer, the data management and information processing layer comes into play. In this stage, the raw data received from the sensors are organized, cleaned, and integrated to prepare them for analysis. This process is especially important in systems that use a large number of sensors, since sensor data often contains incomplete values or noise. Data processing platforms can extract useful information from raw data using various data analysis methods and provide it to analytical models. This step actually creates a bridge between environmental data and the digital model (Nozari, 2026b).

The next layer in the integrated digital twin framework is the modeling and simulation environment. In this layer, a virtual model of the house is created that is able to represent the behavior of the physical system. The digital model can hold information about the building structure, equipment performance, and environmental conditions and is continuously updated using real-world data. This feature allows the digital model to display the current state of the system with high accuracy and at the same time allows for the analysis of different scenarios. For example, the effect of changes in ambient temperature, increased energy consumption, or equipment failure can be examined in the digital model and its results used for management decision-making (Abdallah-Ou-Moussa et al., 2026).

Finally, the decision-making and control layer is in this framework, which is responsible for converting the results of the analysis into practical actions. At this stage, the output of the analytical models and digital simulations can be transferred to the physical system in the form of management recommendations or control commands. For example, if the data analysis shows that energy consumption in a part of the house has increased too much, the system can automatically change the settings of the relevant equipment. This process allows a feedback loop to be formed between the digital and physical spaces, and the system can dynamically adapt to environmental conditions. The general structure of this interaction between the different layers is presented in Figure 3.

Figure 3. Integrated conceptual framework of digital twin–enabled resilient smart homes

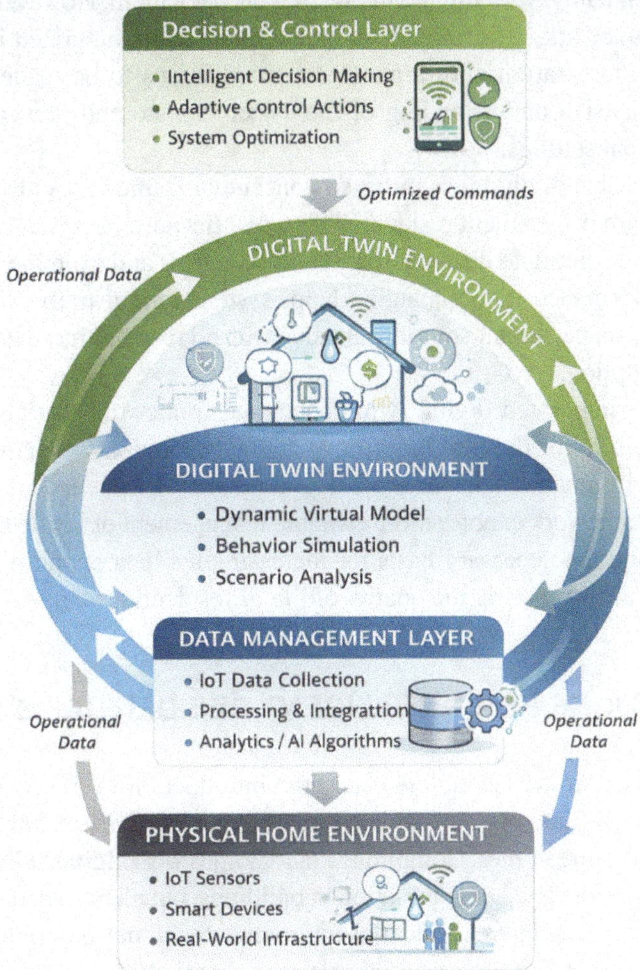

In this conceptual framework, four main layers interact with each other in an integrated cycle. The physical layer of the house generates environmental and operational data, while the data management layer is responsible for collecting and organizing this information. The data is then transferred to the digital modeling environment to analyze and simulate the system's behavior. Finally, the analysis results are translated into control actions in the decision-making layer, which in turn affect the performance of the physical system. This continuous cycle allows the home to continuously assess its state and respond appropriately to environmental changes.

One of the important advantages of such a framework is the coordination between different parts of the residential system. In many smart homes, different devices operate independently, with limited interaction between them. However, in a digital twin-based framework, all data and analytical models are integrated into a single structure. This integration allows management decisions to be made based on a more comprehensive understanding of the system's state and prevents conflicts between different settings.

On the other hand, digital twin-based conceptual frameworks also enable the development of more resilient residential systems. Because in such a structure, the digital model can simulate different disruption scenarios and examine their impact on system performance. This capability helps system administrators to design appropriate crisis management solutions in advance and respond more quickly in the event of a disruption.

Overall, the integrated digital twin conceptual framework can be considered as an infrastructure for the development of a new generation of smart homes. By combining environmental data, analytical models, and intelligent decision-making systems, this framework enables more efficient management of residential environments, providing the necessary basis for increasing resilience, optimizing energy consumption, and improving the quality of life of residents.

8. IMPLICATIONS FOR INTELLIGENT RESIDENTIAL SYSTEMS

The spread of digital technologies and the introduction of advanced analytics in residential environments have opened up new perspectives for the design and management of future homes. Combining smart homes with digital twin technology not only increases the level of automation in buildings, but also enables the creation of smarter and more adaptive residential environments. In such environments, management decisions are no longer based solely on fixed settings or simple controls, but are based on real-time data analysis and predictive models. This development can play an important role in improving the quality of life of residents, reducing resource consumption, and increasing the sustainability of residential systems (Nozari & Yordanova, 2026).

One of the most important consequences of the use of digital twins in smart homes is the change in the way energy is managed in residential environments. In many buildings, energy consumption is managed statically, without taking into account the actual conditions of the environment or the behavior of residents. But in digital twin-based homes, energy consumption data is continuously collected and analyzed, and based on this analysis, consumption patterns can be identified. This information allows the system to dynamically change the settings of heating, cooling, and lighting equipment and prevent unnecessary energy consumption. Such an approach can help reduce energy costs and increase the efficiency of home systems.

On the other hand, the use of digital twins in residential environments can play an important role in improving the safety and security of homes. In this structure, data related to the status of equipment, environmental conditions, and activities inside the home are recorded in an integrated digital model. Analysis of this data can help identify abnormal patterns or dangerous conditions. For example, unexpected changes in ambient temperature, increased levels of pollutant gases, or malfunctions in safety equipment can be quickly identified in the digital model, and the necessary warnings can be sent to residents or control systems. Such a capability enables the residential system to respond appropriately before a minor problem becomes a crisis (Tubis & Werbińska-Wojciechowska, 2026).

Another important consequence of the development of smart residential systems is the possibility of creating residential environments that adapt to the behavior of residents. Traditional homes usually have fixed settings for various systems, while digital twin-based homes can analyze the behavioral patterns of residents and adapt the indoor environment to these patterns. For example, the system can identify the usual times of residents' presence at home, their equipment usage patterns, or their temperature preferences, and provide appropriate environmental conditions based on this information. Such a capability not only helps to increase the comfort of residents, but can also improve the efficiency of home systems.

In addition, digital twins can play an important role in managing the maintenance and repair of home equipment. In many homes, equipment repairs are usually carried out after a breakdown occurs, which can lead to additional costs or disruption of system performance. But in digital twin-based homes, the performance data of the equipment is continuously analyzed and early signs of failure can be identified. This information allows the system to suggest preventive maintenance measures and prevent sudden failures. Such an approach can extend the lifespan of the equipment and reduce maintenance costs.

From a larger perspective, the development of smart homes based on digital twins can also influence the way urban infrastructure is designed and managed. When a large number of homes are equipped with such systems, the data generated at the neighborhood or city level can be used to analyze energy consumption patterns, manage network loads, and plan urban planning. This can help develop smart cities and enable better coordination between residential systems and urban infrastructure (Sompolgrunk et al., 2026).

Overall, the application of digital twins in residential environments has broad implications for the design and management of future homes. By combining environmental data, analytical models, and intelligent decision-making systems, this technology can provide a platform where homes become not only smarter, but also more sustainable and resilient. Such developments could play an important role in shaping a new generation of residential environments that are able to adapt to changing environmental conditions and the evolving needs of residents.

9. CONCLUSION

Rapid developments in digital technologies, especially in the field of Internet of Things, data analytics and cyber-physical systems, have paved the way for the formation of a new generation of residential environments in which houses are no longer just spaces for residence, but are transformed into intelligent, dynamic and data-driven systems. In the meantime, the concept of digital twin, as one of the emerging approaches, has been able to offer a new perspective for the management and analysis of residential environments. By creating a virtual representation of the physical system of the house and continuously synchronizing it with real data, the digital twin enables precise monitoring of the state of the environment, analysis of system behavior and making data-based management decisions.

In this chapter, an attempt was made to systematically examine the conceptual foundations of resilient smart homes based on digital twins. First, the concept of smart homes and their role in the transformation of residential environments were explained, and then the concept of resilience in residential systems was considered. It was then shown how combining smart technologies with advanced analytics can lead to environments that are not only more efficient but also resilient to environmental and technical disruptions. An examination of the digital twin paradigm in built environments also showed that this technology can act as a bridge between the physical and digital spaces, enabling analysis and simulation of the behavior of residential systems.

Next, the main components of digital twin-based smart homes were examined, and the roles of different layers of the system, including sensors, communication infrastructure, data processing platforms, and digital modeling environments, were explained. Together, these components form a multi-layered architecture in which data is collected from the physical environment and, after analysis in the digital model, the results are fed back to the system in the form of control actions or management recommendations. Such a cycle allows for a dynamic decision-making process in which the residential system can continuously assess its state and make adjustments as needed.

This chapter also discusses the role of data intelligence and dynamic modeling in the performance of digital twins. Analyzing environmental data and using predictive models can help identify patterns of system behavior and predict future changes. These capabilities enable better energy management, improved equipment performance, and increased safety in residential environments. In addition, examining resilience mechanisms showed that the digital twin can help maintain stable home performance in changing environmental conditions by continuously monitoring data, simulating different scenarios, and providing adaptive solutions.

One of the most important achievements of this study was the provision of an integrated conceptual framework for resilient smart homes based on digital twins. This framework shows how the interaction between physical, data, and analytics layers can lead to the creation of a smart and adaptive residential system. In such a structure, environmental data is continuously collected and analyzed, and the results of these analyses are used in management decisions. This approach can help improve the efficiency of residential systems, reduce resource consumption, and increase resilience to disruptions.

In summary, the development of smart homes based on digital twins can be considered one of the important paths in the transformation of future residential environments. By combining real-world data, analytical models, and intelligent decision-making systems, this technology enables the creation of environments that are able to dynamically adapt to environmental conditions and the needs of residents. Such developments can play an important role in improving the sustainability, efficiency, and quality of life in residential environments and pave the way for the formation of a new generation of smart and resilient homes in future cities.

REFERENCES

Abdallah-Ou-Moussa, S., Wynn, M., & Rouaine, Z. (2026). Leveraging Blockchain and Digital Twins for Low-Carbon, Circular Supply Chains: Evidence from the Moroccan Manufacturing Sector. *Sustainability (Basel)*, *18*(2), 991. DOI: 10.3390/su18020991

Adibi, S., Rajabifard, A., Shojaei, D., & Wickramasinghe, N. (2024). Enhancing healthcare through sensor-enabled digital twins in smart environments: A comprehensive analysis. *Sensors (Basel)*, *24*(9), 2793. DOI: 10.3390/s24092793 PMID: 38732899

Afolabi, A., Ogunrinde, O., & Zabihollah, A. (2025). Digital Twin and AI Models for Infrastructure Resilience: A Systematic Knowledge Mapping. *Applied Sciences (Basel, Switzerland)*, *15*(24), 13135. DOI: 10.3390/app152413135

Al-Ali, A. R., Gupta, R., Zaman Batool, T., Landolsi, T., Aloul, F., & Al Nabulsi, A. (2020). Digital twin conceptual model within the context of internet of things. *Future Internet*, *12*(10), 163. DOI: 10.3390/fi12100163

Bedi, J., & Wani, N. A. (2026). Artificial intelligence for home digital twins. In *Home Digital Twins* (pp. 227–242). Elsevier.

Divya, A., Ramesh, S., Chakrasali, S. V., Kumar, M. K., Taj, M., Ganiya, R. K., Nageswara Rao, J., K, B., & Qamar, S. (2026). Federated Digital-Twin–Empowered Blockchain Architecture for Autonomous and Resilient Smart Water Management in Sustainable Cities. *Sustainable Computing : Informatics and Systems*, *50*, 101303. DOI: 10.1016/j.suscom.2026.101303

Hossain, M. I., Hossan, M. R., Shaon, Z. H., & Ferdous, M. N. (2026). Linking digital twin paradigm for urban heat monitoring and policy integration to building smart city climate resilience. *Discover Cities*, *3*(1), 1. DOI: 10.1007/s44327-025-00179-8

Hossain, M. I., Hossan, M. R., Shaon, Z. H., & Ferdous, M. N. (2026). Linking digital twin paradigm for urban heat monitoring and policy integration to building smart city climate resilience. *Discover Cities*, *3*(1), 1. DOI: 10.1007/s44327-025-00179-8

Li, Y., Zhang, C., Guo, Z., Yang, N., Yuan, D., & Ding, W. (2026). SafeTwinNet: Secure and Adaptive Learning Ecosystems for Digital Twins. IEEE Communications Standards Magazine.

Nozari, H. (2025). Integrating Neuromorphic Intelligence with Fuzzy Multi-Objective Models in Resilient Green Supply Chain Design. Process Integration and Optimization for Sustainability, 1-14.

Nozari, H. (2026). Digital twin-based predictive maintenance in cold chain logistics. *Operations Research, 60*(1), 481–499. DOI: 10.1051/ro/2025167

Nozari, H. (2026). Neuromorphic-driven energy optimization in smart urban logistics. In *Energy-Efficient Transformative Technologies for Data-Driven Smart Cities* (pp. 211–228). Elsevier. DOI: 10.1016/B978-0-443-27618-7.00015-7

Nozari, H., Abdi, H., Szmelter-Jarosz, A., & Rahmaty, M. (2025). A digital twin-based neutrosophic multi-objective optimization model for smart supply chains in the food industry. *Journal of Intelligent Manufacturing*, 1–15. DOI: 10.1007/s10845-025-02689-z

Nozari, H., & Szmelter-Jarosz, A. (2025). A Predictive Maintenance Approach for Composting Plants Based on ERP and Digital Twin Integration. *Machines (Basel), 13*(12), 1123. DOI: 10.3390/machines13121123

Nozari, H., Szmelter-Jarosz, A., & Weiland, D. (2025). A Fuzzy Multi-Objective Sustainable and Agile Supply Chain Model Based on Digital Twin and Internet of Things with Adaptive Learning Under Environmental Uncertainty. *Applied Sciences (Basel, Switzerland), 15*(19), 10399. DOI: 10.3390/app151910399

Nozari, H., & Yordanova, Z. (2025). An intelligent digital twin approach for optimizing multi-channel supply chains in uncertain environments. Green Technologies and Sustainability, 100309.

Nozari, H., & Yordanova, Z. (2025). Hybrid digital twin and quantum AI with fuzzy multiobjective modeling in supply chain management. *Edelweiss Applied Science and Technology, 9*(10), 609–628. DOI: 10.55214/2576-8484.v9i10.10482

Nozari, H., & Yordanova, Z. (2025). A fuzzy–digital twin optimization framework for simultaneous management of waste and energy consumption in sustainable manufacturing. An International Journal of Optimization and Control: Theories & Applications, 25360152.

Nozari, H., & Yordanova, Z. (2026). An Adaptive Fuzzy Multi-Objective Digital Twin Framework for Multi-Depot Cold-Chain Vehicle Routing in Agri-Biotech Supply Networks. *Logistics (Basel), 10*(2), 27. DOI: 10.3390/logistics10020027

O. Okimi, T., Adeoye, O., & Omoboye, O. (2026). Integrating digital twins and decision support systems for predictive risk management in urban renewal. Digital Twin, 2612464.

Patel, U. R., Ghaffarianhoseini, A., GhaffarianHoseini, A., & Burgess, A. (2026). Towards a well-being-oriented framework for urban digital twins. *Cities (London, England), 169*, 106579. DOI: 10.1016/j.cities.2025.106579

Piras, G., Agostinelli, S., & Muzi, F. (2024). Digital twin framework for built environment: A review of key enablers. *Energies, 17*(2), 436. DOI: 10.3390/en17020436

Ristvej, J., Halúsková, B., Nováková, K., & Chovanec, D. (2026). Implementation Maturity Levels of Digital Twin Technology and Data Content Design for Flood Digital Twin. *Smart Cities*, *9*(2), 28. DOI: 10.3390/smartcities9020028

Rizo-Maestre, C., Flores-Moreno, J. M., Nebot-Sanz, A., & Huesca-Tortosa, J. A. (2026). Multi-Agent Systems and Digital Twins as a Basis for Smart Buildings with Integrated Sustainable Efficient Ventilation. *Buildings (Basel, Switzerland), 16*(5), 1026. DOI: 10.3390/buildings16051026

Sompolgrunk, A., Golzad, H., Liu, H., Banihashemi, S., & Najafi, M. (2026). The evolution of digital twins within the big BIM data landscape. Smart and Sustainable Built Environment, 1-28.

Trivedi, A., & Juyal, S. A. (2026). Smart Cities and Urban Circularity: Advancing Sustainable Transformation through Digital Twins, AI, and Capacity Building. In *Green Horizons: Advancing Climate Solutions Through Green Finance, Circular Economy, and ESG* (pp. 339–356). Springer Nature Switzerland. DOI: 10.1007/978-3-032-03484-7_18

Tubis, A. A., & Werbińska-Wojciechowska, S. (2026). House of resilience for energy supply chains: A digitalization-based approach to enhancing supply chain robustness. *Environment Systems & Decisions*, *46*(1), 1. DOI: 10.1007/s10669-025-10054-x

Widanage, C., & Kim, K. P. (2026). Digital Twin for Sustainable Social Housing: Integrating BIM and MMC Towards Industry 5.0. *Encyclopedia*, *6*(2), 30. DOI: 10.3390/encyclopedia6020030

Yordanova, Z., & Nozari, H. (2025). Harnessing cyber-physical systems and digital twins for industrial innovation and economic resilience. In *Dynamic and Safe Economy in the Age of Smart Technologies* (pp. 33–48). IGI Global Scientific Publishing. DOI: 10.4018/979-8-3693-4369-2.ch003

Yordanova, Z., & Nozari, H. (2026). Cross-layer optimization for energy efficiency in super-smart cities of the future. In *Energy-Efficient Transformative Technologies for Data-Driven Smart Cities* (pp. 125–144). Elsevier. DOI: 10.1016/B978-0-443-27618-7.00013-3

Yordanova, Z., & Nozari, H. (2026). An Agentic Digital Twin Framework for Fuzzy Multi-Objective Optimization in Dynamic Humanitarian Logistics. *Algorithms*, *19*(3), 198. DOI: 10.3390/a19030198

Chapter 5
Resilient Smart Buildings Using Adaptive Digital Twin Frameworks

Sophia Weber
PwC, Germany

Valdrin Dervishaj
http://orcid.org/0000-0002-1677-9552
Berlin College of Business and Technology, Germany

ABSTRACT

This research presents a conceptual and analytical framework for the development of resilient smart buildings based on adaptive digital twins. The main focus is on integrating real-time monitoring, predictive analytics, adaptive control, and multi-objective optimization to improve energy efficiency, maintain user comfort, and increase the functional resilience of the building. In this framework, the digital twin acts as a decision-making core and enables intelligent response to environmental disturbances and changes through continuous communication between the physical system and the digital model. Also, the role of resilience indicators, mathematical modeling, and connection to smart energy infrastructures in shaping a new generation of sustainable, adaptive, and reliable buildings is explained.

1. INTRODUCTION

In recent years, buildings have transformed from mere spaces for habitation and activity into complex, dynamic, and highly technology-dependent systems. Urbanization, increased energy consumption, climate change, and unpredictable

DOI: 10.4018/979-8-2600-1669-5.ch005

events have all made the traditional approach to building design and operation no longer meet today's needs. In the meantime, the concept of "smart building" was proposed as an initial response to these challenges; but experience has shown that intelligence alone is not enough. What has become more important than ever is the ability of a building to adapt, continue to function, and recover quickly in the face of changing conditions and crises; a concept known as "resilience."

Resilience in the field of building does not simply mean resistance to a shock or disruption, but also includes the ability to predict, adapt, and return to favorable conditions. For example, a resilient building should be able to intelligently adjust its performance and prevent degradation of service quality in the face of sudden temperature increases, fluctuations in the power grid, or even changes in user behavior. This level of responsiveness requires a deep understanding of the current state of the system as well as the prediction of its future behavior; a matter that is difficult to achieve with traditional building control and management methods (Hossain et al., 2026).

In this regard, digital twin technology has attracted widespread attention as one of the most innovative approaches in managing complex systems. A digital twin is a digital and dynamic representation of a physical system that simulates the current and future state of the system using real-time data, analytical models, and intelligent algorithms. This technology allows for the observation, analysis, and optimization of building performance at a level that was previously unattainable. However, many current applications of digital twins in the building sector still remain at the level of monitoring and analysis and have not moved towards adaptive and automated decision-making.

The main challenge here is the transition from static digital twins to dynamic and adaptive frameworks that can actively play a role in the process of building control and optimization. An adaptive digital twin not only reflects the state of the system, but also continuously improves the performance of the building by learning from data, continuously updating models and applying intelligent decisions. This approach is doubly important, especially in conditions of high uncertainty and rapid change (Yorucu et al., 2026).

On the other hand, the integration of the concept of resilience with the digital twin has opened up new horizons in the design and operation of smart buildings. In this framework, resilience is considered not only as a qualitative characteristic, but also as a measurable and optimized objective. In other words, the level of resilience of a building can be optimized using mathematical models and optimization algorithms, along with criteria such as energy consumption, thermal comfort and operating costs. This multi-objective perspective allows for a balance between sometimes conflicting objectives and leads to more accurate and efficient decision-making (Nozari & Yordanova, 2026).

In addition, advances in areas such as the Internet of Things, machine learning and big data analytics have provided the necessary platform for the implementation of adaptive digital twins. Smart sensors can collect accurate and real-time information on various building parameters such as temperature, humidity, air quality and energy consumption. After processing and analysis, this data is transferred to digital models and becomes the basis for automated decision-making. As a result, the building becomes a system that not only perceives the environment, but also actively responds to it.

In light of the above, this chapter aims to provide a comprehensive framework for the development of smart, resilient buildings based on adaptive digital twins. The main focus is on explaining key concepts, introducing appropriate architectures, and providing analytical and optimization models that can pave the way for the design of a new generation of buildings. In this approach, an attempt has been made to establish a link between theoretical concepts and practical tools and to clearly outline the path from passive systems to smart, adaptive, and resilient systems (Khdhir et al., 2026).

Finally, it can be said that the combination of digital twins with adaptive mechanisms is a fundamental step towards realizing buildings that are not only efficient and sustainable, but also capable of demonstrating stable and reliable performance in the face of future uncertainties and challenges. This transformation will play a key role in the formation of smart and sustainable cities and will pave the way for fundamental changes in the way humans interact with built environments.

2. CONCEPTUAL FOUNDATIONS OF RESILIENT SMART BUILDINGS

The concept of smart resilient buildings is the result of the convergence of several important developments in the fields of building engineering, smart technologies, energy management and cyber-physical systems. In traditional approaches, buildings were mainly considered as static structures whose main purpose was to provide space, physical safety and, in some cases, reduce operating costs. However, the growth of urbanization, the increase in the intensity of climate change, fluctuations in energy consumption patterns and the emergence of various disturbances have presented this approach with serious limitations. In such circumstances, it is no longer enough for a building to be smart, but the building must also be able to perform stably and reliably in the face of changes, pressures and crises (Nguyen & Adhikari, 2026).

In this context, the concept of resilience has become one of the key components in the new generation of buildings. Resilience in a building does not only mean structural resistance to physical shocks, but also refers to the ability of the system to maintain performance, adapt to new conditions and recover quickly after a disturbance. In other words, a resilient building should be able to maintain an acceptable level of performance even under abnormal conditions and, in the event of a performance degradation, return to the desired state in the shortest possible time. This understanding elevates resilience from a purely technical characteristic to a functional and management capability (Yordanova & Nozari, 2026).

Conceptually, three main components of resilience in smart buildings can be considered. The first component is performance stability, which refers to the ability of the system to maintain proper functioning in the face of disturbances. The second component is adaptability; that is, the building can dynamically change its settings based on environmental, operational, and behavioral conditions. The third component is recoverability, which indicates the speed and quality of the system's return to normal after a disturbance. Together, these three components form the theoretical foundation of resilient smart buildings and distinguish them from purely smart buildings (Nozari, 2025).

To better understand this distinction, it is useful to compare traditional buildings, smart buildings, and resilient smart buildings. As shown in Table 1, the differences between these three generations are not only in the level of technology used, but also in the type of control, the level of data utilization, the ability to predict, the way to respond to disturbances, and the level of integration of the systems.

Table 1. Comparison of Traditional, Smart, and Resilient Smart Buildings

Features / Building Type	Traditional Buildings	Smart Buildings	Resilient Smart Buildings
Control Type	Manual or basic	Automated (Rule-based)	Adaptive and intelligent
Data Utilization	Very limited	Moderate (Real-time monitoring)	Advanced (Real-time + Predictive analytics)
Prediction Capability	None	Limited	High (AI-driven)
Response to Disturbances	Slow and unsystematic	Semi-automated	Fast and intelligent
Level of Resilience	Low	Medium	High
System Integration	Low	Moderate	High (Cyber-Physical Integration)
User Involvement	Very high	Moderate	Minimal (Autonomous behavior)

As can be seen in Table 1, traditional buildings rely mainly on manual control or simple mechanisms, and the role of data in them is very limited. In these types of buildings, the response to disturbances is often slow, unsystematic, and dependent on human intervention. In contrast, smart buildings provide a higher level of efficiency and management by using sensors, monitoring systems, and rule-based automated controls. However, in many cases these buildings still remain at the level of reaction to events and have limited capacity for real prediction or adaptation.

In contrast, resilient smart buildings have a more advanced layer of intelligence. These buildings not only receive and process data in real time, but also, using predictive analytics, adaptive models and intelligent decision-making mechanisms, can change their behavior according to the situation. This transition from reactive to predictive and adaptive logic is one of the most important features of the new generation of buildings. In such a framework, the building is no longer a simply controlled environment, but rather a dynamic and responsive system that is able to perceive changes and make decisions based on them.

One important dimension of this transformation is the change in the place of data in building management. In traditional buildings, data plays a marginal role, and decisions are made based more on experience or human intervention. In smart buildings, data is used to monitor conditions and implement predefined controls. In resilient smart buildings, however, data becomes the main element of decision-making. Information from temperature, humidity, air quality, occupancy, energy consumption, and equipment performance can all be analyzed in an integrated system and used as the basis for adaptive actions. This feature paves the way for systems that not only understand the current situation, but also predict future behavior (Nozari, 2024).

Another important point is the reduction of these buildings' dependence on direct user intervention. As shown in Table 1, in traditional buildings, the user's role in regulation and control is very prominent. In smart buildings, part of this role is transferred to automated systems. But in resilient smart buildings, the system can operate autonomously in many situations. This autonomy does not mean eliminating the human role, but rather shifting the focus from direct control to higher-level monitoring, policy-making, and management.

In summary, the conceptual foundation of resilient smart buildings is based on the principle that a building must go beyond everyday efficiency to be able to cope with uncertainty, disruption, and change. Such a building must not only be smart, but must also be able to maintain its performance, adapt its strategies, and recover quickly in the event of a problem. This paradigm shift defines the path from traditional buildings to smart buildings and ultimately to resilient smart buildings, and provides a theoretical foundation for the development of more advanced frameworks, including adaptive digital twins.

3. DIGITAL TWIN ARCHITECTURE FOR ADAPTIVE BUILDINGS

For smart buildings to move beyond pre-defined responses to adaptive and predictive behaviors, a coherent, multi-layered architecture is essential. At the heart of this transformation, the digital twin is more than just a simulation tool. It is an integrated framework that connects the physical world of a building to a dynamic digital model that not only reflects the current state, but also predicts future behavior and contributes to decision-making.

The digital twin architecture in adaptive buildings is typically based on several key layers, each of which plays a specific role in the perception, analysis, and action cycle. These layers operate hierarchically yet interactively, enabling data and decision flow throughout the system. At the lowest level is the physical layer, which includes equipment, sensors, actuators, and all the actual building components. This layer is the main source of data generation and continuously records information such as temperature, humidity, air quality, energy consumption, equipment status, and user presence (Nozari, 2026).

The data generated in the physical layer is transferred to the data layer through communication infrastructure. This layer is responsible for collecting, storing, filtering, and integrating data. The importance of this layer is that it transforms heterogeneous and scattered data into a structured form that can be used for subsequent analysis. At this stage, data quality, time synchronization, and noise removal play a decisive role in the accuracy of upstream models.

Next is the model layer, which is the analytical heart of the digital twin. In this layer, a combination of physical models (such as heat transfer or fluid dynamics models) and data-driven models (such as machine learning algorithms) is used to represent the behavior of the building. These models are continuously updated with real data and therefore, are able to provide a close-to-real picture of the system state. In addition, this layer also allows for the simulation of different scenarios and the prediction of the consequences of decisions.

Above the model layer, there is the decision-making layer, which is responsible for analyzing the output of the models and choosing the best course of action. In this layer, various methods can be used, such as multi-objective optimization, model-based predictive control (MPC), or reinforcement learning algorithms. The main goal at this level is to create a balance between criteria such as energy consumption, user comfort, operating costs, and the level of resilience. The important point is that the decision-making here is not static, but is constantly readjusted according to current conditions and future predictions (Kumar et al., 2026).

Finally, there is the control layer, which applies the commands generated in the decision-making layer to the physical system. These commands can include adjusting the temperature of HVAC systems, changing the lighting intensity, managing airflow, or even shifting energy loads. The loop that starts at the physical layer and returns to the physical layer after passing through the data, model, and decision layers, forms a closed loop that forms the basis of the adaptive performance of the system.

To better understand this structure, the proposed adaptive digital twin architecture for smart buildings is shown in Figure 1.

Figure 1. Multi-Layer Adaptive Digital Twin Architecture for Smart Buildings

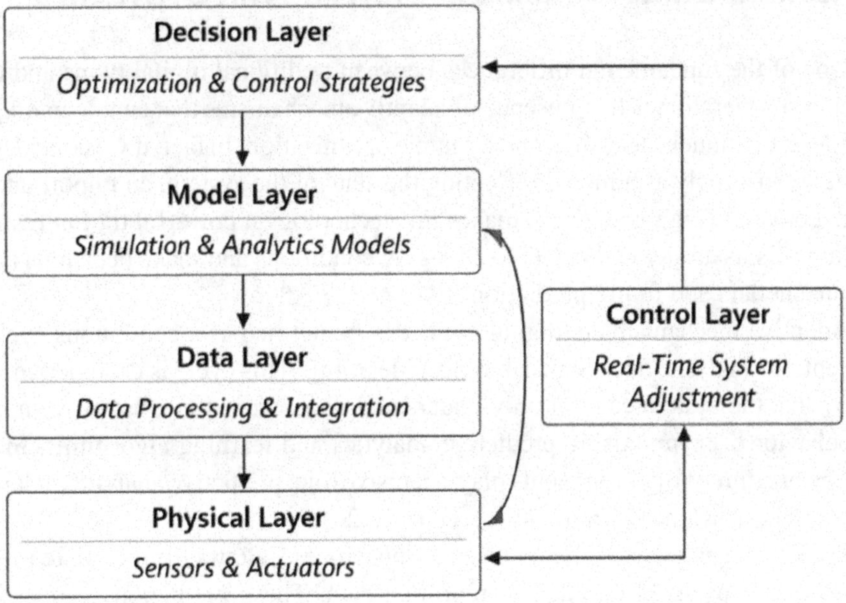

As shown in Figure 1, the data flow starts at the physical layer and, after processing at higher layers, leads to decisions that are then applied back to the physical system. This closed-loop structure allows for a learning and adaptive system that can continuously improve its performance. One of the key points in this architecture is the presence of real-time feedback; this means that any change in the environment or system performance is immediately reflected in the digital model and can influence subsequent decisions.

Another important feature of this architecture is its extensibility and modularity. Each layer can be independently improved or replaced with new technologies without the need to redesign the entire system. For example, simple statistical models can be replaced with advanced deep learning models or control algorithms can be changed based on specific project conditions. This flexibility makes the digital twin a suitable tool for a variety of applications at different scales.

Overall, the digital twin architecture for adaptive buildings provides a platform in which data, models, and decision-making operate in unison. This integration enables the transition from static and reactive systems to dynamic, predictive, and resilient systems. In other words, the building transforms from a passive consumer of resources to an intelligent, self-regulating system that can choose the best possible response in the face of change.

4. ADAPTIVE MECHANISMS IN DIGITAL TWIN SYSTEMS

One of the fundamental differences between traditional digital twins and their advanced generation is the presence of adaptive mechanisms that enable intelligent response, continuous learning, and dynamic optimization. In fact, if we consider the digital twin simply as a model reflecting the state of the system, an important part of its potential is ignored. What makes this technology a powerful tool in resilient buildings is its ability to "adapt" to changing conditions and make decisions based on current data and future predictions.

Adaptive mechanisms in digital twins can be defined as a continuous cycle of perception, analysis, decision-making, and learning. This cycle is designed in such a way that the system can not only understand the current state, but also improve its behavior over time using predictive analytics and learning algorithms. In this framework, three key components play a decisive role: predictive analytics, adaptive control, and machine learning (Wang et al., 2026).

Predictive analytics, as the first step in this process, allows the system to predict future trends using historical and real-time data. These predictions can include temperature changes, energy consumption patterns, user behavior, or even the occurrence of anomalies in equipment performance. The importance of this section is that it moves decision-making from a reactive mode to a proactive one. In other words, the system can take the necessary measures to prevent or mitigate a problem before it occurs.

The next step is adaptive control, which applies the necessary measures to the system based on the output of predictive analytics. This control can take the form of adjusting HVAC system parameters, managing energy loads, or changing operating strategies. The main feature of this type of control is its flexibility in the face of changing conditions. Unlike controls based on fixed rules, adaptive control continuously adapts to new conditions and tries to provide the best possible performance at any given moment (Ristvej et al., 2026).

The third component is automatic learning, which goes beyond the momentary adjustments of the system. In this part, the digital twin learns from its past experiences and improves decision-making policies using machine learning or reinforcement learning algorithms. This learning can include identifying hidden patterns, optimizing control strategies, or even redefining the system's goals. In this way, the system becomes smarter over time and its performance gradually improves.

To better understand the interaction between these three components, the general structure of the adaptive mechanisms in the digital twin is shown in Figure 2.

Figure 2. Key Adaptive Mechanisms in Digital Twin Systems

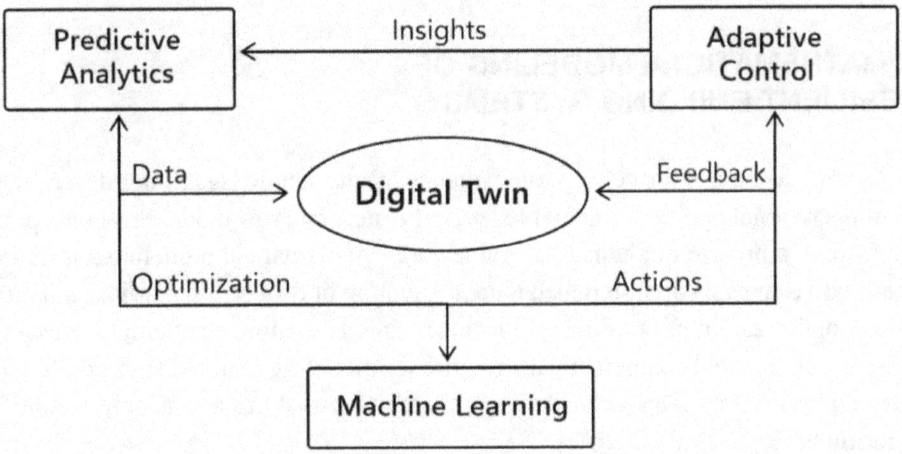

As shown in Figure 2, the digital twin is at the center of an interactive cycle that connects three main components. Predictive analytics turns data into insights; adaptive control turns these insights into actionable actions; and machine learning evaluates the results to improve future policies. This cycle is repeated continuously, each time bringing the system to a higher level of efficiency and adaptability.

One of the important points in this structure is the presence of multiple feedback loops. Unlike traditional control systems, which usually have a simple feedback path, in an adaptive digital twin, feedback occurs at multiple levels. For example, the results of control actions not only feed back to the data layer, but also directly affect the learning process. This feature allows for continuous modification of models and strategies, allowing the system to adapt to complex and nonlinear conditions (Mammone et al., 2026).

On the other hand, these mechanisms are designed to balance different goals. In a resilient smart building, goals such as reducing energy consumption, increasing user comfort, and maintaining system stability are usually considered simultaneously. Adaptive mechanisms, using multi-objective analysis and continuous learning, are able to consider these goals simultaneously and make decisions that establish the best balance between them.

Finally, it can be said that adaptive mechanisms transform the digital twin from a monitoring tool to a decision-support and even decision-making system. This transformation plays a key role in the realization of resilient buildings, because such buildings require systems that can act intelligently and dynamically in the face of uncertainty. In this framework, the digital twin not only reflects reality, but also acts as an active entity in managing and optimizing the system; an entity that can learn, adapt, and ultimately ensure stable and reliable performance.

5. MATHEMATICAL MODELING OF RESILIENT BUILDING SYSTEMS

In order to elevate the concept of resilience in smart buildings from a descriptive to an operational and implementable level, it is necessary to model this concept in the form of a precise mathematical framework. Mathematical modeling allows for quantitative analysis of system behavior, evaluation of different scenarios, and ultimately optimization of building performance. In this section, an attempt is made to present a conceptual-mathematical structure for modeling resilient smart buildings that can be used as a basis for developing decision-making and adaptive control algorithms (Chen et al., 2026).

In the simplest case, a smart building can be considered as a dynamic system that has specific inputs, states, and outputs. Inputs include parameters such as environmental conditions (temperature, humidity, solar radiation), user presence, and control signals. System states can include indoor temperature, air quality, equipment status, and stored energy levels. Outputs also typically include metrics such as thermal comfort, energy consumption, and overall system performance.

To model this system, first the sets, parameters, and decision variables are defined, which form the basic framework of the model. These components are presented in Table 2.

Table 2. Sets, Parameters, and Decision Variables in Resilient Building Modeling

Category	Symbol	Description
Set	$t \in T$	Set of discrete time periods
Set	$z \in Z$	Set of building zones
Parameter	T_t^{out}	Outdoor temperature at time t
Parameter	$O_{z,t}$	Occupancy level in zone z at time t
Parameter	E^{max}	Maximum allowable energy consumption
Parameter	T^{min}, T^{max}	Acceptable indoor temperature bounds
Decision Variable	$u_{z,t}$	Control input (e.g., HVAC level) in zone z at time t

continued on following page

Table 2. Continued

Category	Symbol	Description
State Variable	$T_{z,t}$	Indoor temperature in zone z at time t
Output Variable	E_t	Total energy consumption at time t
Output Variable	R_t	Resilience index at time t

Based on the definitions provided in Table 2, the dynamic behavior of the building can be modeled using state equations. For instance, the evolution of indoor temperature in a given zone can be expressed as a function of outdoor temperature, internal heat gains due to occupancy, and the control input of the HVAC system:

$$T_{z,t+1} = f\left(T_{z,t}, T_t^{\text{out}}, O_{z,t}, u_{z,t}\right) \tag{1}$$

This equation indicates that the future state of the system depends on its current state, environmental conditions, and control decisions. Such a formulation enables forecasting of system behavior over future time steps and forms the basis for predictive decision-making.

To evaluate system performance, one or more objective functions must be defined. In resilient smart buildings, these objectives are typically multiple and potentially conflicting. For example, minimizing energy consumption may conflict with maximizing occupant comfort. In addition, resilience is introduced as a key performance objective. Therefore, the problem is formulated as a multi-objective optimization problem:

$$\min\left(E_t, D_t, -R_t\right) \tag{2}$$

where:

- E_t: energy consumption
- D_t: thermal discomfort index
- R_t: resilience index

The resilience index can be defined as a composite measure incorporating several dimensions, such as system stability, adaptability, and recovery capability. For example:

$$R_t = \alpha \cdot S_t + \beta \cdot A_t + \gamma \cdot \text{Rec}_t \tag{3}$$

where:

- S_t: system stability
- A_t: adaptability
- Rec_t: recovery performance
- α, β, γ: weighting coefficients

The model must also include a set of constraints that define the operational limits of the system. These constraints may represent physical, comfort-related, and energy-related boundaries:

$$\begin{aligned} T^{\min} \leq T_{z,t} \leq T^{\max} \\ E_t \leq E^{\max} \\ u_{z,t} \in [u^{\min}, u^{\max}] \end{aligned} \tag{4}$$

These constraints ensure that control decisions remain within safe and acceptable operating ranges.

A key feature of this modeling framework is its dynamic nature. Unlike static models, decisions are updated iteratively at each time step based on newly available information. This characteristic enables seamless integration with digital twin systems, where real-time data continuously feeds the model and supports adaptive decision-making.

Moreover, this mathematical structure can be combined with advanced optimization algorithms such as NSGA-II or reinforcement learning techniques. In such cases, the system can autonomously learn optimal control policies under varying conditions and progressively improve its performance. This capability is particularly valuable in complex and uncertain environments.

6. MULTI-OBJECTIVE OPTIMIZATION FOR ADAPTIVE RESILIENCE

In smart resilient buildings, decision-making is not based solely on a single objective, but rather a set of simultaneous and sometimes conflicting objectives must be considered. For example, reducing energy consumption can conflict with maintaining thermal comfort for users, or increasing the level of resilience may lead to increased operating costs. Therefore, multi-objective optimization frameworks are proposed as one of the main pillars in the design and operation of this type of buildings. These frameworks enable the simultaneous analysis and management of multiple performance criteria and allow the decision-maker to choose the best balance between different objectives (Al-Jamal et al., 2026).

In this context, the optimization problem can be defined as a multi-objective model in which several objective functions are optimized simultaneously. These objectives usually include minimizing energy consumption, minimizing thermal dissatisfaction, and maximizing the resilience index (Divya et al., 2026).

A key feature of this type of problem is that there is usually no single solution that optimizes all objectives simultaneously. Instead, a set of optimal solutions, called "Pareto", is formed, each of which represents a specific trade-off between the objectives. The final choice among this set depends on the decision-maker's preferences or operational policies.

In adaptive buildings based on digital twins, multi-objective optimization is not performed statically, but is a dynamic and real-time process. This means that as new data from the real environment enters, the model is updated and the optimization problem is solved again. This feature allows the system to adapt to changing conditions and ensures that decisions are always appropriate to the current situation and future predictions. In this framework, the digital twin plays a key role in performing optimization as a computational and analytical platform (Vikram & Manjokumar, 2026).

To solve multi-objective optimization problems in such environments, various methods are used. One of the most common of these methods is meta-heuristic algorithms such as NSGA-II, which are capable of generating a diverse set of Pareto solutions. These algorithms perform well, especially in complex and nonlinear problems with multiple decision variables and constraints. In addition, machine learning and reinforcement learning-based methods have also emerged as new tools in this field that can learn optimal policies through interaction with the environment.

Another important point is how to define and weight different objectives. In many applications, it is necessary to determine the relative importance of objectives in order to select the appropriate option from among the Pareto solutions. This weighting can be done statically or dynamically. In the dynamic mode, the weights change based on environmental conditions or operational priorities. For example, during peak energy consumption conditions, energy reduction may be a priority, while during critical conditions, increasing resilience becomes more important. This flexibility is one of the important advantages of using adaptive frameworks alongside multi-objective optimization (Farid et al., 2026).

In addition, it is important to consider uncertainties in the optimization process. Parameters such as user behavior, weather conditions, or equipment performance may not be precisely predictable. In these cases, the use of robust or stochastic optimization methods can help improve system performance. These methods attempt to provide solutions that perform reasonably well under a wide range of possible conditions.

Another important aspect of multi-objective optimization is its relationship with control mechanisms. The output of the optimization process is usually applied to the system in the form of a set of control decisions (such as HVAC settings or energy load management). These decisions should not only meet short-term goals, but also contribute to the sustainability and resilience of the system in the long term. Here, combining optimization with predictive control (MPC) can be an effective solution, as this method allows for the consideration of the future time horizon in the decision-making process.

Overall, multi-objective optimization is considered a key tool in the design and operation of resilient smart buildings. This approach allows for the simultaneous management of different and conflicting goals and allows the system to provide the best possible performance in changing conditions. Combined with digital twins and adaptive mechanisms, this type of optimization becomes a dynamic, learning and intelligent process that can play a decisive role in the realization of the buildings of the future.

7. RESILIENCE METRICS AND PERFORMANCE EVALUATION

In order for the concept of resilience in smart buildings to move from a theoretical idea to a measurable and manageable criterion, it is necessary to define a set of quantitative indicators that can assess the performance of the system under different conditions. Without these indicators, it is neither possible to compare different solutions nor to accurately assess the improvement achieved by the use of technologies such as digital twins or adaptive control. Therefore, defining and measuring resilience indicators is a key component in the design and operation frameworks of resilient smart buildings.

Resilience is inherently a multidimensional concept and cannot be described by a single indicator. Instead, it is necessary to consider a set of indicators, each of which reflects an aspect of the system's behavior in the face of disturbances. In general, these metrics can be classified into three main categories: performance stability metrics, adaptability metrics, and recovery metrics.

The first category, performance stability metrics, refers to the ability of a system to maintain its performance level in the event of a disturbance. For example, it can be examined to what extent the thermal comfort quality or system performance level is maintained during a sudden increase in external temperature or a temporary power outage. These metrics are usually defined as a percentage of the desired performance and indicate how resilient the system is to stress.

The second category, adaptability metrics, measures the ability of the system to change its behavior to suit new conditions. These metrics can include the speed of response to changes, the rate of change in control policies, or the effectiveness of new settings. For example, if a building can adjust its ventilation settings to maintain the desired temperature within a short period of time after a thermal load increase, it can be said to have a high level of adaptability.

The third category is recovery indicators, which are related to the system's ability to return to normal conditions after a disturbance. One of the most important criteria in this regard is the recovery time; that is, the time it takes for the system to return from an unfavorable state to a stable state. In addition, the amount of performance loss during the recovery period can be considered, as well as the costs associated with it. These indicators are especially important in assessing the performance of the system in the face of serious crises.

To combine these different dimensions, an overall index called the "resilience index" is usually defined, which is calculated as a combination of the above criteria. This index can be defined as a weight of different components, in such a way that the relative importance of each dimension is taken into account. For example, in a hospital building, performance sustainability may be more important than energy efficiency, while in an office building, these priorities will be different. Therefore, flexibility in defining indicators is an important feature in this area.

One common method for assessing resilience is to analyze the behavior of the system over time, especially when faced with a specific disturbance. In this method, the system performance is considered as a function of time and the amount of performance degradation, the duration of the failure, and the speed of recovery are analyzed. This type of analysis can provide a comprehensive picture of how the system responds to disturbances and reveal its weaknesses.

In addition to resilience indicators, other performance metrics need to be evaluated simultaneously. These metrics include energy consumption, user comfort levels, indoor air quality, and operating costs. In fact, performance evaluation in resilient smart buildings is a multi-criteria process in which different objectives must be balanced. This doubles the importance of using integrated evaluation frameworks.

Digital twin plays an important role in this evaluation process. Using real-time data and predictive models, various indicators can be calculated and monitored continuously. This capability enables the dynamic evaluation of system performance and allows building managers to be aware of the resilience status in real time. In addition, different scenarios can be simulated in the digital twin environment and the impact of different decisions on performance indicators can be examined.

Another important point is the relationship between performance evaluation and the optimization process. The indicators defined in this section are usually used as objective functions or constraints in optimization models. In other words, what is measured is what will be optimized. Therefore, the accuracy and comprehensiveness of the definition of indicators have a direct impact on the quality of the final decisions.

Finally, it can be said that the definition of resilience indicators and performance evaluation is an integral part of the design of adaptive smart buildings. These indicators not only provide the possibility of measuring the current state of the system, but also act as tools to guide control and optimization processes. By using these frameworks, it is possible to design buildings that are not only efficient under normal conditions, but also demonstrate stable, reliable and intelligent performance in the face of challenges and uncertainties.

8. INTEGRATION WITH SMART INFRASTRUCTURE AND ENERGY SYSTEMS

One of the most important aspects in the development of resilient smart buildings is how they interact with smart city infrastructure and energy systems. In the past, buildings were considered as relatively independent units that were powered only by the electricity grid or other energy sources. However, in modern approaches, the building is defined as an active node in a larger ecosystem that includes smart power grids, renewable energy sources, energy storage systems and other urban infrastructure. This change in attitude elevates the role of the building from a passive consumer to an active and contributing element in energy management and system sustainability.

In this framework, the digital twin plays a key role in creating this integration. The digital twin can act as an intermediary layer that collects, analyzes and synchronizes data related to the building, the energy grid and other infrastructure. This layer not only allows for system status monitoring, but also provides a platform for coordinated decision-making between different components. In other words, the digital twin allows the building to operate at a level beyond its physical boundaries and interact with its surrounding environment in a more optimal manner.

One of the most important components of this integration is the connection of the building to the Smart Grid. In such a system, the building can receive information about the current energy price, the network load status, and even demand forecasts and make decisions based on them. For example, during times of high energy demand and high electricity prices, the building can reduce its consumption or use stored resources. Conversely, during times when energy is cheaper or more abundant, high-consuming operations can be transferred to that time period. This type of interaction, in addition to reducing costs, also contributes to the sustainability of the entire grid.

In addition, the integration of renewable energy sources such as solar panels or wind systems into the building plays an important role in increasing resilience. These sources can provide part of the building's energy needs independently and reduce grid dependency. However, the intermittent nature of these resources requires intelligent and predictive management. Here, the digital twin can maximize system efficiency by predicting energy production and adapting it to consumption patterns.

Energy storage systems are another important component in this ecosystem. Batteries or other storage technologies can store excess energy during production and use it when needed. This capability is particularly important in critical situations or power outages and can play a decisive role in maintaining the building's performance. Optimal management of these resources requires accurate and real-time decision-making, which is made possible through the digital twin and advanced control algorithms.

To better understand these interactions, the architecture of integrating smart buildings with energy infrastructure is shown in Figure 3.

Figure 3. Integration of Digital Twin-Enabled Smart Buildings with Energy and Infrastructure Systems

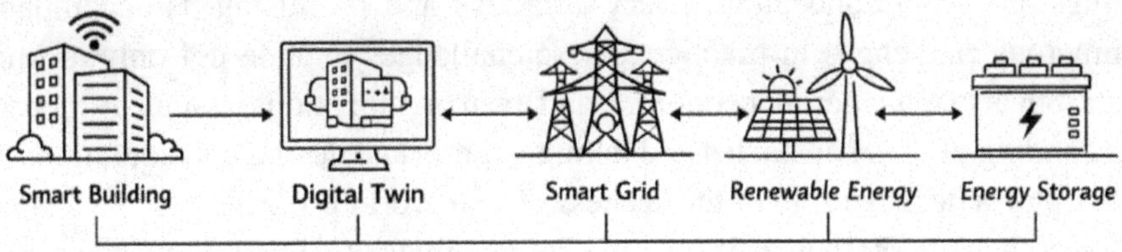

As can be seen in Figure 3, the digital twin is at the center of this ecosystem and acts as a communication hub between the building and other components. Data is collected from various sources such as the power grid, energy generation systems, and internal building equipment and processed in the digital twin. Then, optimal decisions are sent to the various components in the form of control signals. This structure creates a closed and coordinated loop in which all components operate in an integrated manner (Tlili et al., 2026).

One of the main advantages of this approach is increased system resilience to disturbances. For example, in the event of a power outage from the grid, the building can automatically switch to storage or internal generation resources. Also, in case of fluctuations in renewable energy production, the system can maintain balance by adjusting consumption or using storage. This level of coordination is only possible if information is exchanged accurately and in real time between components and decisions are made intelligently (Li et al., 2026).

Another important point is the role of this integration in multi-objective optimization. As mentioned in the previous sections, objectives such as reducing energy consumption, increasing comfort and improving resilience must be considered simultaneously. Connecting the building to smart infrastructure provides access to more options for decision-making and, as a result, the solution space is expanded. This can lead to more optimal and sustainable solutions.

Finally, it can be said that integrating smart buildings with energy and urban infrastructure is a fundamental step towards realizing sustainable and resilient systems. Meanwhile, the digital twin, as a key element, plays a coordinating and decision-making role, enabling optimal resource utilization. This approach not only improves building performance, but also contributes to the sustainability and efficiency of the entire urban system.

9. IMPLEMENTATION CHALLENGES AND DESIGN CONSIDERATIONS

The implementation of adaptive digital twin-based frameworks in resilient smart buildings, although conceptually very attractive and promising, is accompanied by numerous challenges in practice. These challenges include not only technical aspects, but also managerial, economic, and even organizational issues. A thorough understanding of these obstacles and taking appropriate design considerations into account play a decisive role in the success of real-world projects.

One of the first challenges is data quality and integrity. Digital twins are highly dependent on real-time and accurate data from the physical environment. In practice, data collected from sensors may be subject to noise, latency, or even measurement errors. Furthermore, data is often collected from different sources and in heterogeneous formats, making it difficult to integrate them. Without a reliable data infrastructure, analytical and decision-making models cannot perform optimally. Therefore, designing data management systems that include filtering, time synchronization, and data validation is a fundamental requirement in this area (Othman et al., 2026).

The second challenge is related to the accuracy and complexity of the models. The models used in the digital twin must be able to represent the actual behavior of the building with acceptable accuracy. However, increasing model accuracy is usually accompanied by increasing computational complexity. This is especially important in real-time applications that require fast decision-making. Therefore, one of the design considerations is to strike a balance between model accuracy and computational efficiency. In many cases, the use of hybrid models that include physical and data-driven parts can be a suitable solution.

Another issue is real-time processing and system scalability. Modern smart buildings generate huge volumes of data in short time intervals. Processing this data and executing analytical and optimization algorithms in a timely manner requires powerful computing infrastructure. The use of cloud computing, edge computing, and distributed architectures can help reduce latency and increase system efficiency. At the same time, the system design must be such that its performance does not degrade as it scales (for example, at the level of a complex or smart city).

Another important challenge is cybersecurity and data privacy. Since the digital twin is directly connected to the physical and control infrastructure, any intrusion or cyber attack can have serious consequences. In addition, the collected data may contain sensitive information related to user behavior or consumption patterns. Therefore, the system design must include strong security mechanisms such as encryption, multi-factor authentication, and access monitoring. Attention to this issue is essential not only to maintain system security, but also to gain user trust.

From a design perspective, one of the key principles is the modularity of the system. The digital twin architecture should be designed in such a way that its different components (such as the data layer, model or control) can be developed, replaced or upgraded independently. This feature allows for gradual system updates and adaptation to new technologies. Otherwise, any small change may require a redesign of the entire system, which is not economically and operationally viable.

Interoperability is another important consideration. Smart buildings usually include equipment and systems from different manufacturers that use different protocols and standards. The lack of unified standards can create a serious obstacle to system integration. The use of open standards and the design of flexible communication interfaces can greatly reduce this problem.

On the other hand, implementation and operation costs must also be considered. Deploying an adaptive digital twin requires investments in areas such as sensors, communication infrastructure, analytical software, and computing resources. In addition, the costs of maintaining and updating the system must also be considered. Therefore, cost-benefit analysis and the selection of economic solutions play an important role in making decisions for implementation.

Another important point is organizational acceptance and user readiness. Even the most advanced systems will not be used properly if they are not accepted by users or managers. Therefore, designing simple user interfaces, providing appropriate training, and creating transparency in system performance are among the factors that can help to better accept this technology (Kumar et al., 2026).

Finally, it should be noted that implementing digital twins in resilient buildings is a gradual and evolutionary process. Rather than trying to implement a complete system in one go, a phased approach can be used, where the system is gradually expanded and new capabilities are added. This approach not only reduces implementation risks, but also allows for continuous learning and improvement.

Overall, the challenges and design considerations in implementing adaptive digital twins demonstrate the complexity of the field and the importance of systematic and integrated approaches. By considering these factors, we can pave the way for the development of resilient smart buildings in a way that is not only technically efficient, but also economically, operationally, and socially sustainable.

10. CONCLUSION

The transformation in the way buildings are designed, operated, and managed has accelerated dramatically in recent decades. In the meantime, the combination of the concepts of smart building, resilience, and digital twin has opened a new horizon in the development of sustainable and adaptive infrastructure. What this research emphasizes is the transition from buildings with static and reactive performance to dynamic, learning, and predictive systems that can provide reliable and sustainable performance in the face of uncertainties and environmental changes.

In this framework, the conceptual foundations of resilient smart buildings were first examined and it was shown that resilience goes beyond resistance to disturbances and includes dimensions such as performance stability, adaptability, and rapid recovery. This perspective paves the way for redefining the role of the building from a passive consumer to an active and responsive system. Subsequently, the digital twin architecture was introduced as the main infrastructure of this transformation; An architecture that enables real-time monitoring, analysis, and decision-making by creating a continuous link between the physical world and the digital model.

One of the most important achievements of this research is the explanation of the role of adaptive mechanisms in improving the performance of the digital twin. It was shown that the combination of predictive analytics, adaptive control, and machine learning transforms the digital twin from a purely analytical tool into an intelligent decision-support system. This transformation enables rapid and effective response to changes and paves the way for continuous improvement of system performance over time.

In the mathematical modeling section, a framework for quantitatively expressing the behavior of resilient buildings was presented, in which objectives such as energy consumption, user comfort, and resilience index were considered simultaneously. These models showed that resilience can be introduced into decision-making processes as a measurable objective and the best balance between different objectives can be achieved through multi-objective optimization methods. In this regard, the importance of using advanced algorithms and adaptive mechanisms in solving complex and dynamic problems was highlighted.

Also, the role of integrating buildings with smart energy infrastructures was considered. Connecting to smart grids, utilizing renewable energy sources, and using storage systems not only helps improve energy efficiency, but also significantly increases the level of system resilience. In the meantime, the digital twin, as a coordinating core, enables integrated management of these components and enables decision-making at a larger level.

On the other hand, the challenges of implementing these technologies were also examined. Issues such as data quality, model complexity, the need for real-time processing, cybersecurity, and implementation costs are among the factors that must be considered in the design and deployment of real systems. Addressing these challenges and incorporating appropriate design considerations plays a critical role in the success of digital twin-based projects.

Overall, the results of this research show that adaptive digital twin can be used as a comprehensive framework for developing resilient smart buildings. By combining data, models, optimization, and learning, this framework enables the creation of systems that are not only efficient and sustainable, but also capable of performing reliably in variable and unpredictable conditions. Such systems are an important step towards realizing smart, sustainable, and human-centered cities.

Ultimately, it can be said that the future of buildings is moving towards systems that continuously learn from their environment, make decisions, and adapt to new conditions. In this direction, adaptive digital twin is not only a technological tool, but also a new paradigm in building engineering and management thinking; A paradigm that can play a decisive role in shaping the smart and resilient infrastructure of the future.

REFERENCES

Al-Jamal, M. Q., Alsarhan, A., Aljamal, Q., AlJamal, M., Khassawneh, B. S., Al Nuaim, A., & Al Nuaim, A. (2026). AI-Enabled System-of-Systems Decision Support: BIM-Integrated AI-LCA for Resilient and Sustainable Fiber-Reinforced Façade Design. *Information (Basel), 17*(2), 126. DOI: 10.3390/info17020126

Chen, Q., Wang, X., & Liu, L. (2026). Next-Generation AI-Driven Digital Twin Framework for Market Volatility Warning and Risk Detection in Global Logistics Supply Chains. *IEEE Communications Standards Magazine*, 1–10. DOI: 10.1109/MCOMSTD.2025.3641867

Divya, A., Ramesh, S., Chakrasali, S. V., Kumar, M. K., Taj, M., Ganiya, R. K., Nageswara Rao, J., K, B., & Qamar, S. (2026). Federated Digital-Twin–Empowered Blockchain Architecture for Autonomous and Resilient Smart Water Management in Sustainable Cities. *Sustainable Computing : Informatics and Systems, 50*, 101303. DOI: 10.1016/j.suscom.2026.101303

Farid, H. A., Rasti, M., Pongracz, E., & Anvari-Moghaddam, A. (2026). Digital Twin Assisted Real-Time Energy Management System For Smart Homes. *Energy and Building*, ●●●, 116978.

Hossain, M. I., Hossan, M. R., Shaon, Z. H., & Ferdous, M. N. (2026). Linking digital twin paradigm for urban heat monitoring and policy integration to building smart city climate resilience. *Discover Cities, 3*(1), 1. DOI: 10.1007/s44327-025-00179-8

Khdhir, R., Magdich, A., Elloumi, M., Asklany, S., Hamdaoui, R., Abbas, G., & Elrashidi, A. (2026). Building resilient national critical infrastructure: A digital twin-based framework for comprehensive insider and external threat detection. *Journal of King Saud University. Computer and Information Sciences, 38*(3), 118. DOI: 10.1007/s44443-026-00464-5

Kumar, M., Saranya, G., Radder, U., Sundaramurthy, V., Divya, A., & Masih, R. K. (2026). Reinforcement-led digital twin optimization for secure operation in cyber-physical grids. *Electric Power Systems Research, 253*, 112423. DOI: 10.1016/j.epsr.2025.112423

Kumar, M., Saranya, G., Radder, U., Sundaramurthy, V., Divya, A., & Masih, R. K. (2026). Reinforcement-led digital twin optimization for secure operation in cyber-physical grids. *Electric Power Systems Research, 253*, 112423. DOI: 10.1016/j.epsr.2025.112423

Li, M., Yang, C. M., Lo, W., & Kao, Y. W. (2026). A Digital-Twin-Enabled AI-Driven Adaptive Planning Platform for Sustainable and Reliable Manufacturing. *Machines (Basel), 14*(2), 197. DOI: 10.3390/machines14020197

Mammone, V., Binetti, M. S., & Massarelli, C. (2026). Digital Twin in Territorial Planning: Comparative Analysis for the Development of Adaptive Cities. *Urban Science (Basel, Switzerland), 10*(2), 80. DOI: 10.3390/urbansci10020080

Nguyen, T. D., & Adhikari, S. (2026). Advancing Digital Twins for Building Lifecycle Management in Construction: A Systematic Literature Review. *Buildings (Basel, Switzerland), 16*(6), 1151. DOI: 10.3390/buildings16061151

Nozari, H. (2024). Green supply chain management based on artificial intelligence of everything. *Journal of Economics and Management, 46*(1), 171–188. DOI: 10.22367/jem.2024.46.07

Nozari, H. (2025). Integrating Neuromorphic Intelligence with Fuzzy Multi-Objective Models in Resilient Green Supply Chain Design. Process Integration and Optimization for Sustainability, 1-14.

Nozari, H. (2026). Digital twin-based predictive maintenance in cold chain logistics. *Operations Research, 60*(1), 481–499. DOI: 10.1051/ro/2025167

Nozari, H., & Yordanova, Z. (2026). A Self-Deciding Adaptive Digital Twin Framework Using Agentic AI for Fuzzy Multi-Objective Optimization of Food Logistics. *Algorithms, 19*(3), 218. DOI: 10.3390/a19030218

Othman, A. A. E., & Mohamed Abdrabou, N. (2026). A digital twin framework for enhancing building maintainability during the design process in architectural design firms in Egypt. *International Journal of Construction Management*, 1–18. DOI: 10.1080/15623599.2026.2624416

Ristvej, J., Halúsková, B., Nováková, K., & Chovanec, D. (2026). Implementation Maturity Levels of Digital Twin Technology and Data Content Design for Flood Digital Twin. *Smart Cities, 9*(2), 28. DOI: 10.3390/smartcities9020028

Tlili, F., Ayed, S., & Fourati, L. C. H. A. A. R. I. (2026). Integration paradigm of intelligent digital twin into UAVs systems. *Soft Computing, 30*(4), 1–18. DOI: 10.1007/s00500-025-11065-1

Vikram, C., & Manojkumar, R. (2026). AI-Digital Twin Synergy for Smart and Autonomous Wind Energy Systems. In Advanced Wind Energy Systems: Grid Integration, Markets, and Sustainable Infrastructure (pp. 137-174). IGI Global Scientific Publishing.

Wang, L., Zhang, X., & He, W. (2026). The construction and optimization of resilient community living sphere driven by digital twins. *Scientific Reports*, *16*(1), 4268. DOI: 10.1038/s41598-025-34455-9 PMID: 41547970

Yordanova, Z., & Nozari, H. (2026). Enhancing Innovation and Resilience in Entrepreneurial Ecosystems Using Digital Twins and Fuzzy Optimization. *Digital*, *6*(1), 25. DOI: 10.3390/digital6010025

Yorucu, V., Bekun, F. V., & Yitmen, I. (2026). Interdisciplinary Studies on Urban Resilience: Urban Digital Twins and Construction 5.0 for Resilient, Sustainable and Smart Cities. In *Modern Approach of Resilient and Sustainable Smart Cities* (pp. 141–155). Springer Nature Switzerland. DOI: 10.1007/978-3-032-05715-0_9

Chapter 6
Semantic Twins for Enhanced Intellectual Asset Management:
Assessing Patent Feasibility for Industrial Applications

Miglena Molhova-Vladova
http://orcid.org/0000-0002-2367-8549
University of National and World Economy, Bulgaria

Teodora Nikolova
http://orcid.org/0009-0003-9945-7173
University of Library Studies and Information Technologies, Bulgaria

ABSTRACT

Intellectual asset management in technology-intensive industries has traditionally used bibliometric and citation-based methods to provide insight into the structure of the patent landscape but does not provide much information on the content of a particular claim. These approaches fail to determine if the technical and methodological content of a particular patent actually applies to the knowledge structures and operational limitations and regulatory demands of a specific industrial environment. The chapter proposes a conceptual and operational answer to that question, namely the semantic twin, a formally and machine-understandable model of a body of knowledge that is a virtual counterpart to patent claims and can be assessed, rated, and understood in terms of their practical applicability in the real world.

DOI: 10.4018/979-8-2600-1669-5.ch006

1. INTRODUCTION

Over the last few decades, patents have played a paradoxical role in corporate strategy of innovation. On the one hand, they constitute the main legal tools by which companies claim ownership of technological knowledge, and their accretion is generally viewed as an indicator of the output of innovation. On the other hand, the volume of the total world patent stock, which is now well over ninety million active documents, makes the task of evaluating them for strategic purposes a very resourceful one. The mismatch of the quantity of potentially relevant patents and capacity of organizations to evaluate them meaningfully represents a structural gap in the intellectual asset management process – one that shapes technology management and policy developing, research and development (R&D) investment allocations, freedom-to-operate determinations in ways that are often opaque to the organizations involved (Van Zeebroeck & Graham, 2010; Hall et al., 2014).

The current methods of patent analysis can be divided into two groups. The first one is bibliometric and network-analytic in nature: it focuses on citation patterns, filing paths, technology categories, and co-inventor networks to chart the structural characteristics of patent landscapes. These approaches assume that the semantic content of patent claims is opaque: a patent text is a node in a network, and its value is represented by its location in that network, and not by its actual analysis of what the patent actually declares to be doing (Trajtenberg, 1990; Lanjouw & Schankerman, 2001). The second dimension of approaches uses natural language processing, topic modelling and machine learning classification to derive features in patent text (Tseng et al., 2007). These techniques can determine the nature of a patent, but they cannot judge whether the claims made are coherent to the formal ontological representation of a target domain, whether the axioms of the patent are logically satisfiable by the domain model, or what targeted adjustments would be needed to align the patent with an existing technological infrastructure or simply put - is the patent feasible at all at an industry level.

This paper introduces the semantic twin concept that offers such a tool. The semantic twin is based on the digital twin paradigm which has disrupted the practice of monitoring and maintenance in engineering fields by establishing virtual copies of real systems that may be simulated, questioned and updated in real time (Grieves, 2014; Tao et al., 2019). While a digital twin is a model of a system behavior in particular conditions, a semantic twin expresses what a body of knowledge is, what the logical connections are between concepts, what the rules are that govern valid knowledge states, and what requirements must be met before a given claim can be said to make sense and be relevant to a given body of knowledge.

2. LITERATURE REVIEW

2.1. Digital Twins and the Semantic Dimension of Knowledge Representation

The concept of digital twins and the semantic dimension of knowledge representation is systematically presented in Grieves (2014) who describes digital twin as a virtual version of a tangible product, system, or process, to which real-time data is provided by its physical counterpart, allowing it to be simulated, monitored, and proactively maintained throughout the operation lifecycle. Tao et al. (2019) described the architecture of digital twins, which consists of three constitutive functions: data acquisition, model construction, and service delivery, thus defining the difference between twins and simpler simulation settings. Rasheed et al. (2020) state that the practical usefulness of digital twin directly depends on the semantic richness of the underlying model.

The development of semantic-specific twin architectures has attracted a great amount of academic interest over the past several years. El-Hajj (2026) introduced a systematic ontological analysis of the semantic basis required to implement digital twin implementations and claimed that the ontology is the primary epistemic organization of any twins claim to be about meaning, and not behavior. Jazi et al. (2025) also suggested a semantic-based digital twin framework, combining ontological models with process simulation to facilitate knowledge-based production decisions, in the manufacturing domain, where the semantic and physical aspect of twinning can be integrated to enable more holistic industrial decision-making. Where a digital twin provides the answer to the question of how a given system works, a semantic twin should provide the answer of what the relevant concepts in a field are and how they depend on each other. Arya et al. (2025) designed DTSafe-O, a semantic digital twins ontology for construction safety management, as an example of how the semantic twin can be used as a normative reference model against which construction configurations can be evaluated to be compliant. These contributions tend towards the concept of the semantic twin as a sub-category of the digital twin paradigm, one that has value in its ability to render domain knowledge explicit, formal and computationally accessible.

2.2. Intellectual Asset Management and Patent Feasibility Analysis

Intellectual asset management (IAM) is the strategic process of identifying, protecting, valuing, and exploiting an organization's intangible assets—such as patents, trademarks, know-how, and trade secrets (McConnachie, 1997). It aims to maximize the commercial value of these assets, aligning them with business goals to boost competitiveness and revenue. The IAM of technology-intensive companies involves various activities, such as construction of patent portfolio, licensing strategy, freedom-to-operate analysis, and acquisition of technology. Fischer and Ringler (2007) showed that strategic value of a patent portfolio hinges crucially on the compatibility between the technical content of each patent and the knowledge needs of the innovation path of the firm. Hall et al. (2014) also discovered that despite years of research, there is still consistent finding of evidence that a small percentage of patents bring with them the vast majority of economic value, meaning that the quality of selection and evaluation is much more important than the volume of acquisitions. These results suggest the significance of content-level patent analysis - evaluation based on what patents assert, rather than the number of patents that a company possesses. Bessen and Meurer (2008) studied the precision of patent claims in various technological areas and discovered that often patents have broad, vaguely phrased claims that state what happens but not how it happens, with this especially relevant to the information and communication technology (ICT) domain. The direct effect of this ambiguity is that a patent with semantically underspecified claims conveys less actionable knowledge than a patent with claims that identify the concepts, relationships and logical constraints of described technology with precision. To evaluate the relevance of patent and the possibilities for its implementation one will thus need a formal reference model of what adequate semantic coverage would look like in the target domain. The patent feasibility assessment problem is particularly acute in areas where patent thickets occur - the thicket of overlapping intellectual property rights (IPR) that any innovator must find its way through to commercialize a new technology (Shapiro, 2000). Blind et al. (2006) reported the inhibition of innovation by patent thickets in the ICT industry which is closely applicable to the case study involving software engineering as illustrated in Section 5.

The challenges in the evaluation of patent feasibility are further complicated by the strategic communicative purpose that governs patent drafting. Patent attorneys draft patent claims as broadly as possible while trying to provide limited information about how the patent could be implemented. This creates a systematic gap between the scope of a patent cover and the level of detail needed to implement a patent. Srivastava and Gnyawali (2011) argued that patents do not exist as a legal protection tool only, but as codification tools of knowledge: drafting a patent necessitates tacit knowledge to be made explicit in a manner that can later be transmitted and developed. The semantic twin can answer this very aspect of assessing patent value by offering a formal reference model of the target knowledge domain, which establishes the environment of systematic evaluation of the extent to which a patent disclosure of knowledge matches the knowledge requirements of the recipient organization.

2.3. Research Gap

The above review reveals a convergence research gap – the need to have instruments which can interact with the semantic content of patent claims to conduct content-level feasibility assessment and the conceptual toolkit to create formal domain representations that can be used as evaluation reference models. The study presented in this paper fills this knowledge gap by developing a Logic-Guided Ontology Schema (LOGOS) - based semantic twin, tested for a software defect prioritization system with six actual patents on it, and showing that the resulting concept-mapping and axiom-compliance analysis are structured, reproducible and actionable patent compatibility judgments that could not be generated by any other instrument.

3. THEORETICAL FRAMEWORK

3.1. Defining the Semantic Twin

The semantic twin in the paper is defined as a formal, machine understandable model of the knowledge structure of a field that can be used in testing intellectual assets in general, and a patent in particular, to check conceptual coverage, logical consistency, and practical applicability. This definition differs with the digital twin in both a philosophical and practical way. A digital twin represents the dynamical behavior of a physical or computational system: it contains operational states, input-output relations and time dynamics. A semantic twin represents the constitutive structure of a body of knowledge: it defines what exists, what properties exist, how they relate and what logical constraints make sense of their correct arrangement. The semantic digital twin incorporates semantic technologies such as ontologies, knowledge graphs, and linked data (Hosamo et al., 2022). In the patent context, the semantic twin is useful in the pre-deployment assessment context of patent feasibility, wherein the patent is a description of a technology, which may never have been implemented in the target environment. Second, a semantic twin supports a qualitatively different type of inference: what concept domains are covered by this patent, what logical rules does it obey, and what is the difference between its model of knowledge and the model of the target domain. The semantic twin is realized by the LOGOS protocol, where the ontological representation of a domain is directly represented as a structured prompt to a LLM. This implementation preserves the logical role of ontological evaluation while eliminating the need for standalone ontology infrastructure. The LLM executes these evaluative operations by limited natural language reasoning, generating outputs that are verifiable by the stated rules at each step.

3.2. Ontological Representation and Formal Knowledge Structures

The basic ontological framework differentiates the following structure:

- domain concepts – classes that are central to the target domain.
- attributes – properties that describe the key concepts.
- relations- they describe how the concepts are connected.
- axioms - logical principles which regulate legitimate arrangements of concepts and associations.

All types are expressed as first order logic (FOL) predicates. An important design principle in the ontological representation is the difference between necessary and sufficient conditions of membership of a concept. This difference is relevant in the context of semantic twins to determine partially mapped patent concepts: a patent that covers some, but not all the required conditions of a domain concept is deemed partially compatible, and the omitted conditions are seen as necessary amendments. The use of partial mappings in the LOGOS scoring model as a fractional score of a concept is an accurate realization of this principle: a partial coverage of the defining conditions of a concept will be given a partial credit and the remaining conditions defined as the necessary changes. A principled approach to uncertainty is also represented in the formal knowledge structures. Each concept-mapping assessment is identified to have three types of epistemic status, namely, confirmed, when the treatment of a concept by a patent is clearly and completely coherent to the definition of the domain model; inferred when the concept is not explicitly covered by the patent but can be reasonably determined by the contextual information; and absent when neither the explicit nor reasonable treatment of the concept can be found. The difference between confirmed and inferred mappings is captured in the assessment output and applied to produce confidence-weight scores. The formal knowledge structures also tackle the issue of automated data filling - situations in which a domain attribute may be automatically inferred or assigned, based on the values of other attributes. The explicit definition of automated filling rules as axioms, not as an implicit part of processing logic, makes sure that the LLM is used uniformly on all patents being evaluated and that the filling decisions are represented in the output of the assessment. A patent, which fails to specify its behavior when it needs attribute data on behavior, but it is lacking, has an implicit gap in its domain model, which the LOGOS evaluation can detect and inform.

3.3. The Logic-Guided Ontology Schema (LOGOS) Methodology for Prompt-Based Semantic Twin Construction

The LOGOS methodology is a structured and formally based mechanism of building semantic twins as structured LLM prompts. It was created as a reaction to the empirical finding that the engineering infrastructure needed to implement semantic twins is inaccessible to many organizations that may have a significant need to perform structured patent feasibility studies. LOGOS fills this accessibility gap by organizing domain knowledge into 5 structured layers of the semantic twin architecture, encoded directly withing a single LLM prompt. The methodology integrates Ontology Development 101 (Noy and McGuinness, 2001), MethOntology (Fernandez Lopez et al., 1997), and OntoPrompt (Ye et al., 2022), together with FOL, ensuring consistency of the LLM-based reasoning.

LOGOS architecture is grounded in the Cartesian methodology and comprises five layers. The basis is the Ontological Layer, which defines the main concepts, attributes and relations. All axioms, constraints, error-generation conditions, and automated rules of filling in data are specified on the Logical Layer. The Orchestration Layer operationalizes the ontology by defining the instructions that the LLM should follow. The Computational Layer models quantitative compatibility assessment. The Explanatory Layer asks the LLM to produce natural-language descriptions of each assessment conclusion and to identify the changes required to better align the patent with the domain model.

The directness of the symbolic embedding is a distinction between LOGOS and neuro-symbolic systems (NSS): unlike NSS architectures, LOGOS focuses on embedding the symbolic constraints directly in the prompt. So, LOGOS can be implemented in any organizational setting where the LLM can be accessed (Figure 1).

3.4. The LOGOS Prompt

System Prompt / Role

You are expert in evaluation of the semantic consistency of defect prioritization patents. You do not calculate bug urgency from real defects. You checks whether descriptions and claims in the patents are compatible from semantic point of view with LOGOS framework.

LOGOS Framework Core Concepts, Attributes, and Rules

Defect / Bug

A product deficiency that does not conform to requirements.

Attributes

bug_id: Unique identifier

reporter: The person who reported the defect; Values: User, Product Owner, Team Member

bug_origin: Where the bug was discovered; Values: PROD (production), Internal (development)

functional_area: Affected technical module; Values: Database, API, DevExpress, BusinessCode

quality_attribute: Affected quality attribute; Values: Capacity, Security, FunctionalCorrectness, UserErrorProtection

priority: Business impact level (REQUIRED); Values: High (1), Medium (2), Low (3)

severity: System impact level (REQUIRED); Values: High, Medium, Low

Axioms for Validity and Constraints

During calculation, check the following rules. In case of violation, follow the instructions for handling errors.

Key Terms

LOGOS Axioms

Group 1: Validity of Reporter and Bug Origin

A1: IF Reporter(x, User) AND (Origin(x, Internal) OR Origin(x) = EMPTY)

THEN Origin(x, PROD) AND Warning("Auto-corrected to PROD")

A2: IF Reporter(x, ProductOwner | TeamMember) THEN Origin(x) MUST BE {PROD, Internal}

Group 2: Validity of Functional Area and Quality Attributes

A3: IF Area(x, DB) AND Attr(x, q) THEN (q = Capacity OR q = Security)

A4: IF Area(x, API) AND Attr(x, q) THEN (q = FunctionalCorrectness OR q = Security)

A5: IF Area(x, a) AND Attr(x, q) AND NOT ValidAttr(a, q) THEN Error("Functional area {a} with invalid quality attribute") AND StopCalculation(x)

Group 3: Automated Rules

A6: IF Area(x, DevExpress) AND Attr(x, Empty) THEN Attr(x, UserErrorProtection) AND Error("Functional area {a} with system-assigned quality attribute UserErrorProtection")

A7: IF Area(x, BusinessCode) AND Attr(x, Empty) THEN Attr(x, FunctionalCorrectness) AND Error("Functional area {a} with system-assigned quality attribute FunctionalCorrectness")

A8: ∀x (Area(x, API) ∧ Empty(x, Attr)) → Has(x, Attr, Security) AND Error("Functional area {a} with system-assigned quality attribute Security")

A9: ∀x (Area(x, DB) ∧ Empty(x, Attr)) → Has(x, Attr, Security) AND Error("Functional area {a} with system-assigned quality attribute Security")

Group 4: Required Fields

A10: IF Priority(x, Empty) THEN Error("Priority is required") AND StopCalculation(x)

A11: IF Severity(x, Empty) THEN Error("Severity is required") AND StopCalculation(x)

LOGOS Compatibility Calculation Model

– Score = (Mapped_Concepts / Total_LOGOS_Concepts) × 0.6 + (Satisfied_Axioms / Total_Axioms) × 0.4

Where:

Concepts: severity, priority, reporter, origin, functional area, quality attribute – true/false

Axioms: A1 – A11: Complies (Satisfied Axioms), Partial (Satisfied Axioms), Violates, N/A

– Rating of patents – 0-1

Rating of patents

Processing Instructions

1. For each patent get ID, Title, Description, Claims, Key Words
2. Validity of required fields
   - Map the patent concepts with LOGOS concept
   - When mapping exists, compare the semantic of the concepts
   - When mapping exists, but in the patent the meaning of the concept is missing, create warning; not include in calculation of the scoring
   - When there is no mapping, create an error; not included in calculation of the scoring
   - When mapping exists and the semantic between the concepts is partial, create error, include in calculation of the scoring
   - When mapping and sematic coincide, include is calculation of the scoring
3. Compare the axioms
4. Analyze conflicts
   - Check for semantic conflicts between the concepts when mapping exists; if yes, create error
   - Check for axioms violation; if yes, create error
5. Generate an explanation message for each error and warning
6. Calculate Scoring according to the LOGOS Compatibility Calculation Model
7. Generate recommended actions – reject (0.0 – 0.3)/ integrate with modification of your bug report structure (0.3 – 0.7)/ integrate as it is (Scoring 0.7-1)
8. Example for generated output

Example for generated output

9. Based on LOGOS concepts suggest alternative patents

4. METHODOLOGY

4.1. Research Design

In this paper, a Design Science Research (DSR) methodology is used, already recognized as a suitable solution to work that seeks to generate a new artefact, in this instance, the semantic twin framework and the LOGOS methodology, the usefulness of which can be determined by proving its applicability in solving a well-defined set of problems (Hevner et al., 2004). The dual dedication to rigor, as it makes the artefact have a solid theoretical base, and relevance, as it makes the artefact meet actual practical needs, makes DSR stand out of purely empirical research. The study follows the standard DSR cycle of identifying a problem, its design, construction, evaluation and communication. The cyclic nature of the DSR cycle is also shown in the structure of the paper. The scoping work in the first domain indicated that the competency questions of the software defect prioritizations domain could not be answered satisfactorily without direct reference to the ISO 25010 quality standard - a result which led to a revised domain model to include the Quality Attribute concept that would later turn out to be the most diagnostically important gap in the corpus of evaluated patents. This feedback loop whereby initial design and construction effort brings to light latent requirements that improve the theoretical framework is typical of a productive DSR and one of the main benefits of the methodology over more strictly deductive approaches.

4.2. Data Sources and Materials

The experimental case study employs a dataset, which is a set of six patents found on Google Patents by searching the phrase "software bug report classification" - TR2022020456A2, CN116775871A, CN112905478A, CN114579471A, CN113011179B, and CN117668536A — represent a convenience sample of patents returned by the search query. The semantic twin based on LOGOS was run on Claude Sonnet 4.5 (Anthropic) as the LLM system, and the ontology-based prompt was inputted as the system prompt, and the individual descriptions and claims of the patents were entered as user-turn inputs. The assessment was performed thrice to determine the reproducibility and the standard deviations in the total scores were observed below 0.03 in the three runs.

4.3. Construction Process of Semantic Twin

A LOGOS-based semantic twin is constructed in a five-phase process that is like the ontological engineering cycle of Uschold and Gruninger (1996) with each phase modified to accommodate deployment with LLM. Phase 1 - Domain Scoping - the target domain is defined by eliciting competency questions. In the case of software defect prioritization domain, some of the representative competency questions are: Does this patent consider the relationship between severity of defects and deployment priority? Is the methodology of the patent flexible enough to support the difference between production-origin and development-origin defects? Does the patent offer a means to deal with defects involving incomplete attribute data? Phase 2 Ontological Layer Construction — converts the competency questions to a formal concept specification with the LOGOS Prompt Engineering workflow: domain scoping and competency question elicitation; concept identification and attribute specification; axiom formalization in FOL; processing instruction sequencing; scoring formula specification; and explanation template design. The outputting LOGOS prompt is tested through the input of test cases with known desired results. Phase 3 - Logical Layer Construction - formalizes the restrictions on the valid configurations of concepts as FOL axioms (A1-A11). Phase 4 Orchestration and Computational Layer Construction— converts the specifications into a procedural evaluation protocol and the quantitative scoring model. Phase 5 - Explanatory Layer Construction - creates the output templates and recommendation logic that will inform the creation of assessment reports by the LLM.

4.4. Feasibility Assessment Protocol

The protocol of the feasibility assessment has eight procedural steps that are carried out by the LLM under the guidance of the LOGOS prompt. Step 1- Patent Identification - The LLM is required to identify and record the ID of the patent, its title, abstract, key claims and keywords. Step 2 -Concept Mapping - involves asking the LLM to evaluate, on a concept-by-concept basis, how the patent relates to each concept defined in the Ontological Layer, by whether it is fully, partially, not covered, or missing information. An evaluation of a concept being full mapped occurs when the treatment of the concept by the patent is substantively similar to the definition of the domain model. Step 3-Axiom Compliance Analysis — involves the LLM analyzing each axiom that needs to be applied to the methodology described in the patent.. Step 4 - Conflict Detection - This step involves the LLM detecting semantic and logical conflicts. Step 5 — Error Generation - yields structured error and warning messages on each of the identified gaps or conflicts. Step 6 - Score Calculation - uses the LOGOS Compatibility Calculation Model. Steps 7 to 9 - Recommendation

Generation and Alternative Patent Identification - converts the score and detailed analysis into actionable advice by specifying what changes are necessary to bring partially compatible patents into action; and recommending alternative patents to the concepts that are not covered by the evaluated patents.

5. RESULTS

5.1. Semantic Twin Architecture: Performance in the Case Study

The LOGOS prompt performed as intended across the evaluation of all six software defect prioritization patents. The Orchestration Layer's processing sequence proved robust in handling the heterogenous input format of the evaluated patents - directing the LLM through consistent concept mapping and axiom compliance steps in each case. The principal operational challenge arose in the treatment of implicit patent claims: assertions about a technology's capabilities expressed through the description of a use case rather that stated as explicit claim. The concept definition in the Ontological Layer provided the vocabulary needed to identify such implicit references, but their scoring required conservative treatment within the Computational Layer to avoid overestimating compatibility based on plausible inference rather than demonstrated alignment. When a patent's language could be constructed as addressing a domain concept without doing so explicitly, the partial mapping classification was applied, contributing fractional concept credit rather than full credit, which the Computational Layer handled correctly under the scoring formula.

The most diagnostically significant element of the LOGOS implementation was the encoding of the ISO 25010 Quality Attribute taxonomy as a mandatory concept within the Ontological Layer which made the absence of quality attribute treatment detectable as a formal as rather than merely as a design choice. The Ontological Layer's formal specification of this concept, combined with the Logical Layer's axioms produced the finding that no evaluated patent addressed this dimension – a result that could not have been generated by any purely text-analytic method operating without a formal reference model.

5.2. Framework Validation

The validation evidence that is presented in the case study is based on two complementing mechanisms that combine to deal with the major reliability issues of interest in the context of the tools that are based on LLM. The former is a quantitative mechanism: the three-run reproducibility test confirms that the LOGOS prompt can yield consistent numeric results. The second is the qualitative mechanism: the expert judgement of reasoning outputs: the question of the logical validity of the concept classifications and axiom judgements of the LLM, and not just their internal consistency, is how they compare with the independent judgements of a domain expert. The quantitative reproducibility and qualitative expert validation is a combination that gives grounds to trust in the methodology that could not be brought about by either of the mechanisms. The implementation of the LOGOS is supported by the validation evidence of the reproducibility analysis and the review of the reasoning of the LLM by experts. The maximum difference in the total compatibility scores among three independent evaluation runs of any of the six patents was not more than 0.03, indicating high reproducibility. The clarity of the reasoning outputs of the LLM allowed a domain expert in software quality assurance to screen the outputs line-by-line, and conclude that the concept mappings and axiom assessments of the LLM were logically valid. Another aspect of validation is how sensitive the assessments of the semantic twin are to the wording of the concept definitions of the domain model. Two different versions of the definition of Priority concept were made and sent as part of the LOGOS prompt to a group of three patents. The scores of compatibility, which resulted varied by an average of 0.04, is a small yet not trivial difference that demonstrates the significance of specific standardized definitions of concepts when building semantic twins.

5.3. Case Study: Software Defect Prioritization

The six fundamental LOGOS concepts together with their properties and formal axiom sets that control their relationships in the domain model, are shown in Table 4. The concept specifications of the domain model are an indication of the operational needs of a software engineering team that is required to triage the incoming defect reports in the conditions of incomplete information and time constraints. Table 5 presents the compatibility scores for all six evaluated patents based on the defined ontology and axiom set.

CN112905478A had the highest overall score, with 0.55, due to its explicit treatment of both Priority and Severity. The three patents that had the highest score of 0.35 each were concerned with Priority prediction where the concept alignment was reasonable though not complete Reporter, Origin, and Quality Attribute as part of their methodological approach. CN116775871A and TR2022020456A2 were Incompatible: the former mentioned Priority and Origin in any substantive manner, whereas the latter had the lowest concept score in the corpus (0.05) and did not comply with the axioms.

Axiom compliance analysis of the six patents showed that there was a tendency toward practical significance in that TR2022020456A2, which had scored zero on axiom compliance, failed because it was not interacting with any concept that the rules would be applied to. This is a critical scope miss between the knowledge model of the patent and the domain model, which is fundamentally different than a patent that has the right ideas but uses them in an incorrect way. The former is an indication of a technology that is functioning at a different conceptual register altogether; the latter is an indication of a technology that is in the right register but with logical gaps that could be filled through targeted modifications. The LOGOS assessment procedure identifies these two forms of incompatibility by classifying as inapplicable incompatible axioms, and thus a patent will not be infringed by the axioms applied to the concepts out of scope of the stated scope. This contextual treatment of axiom applicability demands that the LLM make domain-directed judgment over the coverage of the technical claims of each patent, one of the more challenging inference tasks in the LOGOS protocol. The case study illustrated the ability of the LOGOS methodology in identifying alternative patents. Based on the gap in concepts found in the six patents reviewed, the LLM proposed two other patents as candidates that would offer a multi-dimensional scoring foundation that was more in line with the LOGOS domain model: US8898151B2 and US20120078651A1 (Table 6). The automatic identification of alternative candidates, directly out of the knowledge of the LLM, is an example of how the formal specification of concept gaps in the assessment result can be transformed into a patent search strategy without further prompt engineering.

5.4. Discussion of Case Study Findings

The software defect prioritization domain is a certain instructive case study of the semantic twin methodology since it is at the crossroads of two properties that undermine traditional patent analysis: terminological instability and the lack of generally accepted formal standards in patent practice. Terminological instability refers to the fact that the same word, such as urgency, severity, priority, etc., has been applied with varying meanings in various patents, development techniques and in various organizational settings, and keyword based patent searching is therefore not reliable. A lack of formal standards in patent practice implies that there is no standard reference vocabulary that patent authors are supposed to employ, as is the case with other areas of practice. The LOGOS approach satisfies both the characteristics: defining the domain model to ISO 25010 grounds the evaluation vocabulary in a formal international standard; encoding the domain model as an LLM prompt instead of mandating that any patent authors consult it, ensures the evaluation of any patent to the standard regardless of whether the author was aware of it. The fact that the scoring of the six patents assessed indicates that the conceptual space of the patent corpus is heterogeneous. All patents cover one or more of the six concepts in the LOGOS domain model, but none of them attempts to cover all of them. The trend shows that the analyzed patents are a collection of conceptually fragmented spaces where specific methodological inventions are sought individually without considering the larger context of operation that defect triage must operate in. The most important finding of the case study is that no considered patent covered ISO 25010. This fact indicates a systemic gap in standardization: patent methodologies are created without any conceptual standards. A security vulnerability is dealt with in a much different manner compared to a capacity constraint although they both may receive a High priority in the defect tracking system. This lack is visible and measurable by the semantic twin, as it encodes the ISO 25010 as a constitutive element of the defect prioritization ontology, which can be considered the normative reference that this ontology entails. The diagnostic value of the case study was further expanded by the alternative patent identification ability of the LOGOS methodology. The LLM proposed two more patents as candidates to fill the gaps in the concept. Even these more compatible options were observed to miss the Reporter role, Origin environment, and Quality Attribute dimensions, and confirm that the overall multi-dimensional treatment of defect urgency represented in the semantic twin has no published patent analog. This discovery is of strategic importance to an organization that has formulated a defect management architecture that includes quality attribute classification: it defines a patentable conceptual space that is not occupied by any patent in the corpus.

6. CONCLUSION

The paper proposes the implementation of semantic twin construction as the LOGOS methodology, which is the entanglement of structured ontological information concept, attribute, axioms, processing rules, a compatibility scoring formula, and explanatory instructions directly within a LLM using prompts. The framework is tested in the software defect priorities area wherein six patents were examined by Claude Sonnet 4.5 which was provided with the LOGOS prompt. The compatibility scores were between 0.05 and 0.55 with no one marking the Highly Compatible mark. Considerably, the idea of quality attribute as outlined in the ISO 25010 was not found in any of the reviewed patents, indicating a systematic quality failure of standardization in the patent arena of the specified field. The findings prove that the semantic twin can be used as a normative reference model where the patent heterogeneity can be measured, conceptual gaps are determined, and integration trajectories are considered - a sort of content-type patent intelligence that cannot be provided by bibliometric or text-mining methods.

The software defect prioritization case study illustrates the framework in an area where the standard of knowledge in question is authoritative but not wholly implemented in practice, and the normative role of the semantic twin, in setting what comprehensive coverage would appear like, is of particular concern. The case study is a challenging test of the main claims of the framework, not completely representative of the breadth of contexts in which it can be used, but useful evidence that the main claims of the framework can be delivered by a practical implementation of the LOGOS methodology using a practical and easily accessible LLM-prompt interface. The thesis of this paper is that the analysis of patent feasibility needs an entirely new type of analytical tool as compared to the ones that prevail in the market nowadays. The tool for that is called the semantic twin: a formal, machine-readable description of domain knowledge that allows structural, reproducible, explainable compatibility assessment of patent claims. The concept of semantic twin in the paper has been elaborated in terms of theoretical basis, architectural specification and empirically evaluated. The LOGOS approach offers an accessible, concrete implementation route as it directly entails the formalized knowledge of the domain becoming embedded in an LLM via a structured ontological prompt, allowing the semantic twin construction to be accessible to organizations without special knowledge engineering infrastructure.

The semantic twin building process is an analytically productive process by itself, regardless of the specific patent evaluation result. The process of developing the LOGOS domain model of software defect prioritization needed a formalization of the concepts, relationships and logical rules that define best practice in defect prioritization, to bring tacit domain knowledge into explicit form and expose gaps in the current patent environment that would not have been discovered with the same accuracy by informal methods. What the current work has demonstrated is that the concept is valid, architecture is consistent and the empirical findings are promising enough to warrant the larger research program which this paper will inaugurate.

STATEMENT OF AI USE IN THE PAPER

During the preparation of this work the author(s) used Claude Sonnet 4.5. in order to embed the LOGOS prompt in the model for execution, to analyze the submitted patent descriptions and claims as asked in the prompt, and to generate an answer for the user. After using this tool/service, the author(s) reviewed and edited the content as needed and take(s) full responsibility for the content of the published article."

REFERENCES

Arya, V., Colace, F., Lombardi, M., Lorusso, A., Santaniello, D., & Valentino, C. (2025). DTSafe-O: A Semantic Digital Twin Ontology for Construction Safety, International Journal on Semantic Web and Information Systems, 21, 1 36. .DOI: 10.4018/IJSWIS.395999

Bessen, J., & Meurer, M. J. (2008). *Patent Failure: How Judges, Bureaucrats, and Lawyers Put Innovators at Risk.* Princeton University Press.

Blind, K., Edler, J., Frietsch, R., & Schmoch, U. (2006). Motives to patent: Empirical evidence from Germany. *Research Policy*, *35*(5), 655–672. DOI: 10.1016/j.respol.2006.03.002

El-Hajj, M. (2026). Semantic foundations for digital twins: The contribution of ontological analysis. *Frontiers of Computer Science*, *8*, 1757450. Advance online publication. DOI: 10.3389/fcomp.2026.1757450

Fernández López, M., Gómez-Pérez, A., & Juristo, N. (1997). METHONTOLOGY: From Ontological Art Towards Ontological Engineering. Proceeding of the AAAI-97 Spring Symposium on ontological engineering. https://aaai.org/papers/0005-ss97-06-005-methontology-from-ontological-art-towards-ontological-engineering/

Fischer, T., & Ringler, P. (2007). What patents are used for? Empirical evidence from the patent portfolios of Swiss firms. *Journal of Evolutionary Economics*, *17*(5), 553572.

Grieves, M. (2014). *Digital twin: Manufacturing excellence through virtual factory replication*. Florida Institute of Technology.

Hall, B. H., Helmers, C., Rogers, M., & Sena, V. (2014). The choice between formal and informal intellectual property: A review. *Journal of Economic Literature*, *52*(2), 375–423. DOI: 10.1257/jel.52.2.375

Hevner, A. R., March, S. T., Park, J., & Ram, S. (2004). Design science in information systems research. *Management Information Systems Quarterly*, *28*(1), 75–105. DOI: 10.2307/25148625

Hosamo, H. H., Imran, A., Cardenas-Cartagena, J., Svennevig, P. R., Svidt, K., & Nielsen, H. K. (2022). A review of the digital twin technology in the AEC-FM industry. *Advances in Civil Engineering*, *2185170*(1), 1–17. DOI: 10.1155/2022/2185170

Jazi, A. E., Taheri Andani, M., Hasanzadeh, A., & Ameri, F. (2025). A Semantic-Driven Digital Twin Framework for Manufacturing. In Proceedings of the ASME 2025 International Design Engineering Technical Conferences (Vol. 2A, Paper No. DETC2025-169558). ASME. https://doi.org/DOI: 10.1115/DETC2025-169558

Lanjouw, J. O., & Schankerman, M. (2001). Characteristics of patent litigation: A window on competition. *The RAND Journal of Economics*, *32*(1), 129–151. DOI: 10.2307/2696401

McConnachie, G. (1997). The Management of Intellectual Assets: Delivering Value to the Business. *Journal of Knowledge Management*, *1*(1), 56–62. DOI: 10.1108/EUM0000000004580

Noy, N. F., & McGuinness, D. L. (2001). Ontology development 101: A guide to creating your first ontology (Stanford University Knowledge Systems Laboratory Technical Report KSL-01-05). Stanford University.

Rasheed, A., San, O., & Kvamsdal, T. (2020). Digital Twin: Values, Challenges and Enablers From a Modeling Perspective. *IEEE Access : Practical Innovations, Open Solutions*, *8*, 21980–22012. DOI: 10.1109/ACCESS.2020.2970143

Shapiro, C. (2000). Navigating the Patent Thicket: Cross Licenses, Patent Pools, and Standard Setting. A. B. Jaffe, J. Lerner, and S. Stern (Eds.) 2000, Innovation policy and the economy (Vol. 1, pp. 119–150). MIT Press.

Srivastava, M. K., & Gnyawali, D. R. (2011). When Do Relational Resources Matter? Leveraging Portfolio Technological Resources for Breakthrough Innovation. *Academy of Management Journal*, *54*(4), 797–810. DOI: 10.5465/amj.2011.64870140

Tao, F., Zhang, H., Liu, A., & Nee, A. Y. C. (2019). Digital twin in industry: State-of-the-art. *IEEE Transactions on Industrial Informatics*, *15*(4), 2405–2415. DOI: 10.1109/TII.2018.2873186

Trajtenberg, M. (1990). A Penny for Your Quotes: Patent Citations and the Value of Innovations. *The RAND Journal of Economics*, *21*(1), 172–187. DOI: 10.2307/2555502

Tseng, Y. H., Lin, C. J., & Lin, Y. I. (2007). Text mining techniques for patent analysis. *Information Processing & Management*, *43*(5), 1216–1230. DOI: 10.1016/j.ipm.2006.11.011

Uschold, M., & Gruninger, M. (1996). Ontologies: Principles, Methods and Applications. *The Knowledge Engineering Review*, *11*(2), 93–136. DOI: 10.1017/S0269888900007797

Van Zeebroeck, N., & Graham, S. J. H. (2010). Comparing Patent Litigation Across Europe: A First Look. *Stanford Technology Law Review : STLR : An Online High-Technology Law Journal from Stanford Law School.*

Ye, H., Zhang, N., Chen, H., & Chen, J. (2022). Ontology-enhanced Prompt-tuning for Few-shot Learning. In *Proceedings of the ACM Web Conference 2022 (WWW '22)*. ACM. DOI: 10.1145/3485447.3511921

APPENDIX

Table 1. Key Terms

Concept	Description	Values
Reporter(x, r)	Role of the person reporting bug x	User, PO (Product Owner), Team Member
Origin(x, o)	Where bug x was observed	PROD, Internal
Area(x, a)	Software area in which the defect occurs	DB, API, DevExpress, BusinessCode
Attr(x, q)	Quality attribute related to the bug	Capacity, Security, FunctionalCorrectness, UserErrorProtection
Priority(x, p)	Bug priority level	{Value} or Empty
Severity(x, s)	Bug severity level	{Value} or Empty
Error(msg)	System error message output	Text string
StopCalculation(x)	Flag to halt calculation	-

Table 2. Rating of patents

Range	Compatibility Level	Description
0.0 – 0.3	Incompatible	Fundamental conflicts
0.3 – 0.7	Partially Compatible	Gaps
0.7 – 1	Highly Compatible	LOGOS-aligned

Table 3. Example for generated output

Patent ID	Concept Score	Axiom Score	Total Score	Rating	Recommendation
PT000001A	0.25	0.50	0.75	Highly Compatible	LOGOS-aligned

Table 4. LOGOS Semantic Twin: Core Concepts, Attributes, and Formal Axiom Groups

LOGOS Concept	Key Attributes	Permitted Values	Governing Axiom Group
Severity	System impact level (REQUIRED)	High, Medium, Low	A11: severity must be present
Priority	Business impact level (REQUIRED)	High (1), Medium (2), Low (3)	A10: priority must be present
Reporter	Role of reporting agent	User, Product Owner, Team Member	A1–A2: reporter constrains valid origin values
Origin	Discovery environment	PROD (production), Internal (development)	A1–A2: auto-corrected for User reporters
Functional Area	Affected software module	Database, API, DevExpress, Business Code	A3–A5: area constrains valid quality attributes
Quality Attribute	ISO 25010 quality dimension	Capacity, Security, Functional Correctness, User Error Protection	A6–A9: automated assignment when absent

Table 5. LOGOS Compatibility Assessment: Six Software Defect Prioritization Patents

Patent ID	Concept Score	Axiom Score	Total Score	Compatibility Rating	Recommendation
CN112905478A	0.35	0.20	0.55	Partially Compatible	Integrate with modifications
CN117668536A	0.15	0.20	0.35	Partially Compatible	Integrate with modifications
CN113011179B	0.15	0.20	0.35	Partially Compatible	Integrate with modifications
CN114579471A	0.10	0.20	0.30	Partially Compatible	Integrate with modifications
CN116775871A	0.10	0.10	0.20	Incompatible	Reject
TR2022020456A2	0.05	0.00	0.05	Incompatible	Reject

Table 6. Priority 1: Multi-Dimensional Scoring Foundation

Patent ID	Title	LOGOS Alignment	Integration Strategy
US8898151B2	Method and system for scoring software bugs	Uses weighted multi-factor scoring	Adapt formula to include LOGOS-specific dimensions (Reporter, Origin, Functional Area, Quality Attribute)
US20120078651A1	System for Prioritizing and Resolving Defects	Weighted severity + priority + customer impact	Replace "customer impact" with LOGOS Origin (PROD weight = high impact)

Figure 1. The Five-Layer LOGOS Architecture and Its Correspondence to the Cartesian Method

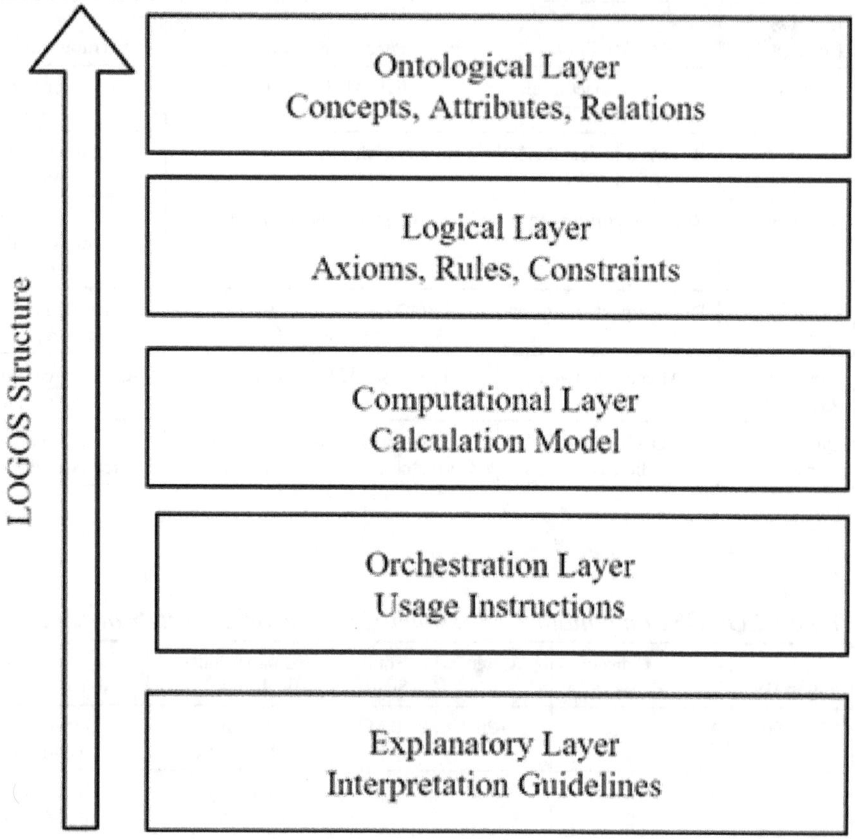

Chapter 7
Smarter Decisions, Faster Results:
The Rise of Cognitive Enterprises

Salam Al E'mari
http://orcid.org/0000-0002-2134-4158
University of Petra, Jordan

Yousef Sanjalawe
http://orcid.org/0000-0002-4442-1865
University of Jordan, Jordan

Fadi Fataftah
http://orcid.org/0000-0002-3299-0972
University Canada West, Canada

ABSTRACT

As organizations confront increasing complexity and data overload, traditional decision-making models are no longer sufficient. This chapter introduces the concept of the cognitive enterprise, an organization that uses Artificial Intelligence (AI) to simulate human reasoning, automate responses, and drive real-time, strategic insights. By leveraging AI technologies such as Natural Language Processing (NLP), Machine Learning Operations (MLOps), predictive analytics, and intelligent automation, cognitive enterprises transform data into action across all business units. The chapter explores how real-time AI enables faster and more accurate decisions in logistics, healthcare, and customer service. It then examines the application of predictive maintenance in industrial settings, where AI-driven models prevent equipment failure, reduce costs, and improve safety. In human resources, AI powers workforce planning through predictive attrition models, resume analysis, and sentiment detection, shifting HR from an administrative role to a strategic partner.

DOI: 10.4018/979-8-2600-1669-5.ch007

1. INTRODUCTION: THE COGNITIVE SHIFT

In the last decade, organizations have registered monumental volumes of data, much more than any person can ever rationally process or respond to within real time. This ought to be a competitive advantage on the surface. However, in reality, sometimes it seems everything is the reverse. The executives are being overwhelmed with dashboards, alerts and reports. Managers are not able to make decisions confidently. And frontline employees, the workers who are directly exposed to the processes, often do not have access to the required information at the moment of necessity. The contemporary enterprise is abundant in information, but lacks wisdom (Shamim et al., 2023).

This paradox has emerged to be one of the most worrying organizational issues in modern-day. Billions of dollars have come out of digital transformation, but so many companies continue to use the old decision-making models that cannot compete with the speed, complexity, and uncertainty of the modern business world. They are experiencing a different type of turbulence, the one that is imposed by swiftly changing markets, customer demands, and the volatility of the supply chain, as well as a pace that can hardly be called linear. This is because in this environment, organizations must be able to gain quicker data but what they really require is smarter decisions (Kalubanga & Gudergan, 2022).

1.1 Why Traditional Decision-Making Is Breaking Down

The basis of making business decisions is becoming outdated. it presupposes that past statistics are a cover-safe indicator of the future, that the work of departments can be compatible, and that leaders have a chance to consider possibilities and then make decisions. However, these fissures in this basis can no longer be overlooked. Inertia is one of the biggest problems. Big organizations tend to act slowly because they are shackled by legacy systems, by outdated processes, and through levels of bureaucracy that are not inclined to adjust. Most of the time, by the time decisions gleaned at the quarterly schedules or meetings, things on the front line have already changed (Allen & Mau, 2025).

And then there is the issue of data silos. In spite of the advancements that technology has brought to enterprise platforms, many organizations have failed to achieve the integration of their data. Thus, marketing data is stored in a completely different platform from supply chain data, while HR data is stored in a third platform. In the absence of integration, it would hardly ever be possible to get a comprehensive view of the business, talk alone of creating patterns and predicting deviations. Human bias is the next invisible factor. Even the best decision-makers may be prone to human bias, which may manifest as decision-making based on gut feelings rather than analysis. Figure 1: Traditional decision-making process vs. that of a cognitive enterprise.

Figure 1. Traditional vs. Cognitive Enterprise Decision Models.

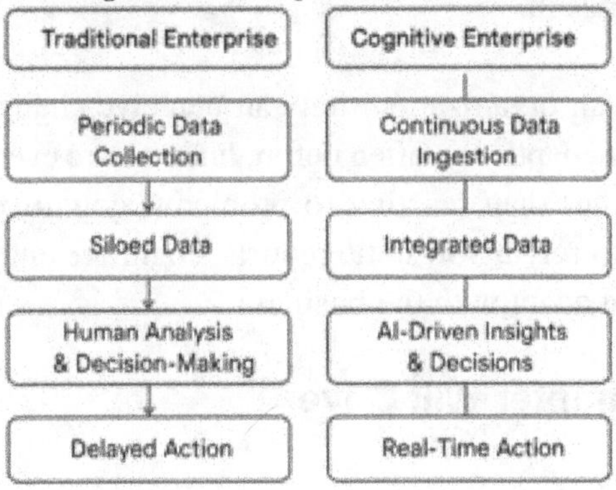

Finally, there is a lack of real-time insight. The world doesn't operate on a quarterly schedule anymore. It operates on minutes, seconds, or milliseconds. However, business decisions are often made using lagging indicators, sales data from yesterday, a financial report from last month, or employee engagement results from last quarter.

1.2 The Rise of the Cognitive Enterprise

However, as a result of the above challenges, a new type of organization is being developed. It is called the cognitive enterprise. A cognitive enterprise is more than a digital business. While a digital business is one that seeks to go from paper to pixels, from spreadsheets to cloud apps, a cognitive enterprise is one that seeks to add intelligence into the core of the business. A cognitive enterprise is one that does not only store data but also interprets it, reasons with it, learns from it, and acts on it in real time (Pilipczuk, 2022).

The heart of all these changes is Artificial Intelligence (AI). Not the highly modern, sci-ific meaning, though. We are talking here about the concept of AI as an instrument, an ensemble of technologies that helps machines mimic human cognitive processes like perception, learning, reasoning, and decision-making (Kowalczuk & Czubenko, 2023). What is revealed in a cognitive enterprise:

- Predictive analytics identifies patterns and predicts what will happen before it happens.
- Intelligent automation removes friction from the workflow to allow for quicker decisions that are also better.

- Natural language processing helps systems understand and respond to spoken language, which makes it possible to extract the intelligence from unstructured information like emails, chats, and documents.
- Adaptive systems are dynamic; they improve with each new piece of information that comes in.

The end result is an organization that can perceive changes, learn from those changes, and respond adaptively, often before humans are even aware that a change has taken place. Rather than reacting to problems, cognitive enterprises predict problems. Rather than relying on static reports, cognitive enterprises use dynamic models that grow and adapt with the business.

1.3 What This Chapter Will Cover

In this chapter, we'll look at how cognitive abilities are revolutionizing the modern enterprise. It's not just about technology, it's about fundamentally reimagining how decisions are made, who makes them, and how fast they can be made. We'll start off with real-time decision-making, where we'll look at how AI is enabling companies to shift from slow and reactive decision-making to fast and data-driven decision-making. We'll look at how immediacy becomes a strategic advantage in decision-making, whether it's in the way that companies are designing their supply chains, their customer service, and so on.

We'll then look at predictive maintenance, one of the most tangible uses of AI in an industrial and manufacturing setting. We'll look at how data is being used not only to monitor performance, but also to prevent breakdowns, reduce costs, and improve safety. We'll then look at human capital intelligence, where machine learning is revolutionizing the way in which companies are recruiting, retaining, and developing their human capital. We'll look at how AI is enabling HR leaders to shift from being administrative to being strategic in their approach to human capital.

Finally, we'll look at the realm of smart finance, where AI is revolutionizing the way financial functions operate, detecting fraud in real time, accurately predicting budgets with unprecedented precision, and even uncovering risks before they become losses. The traditional role of the finance office as a repository of backward-looking reports is giving way to a more forward-looking role. As we move through each of these sections, we'll not only look at the possibilities, but at the necessities. Because becoming a cognitive enterprise is not just about the tools we use, but the habits, roles, and mindsets we create.

Figure 2 displays the analytical architecture for the cognitive enterprise. The data obtained from the operational systems, the Internet of Things (IoT) devices, and the external environment is processed by AI-driven analytics platforms. The processed data is then converted into predictive insights. The predictive insights are used for decision-making in different areas of the business. These areas include supply chain operations, predictive maintenance, human capital management, and financial planning.

Figure 2. Integrated AI Architecture for the Cognitive Enterprise.

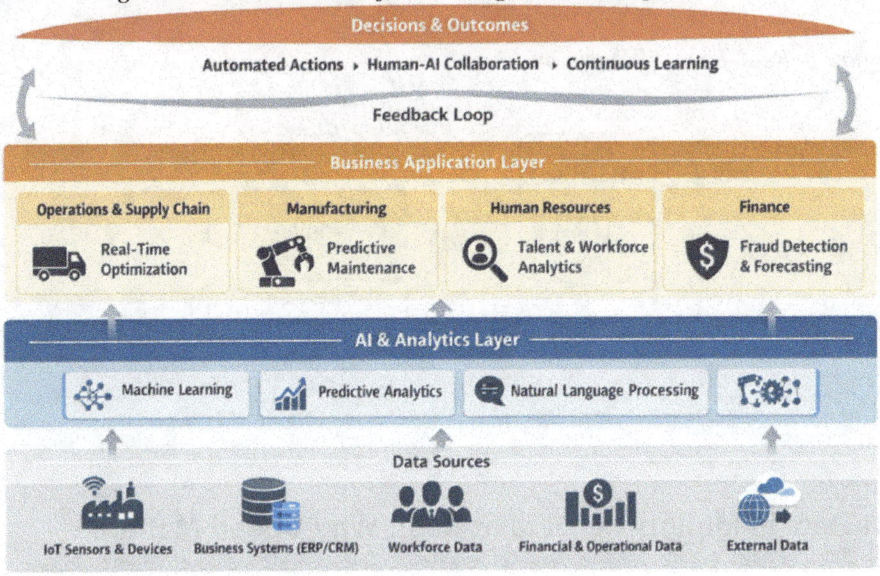

2. FROM INSIGHT TO ACTION: REAL-TIME DECISION-MAKING WITH AI

In the modern world, data is referred to as new oil. However, unlike oil, data does not only need to be extracted and refined. Organizations that are still relying on static data and history are being left behind by their competitors, who are "tapping into real-time data and responding in real-time." The key differences are not only in speed but also in the intelligence. It is not about the data, but about creating a feedback loop between observation, understanding, and action (Stach, 2023).

This is where real-time AI comes in. By processing the data as it is generated, whether from sensors, transactions, or customer interactions, AI helps enterprises transition from hindsight to foresight, and from passive analysis to intelligent action. In this section, we will discuss how the enterprise is moving from static data to data in motion, why this is crucial for enterprise agility, and what enables this transformation. By using real-world examples and looking under the hood of the underlying technologies, we will examine how real-time decision-making is being infused into the daily lives of the enterprise, and what the implications are when the machines think and act faster than us. Figure 3 illustrates the conceptual loop of real-time AI decision-making, with continuous data collection, analysis, and action (Lazaroiu et al., 2022).

Figure 3. Real-Time AI Decision Loop.

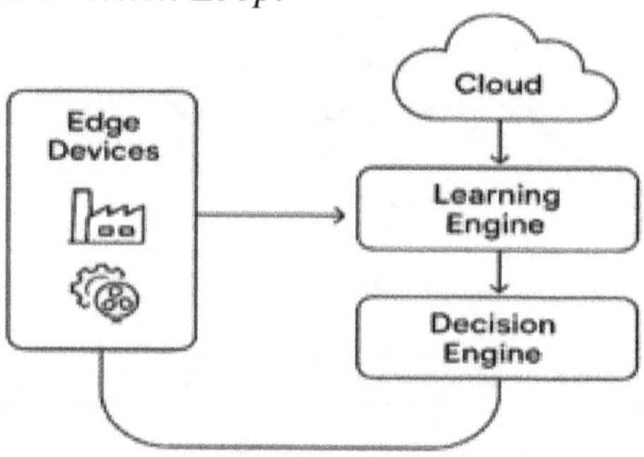

2.1 Data in Motion vs. Data at Rest: Why Speed Matters

Traditionally, businesses have been dealing with data at rest, which refers to the information that is stored in the databases, data warehouses, and reports that are periodically examined. Though this is important, especially with regards to long-term trends, it is not useful where timeliness is a requirement. Data in motion, on the other hand, refers to the data that is currently streaming from the devices, applications, or the systems. It could be the GPS signals from the delivery trucks, the vitals from the patients at the hospitals, or the transactional data from the e-commerce site. When this data is processed in real time, the door is opened to decisions that are contextual and real-time (Parveen & Basit, 2023).

Consider the difference:

- A quarterly report might tell you that customer churn increased over the past three months.
- A real-time churn prediction model can alert your support team the moment a high-value customer begins showing signs of dissatisfaction, giving you time to intervene.

This is not just a technological change. This is a complete paradigm shift in how decisions are made. A world where real-time inputs are the only inputs is not a place where static decision-making cycles apply. A new paradigm of learning and reacting needs to be built into the enterprise, whether it is for optimization, personalization, or preventing impending failures.

2.2 Case Studies: Real-Time AI in Action

2.2.1. Supply Chain Optimization in Retail and Logistics

There are no industries that have faced greater tests of volatility than retail and logistics. Whether it's a pandemic's after-effects, regional issues, or changing customer needs, there has never been a greater need for supply chain intelligence in real-time. The best firms are now using AI-powered platforms that consume and analyze information from hundreds of sources in real-time: inventory levels, supplier updates, weather forecasts, port conditions, and even social media. Not only do these platforms identify issues in supply chains, they also suggest how to resolve them: rerouting goods, order quantities, or supplier allocation (Hrouga & Sbihi, 2023).

The retail and logistics businesses are increasingly using AI-based supply chain control towers that integrate information from various sources such as inventory levels, supplier delivery schedules, weather forecasts, and transportation conditions. For example, AI-based logistics solutions use shipment data and predictive demand information in real-time to route the inventory among the distribution centers and retail stores. Empirical research suggests that AI-based supply chain systems have the potential to reduce the likelihood of delays and inventory outages by enabling businesses to respond rapidly to issues in transportation networks and supplier operations (Shobhana, 2024).

2.2.2 Real-Time Triage in Healthcare

In healthcare, things are even higher. Real-time decision supports can be the life or death of the matter. AI systems that help in the triage of the emergency room are more frequently being used by hospitals. Instead, patient data, such as heart rate, blood oxygen saturation, lab readings, and prior conditions, are constantly compared with the predictive models which were created with thousands of the previous cases. The idea is to give priority to the patients not only depending on the first to reach but based on the urgency of their intervention (Sahoh & Choksuriwong, 2023).

In the health sector, real-time AI technologies aid health practitioners in the continuous analysis of patient data flows, including heart rate, oxygen saturation, lab results, and patient history. These technologies compare patient data with predictive models built from large clinical data sets to identify warning signs of patient deterioration. During the COVID-19 pandemic, hospitals used AI-assisted patient triage technologies that could identify patterns associated with respiratory failure, thus helping in the efficient allocation of intensive care unit (ICU) resources (Nadaf et al., 2025).

2.3 Core Technologies Powering Real-Time AI

Building a real-time AI ecosystem requires more than just algorithms. It involves a tightly integrated stack of technologies that enable data to be collected, interpreted, and acted upon with minimal delay. Three components are especially critical:

2.3.1 Natural Language Processing (NLP)

NLP allows machines to recognize, comprehend and produce human language. During real-time decision-making, NLP is critical in siphoning values out of unstructured information, emails, chat history, social media networks, voice transcriptions, and others. An example would be the AI assistance in customer services that is able to parse messages on a live chat to find nuances of frustration, seriousness, and other information to elevate the need to a human or simple choose a proper resolution within seconds. The same way, in finance, NLP models go through live news feed and earnings reports to pick up signals that might have an impact on trading decisions. Real-time readability produces a strong advantage in the enterprises (Uddin, 2024).

2.3.2 Machine Learning Operations (MLOps)

Machine learning models in the real-time are self-trained, required observation, and tuning. Enter MLOps a collection of practices that essentially makes machine learning and DevOps go together. MLOps is all about ensuring the running state of AI models, which means not only deploying them, but also maintaining, updating, and retraining as new data arrives. An example could be considered a logistics company, which will implement a route optimization model that will factor in the variables such as weather, fuel prices and traffic retrieved on day by day basis. MLOps pipelines are automated, facilitating data consumption, retraining, versioning, and observing performance in a model without a need to intervene manually. In the absence of MLOps, real-time AI soon turns into stale AI, which does not take into account the latest realities of the business environment (Sendas & Rajale, 2025).

2.3.3 Decision Intelligence Systems

The layer of orchestration which combines everything is decision intelligence. These systems integrate predictive analytics, rule-based reasoning, simulation modules and even human feedback arcs to infer and explain decisions. In contrast to traditional business intelligence tools which by themselves merely give information, those based on decision intelligence suggest what to do, and make certain recommendations automatically. They are gaining an application in segments such as fraud detection, credit risk analysis, and management of the operations. It is explainability, not simply automation that is the defining characteristic. In an era when businesses are increasingly dependent upon decisions reached by machines, the trust becomes extremely important. The decisions intelligence systems are usually developed with a form of traceability that facilitates how stakeholders know why a certain decision is taken, and what data was used to inform this decision (Ekundayo, 2024).

2.4 Benefits and Risks of Real-Time AI

2.4.1 Benefits: Speed, Accuracy, and Agility

The advantages of real-time AI are significant and far-reaching (Chataut et al., 2024):

- Speed: Decisions that once took days or weeks can now happen in milliseconds.
- Accuracy: Continuous data ingestion and adaptive learning models improve the precision of forecasts, diagnoses, and recommendations.
- Agility: Real-time systems help organizations pivot quickly in response to changing conditions, whether in customer demand, supply chain dynamics, or market sentiment.

Beyond operational efficiency, real-time AI enables customer intimacy. Businesses can respond to user behavior as it happens, creating more personalized, timely, and relevant experiences. From tailored marketing to dynamic pricing, the possibilities are vast.

2.4.2 Risks: Model Dependence, Opacity, and Ethics

But with power comes responsibility. Indeed, there are concerns over the potential for over-reliance on real-time AI (Patel, 2024):

- Overdependence: As organizations increasingly use real-time AI to make decisions, there may be a problem with over-reliance on such decisions or a failure to detect when errors are occurring.
- Black-box models: Real-time AI uses complex neural network architectures. Can a decision to deny a loan or misdiagnose a patient be explained or contested if the model itself does not explain its reasoning?
- Ethical considerations: Real-time AI may be based on biased or incomplete data. Left unchecked, it may perpetuate inequalities or make unethical decisions for the sake of speed.

For instance, the hiring system might favor applicants over others using the trends of successful hiring in the past, thereby perpetuating gender or racial discrimination. In a real-time system, these processes could occur and go undetected until it is too late. Therefore, to mitigate these dangers, organizations must invest not only in AI but also in governance, which includes thorough testing, auditability, fairness, and ethical review boards to evaluate the impact of the model (Ajunwa, 2020). Table 1 highlights the advantages of real-time AI systems and the associated dangers.

Table 1. Benefits vs. Risks of Real-Time AI.

Benefits	**Risks**
Instantaneous decision-making	Model opacity (black-box decisions)
Improved forecasting	Potential ethical violations
Operational agility	Overdependence on automation
Personalized customer experiences	Bias in training data

2.4 Conclusion: Designing for the Speed of Thought

Real-time AI marks a fundamental shift in the way enterprises operate. No longer are enterprises limited to static, periodic decision-making. Instead, they are flooded with a constant stream of intelligence, a state of perpetual awareness reminiscent of the human mind. Similarly, just as the human mind begins to sense, process, and act on its environment, so too are cognitive enterprises. However, this is not an automatic process. Real-time decision-making requires a solid infrastructure, a well-thought-out strategy, and a cultural ability to trust and challenge real-time AI. The aim is not to replace human judgment with AI, but to enhance it. To give humans the tools they need to act smartly and quickly, without flying blind. In the chapters to come, we will explore these principles in action within particular domains, such as predictive maintenance within industry, intelligent workforce planning within

HR, and smart decision systems within finance. However, the underlying principles are the same: to leverage the power of real-time AI to go from insight to action, and from action to transformation.

3. PREDICTIVE MAINTENANCE: AI IN INDUSTRIAL AND MANUFACTURING CONTEXTS

For decades, the same enemy of operational efficiency has plagued industrial and manufacturing companies across the world. When an essential piece of machinery unexpectedly malfunctions, the effects cascade through the company's schedules, supply chains, and revenue projections, ultimately affecting the end customers. The end result of an unplanned stoppage can be in the hundreds of thousands, or even millions, of dollars per hour, depending on the industry (Yazdi, 2024).

For decades, the only two approaches to maintenance and reducing downtime have been reactive (fix it when it breaks) and preventive (fix it before it breaks, on a schedule). While the latter is an improvement over the former, it still relies on assumption, not actual condition. It's like putting oil in your car every three months, whether it needs it or not. It's wasteful, inefficient, and can still lead to problems being overlooked. That's where predictive maintenance, enabled by AI, is revolutionizing the way industry operates. Now, companies can see problems coming and solve them only when they are absolutely necessary, leading to not only increased efficiency but a smarter, safer, and more sustainable way of doing business (Poór et al., 2019).

3.1 The Problem: The Cost of Downtime and Inefficiency

One of the costliest unplanned events in any industry is unplanned equipment failure. A study by Aberdeen Group and Deloitte found that the average cost of unplanned downtime for any industry can range from $260,000 to as high as $500,000 for every single hour. In any complicated system, such as an oil rig, a manufacturing plant, or an airplane fleet, these figures can add up in a hurry (Chernov et al., 2020).

Beyond direct financial loss, downtime also affects:

- Worker productivity, as staff shift into emergency response mode.
- Supply chain reliability, due to delayed deliveries or broken production promises.
- Customer trust, when commitments are missed or quality suffers.
- Safety and compliance, especially in energy, chemical, and aviation sectors.

Preventive maintenance approaches, though ubiquitous, are not effective. The maintenance schedules are not based on real-world usage or environmental factors. Some equipment is over-maintained, while other equipment fails before the scheduled maintenance date. Maintenance personnel are spending time on non-essential activities, while critical conditions are not being detected. Obviously, a more intelligent approach to asset conditions is needed.

3.2 The AI Solution: From Guesswork to Prediction

Predictive maintenance (PdM) uses a combination of IoT sensors, historical maintenance logs, equipment telemetry, and AI models to detect anomalies, forecast equipment failure, and recommend maintenance interventions just in time. Here's how the process generally works (Dhinakaran et al., 2025):

1. Data Collection: IoT sensors are embedded in critical components, motors, bearings, turbines, compressors, to capture metrics such as temperature, vibration, pressure, humidity, noise, and usage frequency.
2. Data Aggregation and Preprocessing: Raw data is streamed to a centralized platform (on-premise, cloud, or hybrid), where it's cleaned, normalized, and formatted for model input.
3. Model Training and Prediction: Machine learning models, often including anomaly detection, regression, or classification algorithms, are trained on historical failure data to identify patterns associated with asset degradation or imminent failure.
4. Maintenance Recommendation: Based on predicted failure windows, the system suggests optimal times for maintenance or even automatically schedules intervention within enterprise systems.
5. Feedback Loop: Model accuracy improves over time as more data is collected and maintenance outcomes are fed back into the system.

What distinguishes predictive maintenance from older approaches is its precision. Maintenance is neither too early (wasting resources) nor too late (risking failure). It's done when the data, rather than assumptions, says it should be done (Lee et al., 2020).

3.3 Real-World Applications Across Industries

Let's examine how predictive maintenance is transforming some of the world's most asset-intensive sectors:

3.3.1 Aerospace: Predicting Engine Wear and Component Failure

Aircraft engines are among the most sophisticated and costly machinery used today. They are also one of the most data-intensive. A jet engine, for example, sends terabytes of data into the air with each flight, measuring everything from the speed of the turbine to the level of vibrations. Top aerospace companies and airlines use AI models to analyze this data and identify early indicators of wear or fatigue. They use predictive models instead of traditional scheduled inspections. These predictive models identify engines that are operating outside the norm, even if they pass the initial inspection. A leading airline used AI-based predictive maintenance and reduced engine-related delays by 35%. They were able to extend the useful life of key engine components. Best of all, the predictive models help prioritize the maintenance crew's work, so that the most pressing problems are fixed first. (Mohanty, Hazra & Paul, 2020).

3.3.2 Automotive: Smart Assembly and Robotics Maintenance

The automotive industry is a highly automated sector where robotic arms are used in perfect synchronization with conveyor belts. The failure of a robotic arm for a matter of minutes can cause a chain of delays in production. AI-powered predictive maintenance technology is used to monitor robotic equipment in real-time to identify any impending failures. A leading automaker in the world used predictive analytics to monitor over 200 robotic units. By being able to identify impending failures up to five days in advance, they were able to achieve a reduction of 25 percent in unplanned downtime. Predictive maintenance also helps in reducing warranty claims by ensuring that cars are built to precise standards from the outset (Ionescu et al., 2022).

3.3.3 Energy: Grid and Infrastructure Health Monitoring

Energy infrastructure, whether electrical grids, oil pipelines, or wind farms, is expansive, aging, and critical. Monitoring the health of this infrastructure is a massive undertaking. AI has become a cornerstone in the maintenance of uptime, safety, and efficiency. In the case of electrical grids, predictive AI monitors the levels of heat emanating from the transformers, the fluctuating frequencies, and the load imbalance. These predictive models aim to prevent blackouts or fires. In the case of wind farms, predictive models use PdM models to monitor the condition of the turbines. An energy company used AI models to monitor the health of their wind turbines. They were able to predict failures as much as three weeks before they occurred. They were able to reduce their service calls by 30%.

3.4 System Architecture: From Edge to Cloud

Effective predictive maintenance systems rely on a robust technical architecture that allows real-time data collection, processing, and action. Key elements include:

3.4.1 IoT and Machine Learning Model Pipeline

On the front line are IoT devices, which are sensors embedded in physical assets. The devices transmit real-time telemetry data, which is used to feed machine learning models. The data is said to go through a multi-stage process (Shukla, & Sengupta, 2022):

- Edge processing: Initial filtering and threshold-based alerts happen close to the source (on the device or local gateway).
- Data ingestion: Cleaned data is streamed to a centralized system via protocols like MQTT or REST APIs.
- Model inference: AI models trained on historical failure data analyze current conditions and assign risk scores or make predictions.
- Action triggers: Based on confidence thresholds, alerts are generated, or actions (e.g., ticket creation, ordering parts) are initiated automatically.

3.4.2 Cloud-Edge Integration

This predictive maintenance can greatly benefit from the hybrid computing model, where edge computing can be utilized for tasks that require lower latency, and cloud computing can be utilized for tasks that require more computational power to train deep learning models, long-term storage, and visualization. Cloud platforms like AWS, Azure, Google Cloud, etc., may include native services like (Goundar, 2023):

- Time-series databases
- Model training pipelines (AutoML, custom ML)
- Visualization dashboards (Power BI, Tableau)
- Integration with ERP/CMMS systems like SAP or IBM Maximo

This gives the organization the benefits of both worlds: the response time of the edge, combined with the intelligence of the cloud. Figure 4 depicts the hybrid architecture for the predictive maintenance systems, ranging from edge to cloud learning and decision engines.

Figure 4. Predictive Maintenance System Architecture.

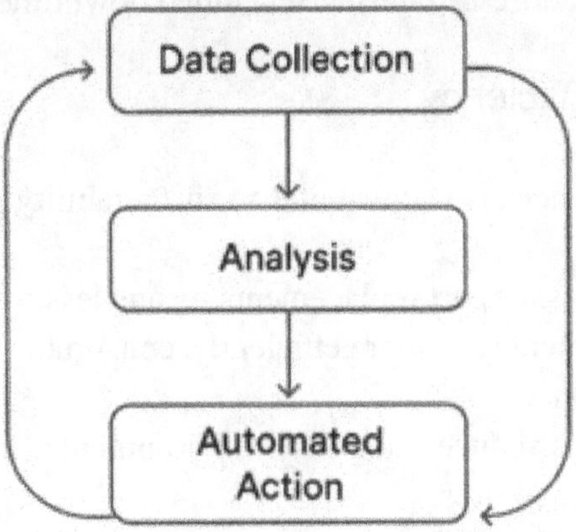

3.5 Business Outcomes: Tangible Value from Prediction

Predictive maintenance, as a concept, is not just a technological upgrade, but a business upgrade. Organizations using PdM techniques are realizing improvements in several areas, as identified by (Asthana, 2025):

1. Cost Savings

- Reduced Downtime: Fewer unexpected failures mean higher uptime and fewer disruptions to production or service delivery.
- Optimized Maintenance: Maintenance is performed only when needed, reducing labor hours and service costs.
- Spare Part Efficiency: Inventory is aligned with actual asset conditions, reducing overstocking or emergency orders.

A global manufacturing firm reported a 40% decrease in maintenance costs after implementing predictive systems across their production lines.

2. Enhanced Worker Safety

Additionally, the working conditions may be hostile, as the maintenance crew may be required to climb tower structures or inspect heavy machinery. With the predictive models, the situation is improved as the models minimize the occurrence of emergency situations. For example, the models may predict overheating of machinery, which is then addressed during scheduled downtime.

3. Environmental Efficiency

Predictive maintenance also contributes to sustainability goals:

- Reducing unnecessary part replacements means less waste.
- Optimized equipment runs more efficiently, consuming less energy and emitting fewer pollutants.
- Fewer breakdowns reduce the risk of environmental hazards like oil spills, leaks, or uncontained fires.

As ESG becomes an essential metric for industrial companies, AI-based maintenance provides both compliance and reputation benefits. Predictive maintenance is one of the most powerful and accessible uses of AI in the industrial world. It changes maintenance from being a cost center to becoming a source of competitive advantage. The effects of predictive maintenance, however, are not limited to the factory floor. It is one of the cornerstones of the cognitive enterprise, an enterprise that does not wait for problems to occur, but one that predicts and responds to them in an intelligent way. It is, in effect, a state of mind, shifting from reactive firefighting to proactive value creation. As we move forward in the coming chapters, we will see how AI drives this revolution, not only in physical assets, but also in human capital and financial systems. When every part of the enterprise thinks, adapts, and responds in real time, the possibilities are no longer limited by yesterday's limitations, but energized by tomorrow's intelligence (Rane, Choudhary & Rane, 2024).

4. HUMAN CAPITAL INTELLIGENCE: WORKFORCE PLANNING MEETS MACHINE LEARNING

The way we work is undergoing a seismic shift. From remote work and generational turnover to the growing requirement for new technical skills, the workforce is in motion. Yet, while technology has revolutionized the way supply chains operate, the way businesses conduct their finances, and the way they run their operations, the HR department is often forced to try and keep up with this motion with outdated tools and approaches. However, this is no longer the case. As more businesses turn to AI-based solutions throughout their operations, a new paradigm is emerging: human capital intelligence, the use of machine learning and advanced analytics to manage the workforce. Just as predictive maintenance allows companies to predict machine failure, AI allows companies to predict turnover risk, identify emerging skills, and even recognize the early stages of employee disengagement. In this chapter, the ways in which AI is revolutionizing the Human Resource function from a supporting function to a strategic powerhouse are explored. It's not simply a matter of hiring more quickly, nor is it simply a matter of evaluating performance more efficiently. Rather, it's about aligning HR strategies with business objectives in an agile, equitable, and highly intelligent way (Kalusivalingam et al., 2020).

4.1 The Talent Crisis: Complexity at the Core of Modern Workforces

Before diving into AI's role, it's important to understand the problem: today's talent landscape is more complex than ever. Organizations face a series of interconnected challenges (Goto, 2021):

- Fast-changing skill needs: Emerging technologies like AI, blockchain, and cybersecurity are driving demand for roles that didn't exist five years ago. Simultaneously, traditional roles are being redefined.
- Hybrid and remote workforces: The post-pandemic era has cemented remote and hybrid models as the norm, requiring new methods of collaboration, performance tracking, and engagement.
- Talent scarcity: Despite economic uncertainties, the global war for talent continues. Certain industries, healthcare, tech, logistics, are facing chronic shortages of qualified workers.
- Employee expectations: Workers today expect more than just compensation. They seek purpose, flexibility, inclusion, and continuous development.

Against this backdrop, HR teams are under pressure to move from administrative management to strategic foresight. But legacy tools, manual spreadsheets, static org charts, and generic engagement surveys, aren't enough. What's needed is a dynamic, data-driven approach to human capital: one that reflects the true complexity and velocity of the modern workplace.

4.2 How AI Helps: Unlocking Human Capital Intelligence

For AI, the same basic value proposition that it offers to any part of an enterprise's operation, pattern recognition at scale, is brought to bear in HR. The interest, and the stakes, are particularly high in HR, though, so let's examine three of the most impactful scenarios.

4.2.1 Resume Parsing and Job-Role Matching Using NLP

An important part of AI technology that has been immediately applied to HR functions is in the recruitment process. Enormous corporations have to process millions of applications each year for job openings. Human recruiters cannot work this quickly, meaning that additional opportunities to fill the openings and any job vacancies take longer to hire. AI is able to process resumes in seconds and determine matches for keywords, extract the required experience, and then evaluate the candidate's profile against the job requirements. These tools also add value as they are able to rank candidates, indicate missing required skills, and then offer job recommendations based on experience. More importantly, tools based on AI are moving further than keyword matching. Modern tools are able to engage in what is known as semantic understanding, which means that they can identify what the candidate meant, even if presented differently. As such, AI can identify the same competency whether or not they describe something as 'data visualization' or 'dashboard creation.' These tools are very useful to address the particularly bias discrimination against 'non-traditional' resumes (Mittal et al., 2020).

4.2.2 Predictive Attrition Models

High employee turnover rates are not only costly, but they are also detrimental to organizational stability. However, most organizations lack effective means of forecasting and predicting possible turnover and the underlying reasons, until it is too late. Using historical data on employees, their tenure, promotions, performance appraisals, and even their interaction with their managers, it is possible to build predictive models that can forecast the chances of an employee quitting in the next 3, 6, or 12 months. These predictive models often employ a variety of classification techniques, such as decision trees, logistic regression, and gradient boost, to generate an attrition score. These models, in conjunction with contextual data, team changes, location, and job market activity, provide early warnings that enable HR teams to initiate interventions, have conversations, and even offer opportunities based on the needs of each employee. Some organizations have been successful in integrating these predictive models into HR dashboards, enabling people managers to proactively act on retaining their talent (Chung et al., 2023).

4.2.3 Employee Sentiment Analysis Through Chat Data and Surveys

Culture is an intangible, yet measurable concept. As communication technologies like Slack, Teams, and emails become more prevalent, employees are inadvertently creating data that can be analyzed to understand their moods, tones, and sentiment. Using NLP and sentiment analysis, AI can automatically scan the aggregate data and identify trends, such as increased frustration in a particular department, reduced morale due to policy changes, or impending burnouts leading up to peak workloads (Ravichandran et al., 2023).

Intelligent survey tools are also leveraging adaptive questioning, where the system adjusts its questioning based on previous answers, to gather more in-depth data on what drives employees' engagement. These are not just data on how employees are feeling, but what they actually care about. While there are obvious ethical concerns that need to be discussed (which we will get to in a bit), organizational leaders are also gaining valuable real-time visibility into the emotional state of their organization, an invaluable advantage in chaotic times.

4.3 Ethical Use of AI in HR: Walking the Line Between Insight and Intrusion

The essence of HR is people, and the application of AI in the HR space is an area that needs to be treated with extraordinary sensitivity. The same AI that can spot high-potential employees or identify risks of employee turnover can, in the wrong hands, be used to reinforce biases or breach privacy (Varma, Dawkins & Chaudhuri, 2023).

4.3.1 Fairness and Bias Mitigation

One of the biggest risks of AI in HR is the amplification of historical biases. If past hiring decisions favored certain demographics, consciously or not, AI models trained on that data may learn to replicate those patterns.

To combat this, organizations must:

- Use bias-auditing tools to detect and mitigate skews in datasets and model predictions.
- Incorporate fairness constraints during model training.
- Regularly retrain models on updated, diversified data.

Some companies are also implementing "ethical AI review boards", cross-functional teams that evaluate high-impact models for potential bias and recommend safeguards.

4.3.2 Transparency and Consent

Employees have a right to know how their data is being used. While AI can yield incredible insights, it must be transparent, explainable, and consensual. This means:

- Clearly disclosing what data is collected and why.
- Giving employees control over certain types of personal data.
- Ensuring that decisions based on AI (such as promotions or performance ratings) are auditable and explainable.

Trust is essential. AI should empower employees and managers, not replace or surveil them.

4.4 Case Examples: Putting Human Capital Intelligence into Practice

Let's look at some real-world applications that highlight how AI is being integrated into core HR processes.

4.4.1 AI in Performance Management

Traditionally, performance reviews are not only infrequent, subjective, and disconnected from day-to-day work. The new AI-based systems leverage continuous feedback loops, where inputs are collected from team members, management, and project data throughout the year. For example, the systems can analyze project management tools (Jira, Asana) and communication channels to identify contributions, productivity bottlenecks, and management styles. These inputs are then compiled to feed more holistic, objective, and more frequent performance reviews. Additionally, AI can help set personalized development objectives, matching employees with internal learning resources or mentoring opportunities against real-time performance metrics (Varma, Pereira & Patel, 2024).

4.4.2 Diversity and Inclusion Modeling

Diversity metrics may be locked away in static reports. However, AI is helping organizations implement dynamic inclusion modeling, where they can monitor diversity in the hiring pipeline, promotions, and even salaries. A company used AI to monitor the patterns of promotions among thousands of their workforce. They found that women and minorities were being promoted at a slower rate, even after adjusting for tenure and performance. Another type of tool allows HR leaders to use scenario modeling. They can use this to test the impact of their proposed policies on diversity before they implement them. For example, they might use this tool to test the impact of flexible work options or changes to benefits. (Barrier et al., 2025).

4.5 Strategic Shift: From Administrative HR to Workforce Intelligence

At the heart of these transformations is a strategic pivot: HR is no longer a back-office function. It is now a driver of competitive advantage. AI gives HR tools to (Kadyan & Singh, 2025):

- Forecast workforce needs based on business strategy and market trends.
- Model talent gaps before they impact performance.
- Design more resilient workforce plans that adapt to change.

This shift is particularly valuable in industries facing disruption. For example:

- A logistics company used AI to forecast a shortage of skilled warehouse labor six months ahead of peak season and proactively launched training programs.
- A tech firm used machine learning to predict what mix of contract, full-time, and gig workers would be needed as new products rolled out.

Instead of reacting to talent shortages or cultural issues, companies using AI in HR can anticipate, plan, and optimize. Human capital becomes a living, evolving asset, measured, understood, and empowered in real time.

4.6 Intelligence That Works for People

The use of AI in human capital is not about replacing people; it's about understanding people more deeply, helping people more effectively, and planning for the future more precisely. As enterprises become more cognitive, they must also realize that intelligence is not only in their systems; it's also in their people. By using machine learning in hiring, retaining, engaging, and developing people, organizations can turn their HR function from a cost center to a strategic engine. Organizations can retain their best people, build a more inclusive culture, and develop a workforce that matches their strategic ambitions. The most successful organizations will not only use machine learning; they will build a culture in which humans and computers collaborate to deliver better work, better workplaces, and better outcomes (Wirtz & George, 2022).

As discussed above, in the next section, it will be shown how this same logic plays out in financial operations. From fraud detection to budgeting, AI is building the next evolution of smart finance – and with it, the next pillar of the cognitive enterprise.

5. SMART FINANCE: AI FOR RISK ANALYSIS, BUDGETING, AND FRAUD DETECTION

In the hyper-dynamic economy of today, financial leaders can no longer afford to be merely historians of the past. Rather, the rapidity of change, whether in the form of disruptions in the global markets or in the demand from customers, calls for financial teams to move from a backward-looking approach to a forward-looking approach. Moreover, the CFO of today needs to be an equal part of an analyst, a strategist, and a technologist. And today, the biggest enabler for the CFO is AI (Saha, 2025).

The revolution that AI is creating in financial management is such that it is helping organizations operate in a faster, smarter, and more secure manner. From risk management to budgeting, from fraud management to compliance, smart financial management systems enabled by AI are helping organizations achieve greater results. This chapter discusses the revolution that AI is creating in financial management, the technology that makes it possible, and the challenges that need to be overcome by organizations.

5.1 The Evolving Role of Finance: From Reporting to Proactive Strategy

Traditionally, the role of the finance function was that of the "scorekeeper" of the business, where they needed to manage the expenses, generate the financial statements, and ensure compliance with the regulatory environment. Although these activities are still relevant, they are no longer sufficient. Today, the finance function is expected to:

- Develop future scenarios based on complex and constantly changing data sets.
- Assist in the decision-making process regarding strategic investments and risks.
- Provide real-time information for decision-making across the enterprise.

The impetus for change is the explosion of data within the enterprise, not only from the customer-facing and operational activities of the company, but also from the overall economic environment. The finance executive is now challenged with using this data as a means of developing the ability to foresee the future. However, the traditional means of managing the enterprise using traditional tools, spreadsheets, and monthly close cycles are not conducive to such an environment. This is where AI enters the picture as the means of developing a proactive, predictive, and intelligent finance operation.

5.2 Core AI Applications in Finance

The finance function offers a fertile ground for AI deployment because of its data richness, need for precision, and highly repeatable processes. Let's examine three areas where AI is already making a profound impact.

5.2.1 Fraud Detection via Anomaly Detection Models

As financial fraud continues to change, so must the solutions to prevent it. Traditional rules-based systems for detecting financial fraud often fail to keep up with the evolving nature of financial fraud, yielding either too many false positives or failing to detect the complexities of financial fraud altogether. Machine learning, and more specifically unsupervised machine learning, performs well at detecting anomalies, even complex ones, that might otherwise go undetected. These types of machine learning algorithms can (Hilal, Gadsden & Yawney, 2022):

- Detect irregular transaction patterns.
- Flag suspicious supplier behavior or procurement anomalies.
- Identify inconsistencies in financial reporting or expense claims.

For example, in the case of credit card fraud, an AI model examines real-time data such as time of day, location, type of store, and frequency. Therefore, if a customer normally buys items in Paris, an AI model will be able to detect if the card is used to buy items in São Paulo, such as electronics. In the case of corporate finance, an AI model will be able to detect internal fraud such as ghost vendors or inflated invoices. This will not only reduce fraud losses for an organization, but it will also improve the ability to comply with anti-fraud regulations.

5.2.2 AI Budgeting Assistants and Scenario Planning

Budgeting is a time-consuming and static process that has traditionally been plagued by the need for manual data entry, long approval cycles, and a lack of transparency. By the time the budget is complete, it is likely obsolete. AI is changing the budgeting process with the ability for dynamic budgeting and modeling. AI assistants can process vast amounts of financial and operational data in order to (Spaniol & Rowland, 2023):

- Suggest budget allocations based on historical patterns.
- Forecast spending and revenue trends.
- Simulate best-case, worst-case, and most likely scenarios.

This allows finance teams to adjust their budgets in near real-time as business conditions change. Take, for instance, a scenario where raw materials prices surge or customer demand changes. AI systems would be able to offer new forecasts and budget changes, thus helping business leaders remain nimble. Some of the more advanced systems even offer natural language interaction, whereby budget owners would be able to pose questions such as, what's our projected cash flow if Q3 sales drop by 10%? and get the answer right away, no spreadsheet wizardry needed.

5.2.3 Credit Scoring Using Alternative Data

Traditional credit scoring systems are greatly dependent on formal financial history, credit reports, repayment of loans, income statements, etc. However, many people and small businesses, especially in developing countries, are not part of the formal credit systems. AI has brought forth opportunities to use alternative credit scoring systems with factors like (Djeundje et al., 2021):

- Mobile phone usage patterns.
- Utility payment history.
- Social media behavior.
- E-commerce purchase history.

Using these non-traditional data sources and machine learning algorithms to analyze them allows financial institutions to gauge creditworthiness with surprising accuracy. Additionally, this will allow for further financial inclusion for underserved segments of the population. Financial technology companies are driving this trend, with some using AI to evaluate creditworthiness and approve loans within a matter of seconds. This has helped to bring many people into the fold for the first time.

5.3 The Tech Stack Behind Smart Finance

Beneath these capabilities lies a robust technology ecosystem that combines automation, analytics, and scalable infrastructure. Two major pillars are worth highlighting:

5.3.1 RPA and AI for Process Automation

Robotic Process Automation (RPA) automates repetitive, rule-based tasks—like invoice processing, reconciliation, or tax calculations. When combined with AI (e.g., NLP and machine learning), RPA becomes far more powerful. This AI-enhanced automation can (Afrin, Roksana & Akram, 2024):

- Read and interpret invoices or contracts (using document intelligence).
- Match payments to open invoices even when formatting errors exist.
- Extract relevant terms from legal documents.

For example, a global accounting team might use an AI-RPA system to process hundreds of invoices daily, flagging only exceptions for human review. The result is a significant reduction in manual workload, improved accuracy, and faster cycle times.

5.3.2 Predictive Analytics for Revenue Forecasting

Traditional revenue forecasting often relies on simple linear models or gut feeling. AI allows for much more nuanced and adaptive forecasting by incorporating a wide range of variables (Ismail et al., 2024):

- Seasonality trends
- Macroeconomic indicators
- Customer behavior and churn rates
- Product-level performance data

Machine learning models such as ARIMA, LSTM neural networks, or gradient boosting machines can forecast revenues with much higher accuracy, especially in volatile markets.

Some organizations integrate these models into interactive dashboards, allowing business units to explore forecast drivers and adjust inputs dynamically. This collaborative, AI-supported planning environment enables finance teams to partner more closely with sales, operations, and marketing in steering the business.

5.4 Challenges: Trust, Transparency, and Compliance

Despite its benefits, the adoption of AI in finance must be tempered with careful consideration. Several critical challenges must be addressed (Chadha et al., 2023).

5.4.1 Regulatory Compliance

Finance is one of the most heavily regulated sectors, and AI must operate within strict compliance frameworks such as:

- GDPR (for data privacy)
- SOX (for financial reporting)
- AML/KYC laws (anti-money laundering and know-your-customer)

Organizations must ensure that AI systems do not violate privacy rights, expose sensitive data, or create blind spots in audit trails. This requires robust data governance, model documentation, and alignment with internal audit and legal teams.

For example, an AI model that analyzes employee expense reports must comply with GDPR by anonymizing data where appropriate and offering data subjects the right to access or delete their data.

5.4.2 Model Interpretability

"Black box" AI models, particularly those employing deep learning techniques, are not transparent in the decision-making process. However, in the finance industry, transparency is a key concern. For instance, in the case of a loan application or a transaction that is considered suspicious by an AI system, the need for transparency is high. Regulators and customers need to understand the basis of the decision-making process. This is where the importance of XAI is recognized. Various finance industry firms are employing hybrid models in the decision-making process, ensuring the trade-off between transparency and accuracy (Djeundje et al., 2021).

5.5 Success Stories: AI at Work in Financial Innovation

Let's explore two compelling examples of how AI is driving real-world success in financial operations (Challoumis, 2024).

5.5.1 Fintech Disruption of Traditional Banks

Fintech startups have embraced AI from day one, using it to:

- Deliver real-time credit approvals.
- Personalize financial advice using robo-advisors.
- Optimize investment portfolios using AI trading algorithms.

A good example is that of a fintech-based lender that used alternative data to facilitate the approval of microloans in under 5 minutes. Because traditional banks were unable to serve the low-income population or the gig economy, the fintech-based lender recorded huge growth as they were able to extend over $500 million in loans over the course of three years while keeping their default rates below the industry benchmarks. What the fintech-based lender did right was not only their use of technology but their capacity to establish trust with their customers through transparency regarding how their creditworthiness was assessed.

5.5.2 AI-Driven Audit Trails in Corporate Finance

Today, large corporations are leveraging AI to revolutionize internal audits and compliance. A global consumer goods company adopted an AI solution that monitored all financial transactions in real-time. The AI system identified unusual transactions, duplicate payments, unusual vendor transactions, or policy infractions and generated auditable trails. The company no longer had to engage in a traditional quarterly audit process. Instead, it adopted a new model of real-time audit assurance. The finance teams could now respond to risks in real-time rather than after months. This new model reduced audit time by 60%. Furthermore, it also allowed the company to achieve SOX compliance in a more efficient way. This allowed internal auditors to focus on high-value tasks (Kummari, 2025).

Table 2. Comparison of AI Applications Across Business Functions

Business Function	AI Application	Key Benefit
Supply Chain	Real-time optimization	Reduced delays
Manufacturing	Predictive maintenance	Minimized downtime
HR	Attrition prediction, resume matching	Strategic hiring
Finance	Fraud detection, scenario planning	Improved risk management

6. CONCLUSION

AI is transforming organizational processes related to information interpretation, decision-making, and response to environmental changes. This chapter aims to explore the concept of a cognitive enterprise, which is a type of organizational form where AI is incorporated into business processes to enable data transformation into real-time intelligence and actions. A cognitive enterprise is one where there is a shift beyond periodic data analysis and reporting, using continuous data, predictive analytics, and intelligent automation to enable faster decision-making processes.

From an analytical perspective, this chapter will demonstrate that the value of AI is not simply about automation but about amplifying the way an organization thinks. Real-time decision technologies enable organizations to transition from backward thinking to forward thinking, helping them respond to Ops hiccups and market forces in real-time. Consider predictive maintenance – using sensors and ML algorithms, you can detect potential failures before they occur, reducing downtime and increasing overall business agility. Or human capital intelligence – using AI-powered analytics, you can forecast future workforce needs, turnover rates, and even support data-driven human capital strategies. And then there's finance – using AI-powered forecasting, fraud detection, and scenario planning, you can transition the finance organization from backward thinking to forward thinking. All these examples demonstrate the value of AI in helping organizations transition from reactive organizations to adaptive organizations that learn and adapt.

For managers and organizational leaders, the transition toward cognitive enterprises carries several important implications. First, the adoption of AI requires integrated data infrastructures and cross-functional collaboration, as the effectiveness of intelligent systems depends on the ability to connect information across operational, financial, and human resource domains. Second, organizations must invest not only in technology but also in organizational capabilities, including data governance, employee training, and cultural readiness for AI-supported decision-making. Third, ethical considerations must be treated as a strategic priority. Issues such as algorithmic bias, transparency, and data privacy require robust governance frameworks to ensure that AI-driven decisions remain trustworthy and socially responsible. Leaders who successfully combine technological innovation with ethical oversight and organizational transformation will be better positioned to realize the full potential of AI-enabled enterprises.

However, the emergence of cognitive enterprises also raises a number of open questions that are worth exploring in more depth. For instance, the potential for integrating AI systems across domains could be explored, such as the potential for decision intelligence platforms to integrate the insights of operations, finance, and human resources all at once. Another potential area of exploration is the governance and ethical use of AI systems. For instance, more research could focus on the potential for increasing the transparency of AI models, auditing the fairness of AI-driven decisions, and increasing the explainability of AI-driven decisions. Another area of research could focus on the organizational and cultural changes required for the use of AI-driven decisions. For instance, more research could focus on the potential for AI-driven decisions in traditional industries. Finally, the long-term impact of human-AI collaboration could also be explored as an area of research.

REFERENCES

Afrin, S., Roksana, S., & Akram, R. (2024). AI-Enhanced Robotic Process Automation: A Review of Intelligent Automation Innovations. *IEEE Access : Practical Innovations, Open Solutions*. Advance online publication. DOI: 10.1109/ACCESS.2024.3513279

Ajunwa, I. (2020). An auditing imperative for automated hiring systems. *Harv. JL & Tech., 34*, 621. https://ssrn.com/abstract=3437631

Allen, K. E., & Mau, T. A. (2025). Breaking down traditional silos: Leading organizations from an interdependent mindset. *Humanistic Management Journal, 10*(1), 125–139. DOI: 10.1007/s41463-024-00198-4

Asthana, R. (2025). Driving User Adoption and Agility in PLM: A Data-Driven Approach to Change Management. https://www.theseus.fi/bitstream/handle/10024/887389/Asthana_Richa.pdf?sequence=2&isAllowed=y

Barrier, K. M., Porche, D. J., Campbell, K. M., Byrd, T. O., Morris, M., Blalack, K. L., Ziglor, C., Tuazon, S., Mouton, C. P., & Southerland, J. H. (2025). Models of Diversity, Equity, Inclusion, and Belonging: Part II—Exploring Models of Inclusion from Other Health Professions for Dentistry. *Dental Clinics of North America, 69*(1), 55–68. DOI: 10.1016/j.cden.2024.08.010 PMID: 39603769

Chadha, P., Gera, R., Khera, G. S., & Sharma, M. (2023). Challenges of Artificial Intelligence Adoption for Financial Inclusion. In *Artificial Intelligence, Fintech, and Financial Inclusion* (pp. 135-160). CRC Press. https://doi.org/DOI: 10.1201/9781003125204-9

Challoumis, C. (2024). the Future of Business-integrating AI Into the Financial Cycle. In *XIV International Scientific Conference* (pp. 44-78).

Chataut, R., Nankya, M., & Akl, R. (2024). 6G networks and the AI revolution—Exploring technologies, applications, and emerging challenges. *Sensors (Basel), 24*(6), 1888. DOI: 10.3390/s24061888 PMID: 38544151

Chernov, D., Sornette, D., Chernov, D., & Sornette, D. (2020). Specific features of risk management in the industrial and agricultural sectors. *Critical Risks of Different Economic Sectors: Based on the Analysis of More Than 500 Incidents, Accidents and Disasters*, 13-145. https://doi.org/DOI: 10.1007/978-3-030-25034-8_2

Chung, D., Yun, J., Lee, J., & Jeon, Y. (2023). Predictive model of employee attrition based on stacking ensemble learning. *Expert Systems with Applications, 215*, 119364. DOI: 10.1016/j.eswa.2022.119364

Dhinakaran, D., Edwin Raja, S., Velselvi, R., & Purushotham, N. (2025). Intelligent IoT-Driven Advanced Predictive Maintenance System for Industrial Applications. *SN Computer Science*, *6*(2), 151. DOI: 10.1007/s42979-025-03695-x

Djeundje, V. B., Crook, J., Calabrese, R., & Hamid, M. (2021). Enhancing credit scoring with alternative data. *Expert Systems with Applications*, *163*, 113766. DOI: 10.1016/j.eswa.2020.113766

Ekundayo, F. (2024). Leveraging AI-Driven Decision Intelligence for Complex Systems Engineering. *Int J Res Publ Rev*, *5*(11), 1–10. DOI: 10.55248/gengpi.5.1124.3343

Goto, M. (2021). Collective professional role identity in the age of artificial intelligence. *Journal of Professions and Organization*, *8*(1), 86–107. DOI: 10.1093/jpo/joab003

Goundar, S. (Ed.). (2023). *Edge Computing: Technology, Management and Integration*. BoD–Books on Demand., DOI: 10.5772/intechopen.105637

Hilal, W., Gadsden, S. A., & Yawney, J. (2022). Financial fraud: A review of anomaly detection techniques and recent advances. *Expert Systems with Applications*, *193*, 116429. DOI: 10.1016/j.eswa.2021.116429

Hrouga, M., & Sbihi, A. (2023). Logistics 4.0 for supply chain performance: Perspectives from a retailing case study. *Business Process Management Journal*, *29*(6), 1892–1919. DOI: 10.1108/BPMJ-03-2023-0183

Ionescu, D., Filipescu, A., Simion, G., Mincă, E., Cernega, D., olea, R., & Filipescu, A. (2022). Communication and control of an assembly, disassembly and repair flexible manufacturing technology on a mechatronics line assisted by an autonomous robotic system. *Inventions (Basel, Switzerland)*, *7*(2), 43. DOI: 10.3390/inventions7020043

Ismail, S., Natarajan, S. K., & Voleti, R. (2024, November). Leveraging Predictive Analytics of Age Demographics for Engagement and Revenue Forecasting. In *2024 2nd International Conference on Advances in Computation, Communication and Information Technology (ICAICCIT)* (Vol. 1, pp. 749-754). IEEE. http://doi.org/ DOI: 10.1109/ICAICCIT64383.2024.10912167

Kadyan, P., & Singh, R. (2025). Smart HRM: Unleashing the Power of Business Intelligence for Workforce Excellence. In *Harnessing Business Intelligence for Modern Talent Management* (pp. 275-308). IGI Global Scientific Publishing. http://doi.org/DOI: 10.4018/979-8-3373-1942-1.ch011

Kalubanga, M., & Gudergan, S. (2022). The impact of dynamic capabilities in disrupted supply chains—The role of turbulence and dependence. *Industrial Marketing Management*, *103*, 154–169. DOI: 10.1016/j.indmarman.2022.03.005

Kalusivalingam, A. K., Sharma, A., Patel, N., & Singh, V. (2020). Optimizing Workforce Planning with AI: Leveraging Machine Learning Algorithms and Predictive Analytics for Enhanced Decision-Making. *International Journal of Artificial Intelligence and Machine Learning, 1*(3). https://cognitivecomputingjournal.com/index.php/IJAIML-V1/article/view/53

Kowalczuk, Z., & Czubenko, M. (2023). Cognitive motivations and foundations for building intelligent decision-making systems. *Artificial Intelligence Review, 56*(4), 3445–3472. DOI: 10.1007/s10462-022-10255-9

Kummari, D. N. (2025). *Advanced Practices in Auditing, Regulatory Compliance, and Smart Manufacturing Systems*. Deep Science Publishing., DOI: 10.70593/978-93-7185-253-1

Lazaroiu, G., Androniceanu, A., Grecu, I., Grecu, G., & Neguri ă, O. (2022). Artificial intelligence-based decision-making algorithms, Internet of Things sensing networks, and sustainable cyber-physical management systems in big data-driven cognitive manufacturing. *Oeconomia Copernicana, 13*(4), 1047–1080. DOI: 10.24136/oc.2022.030

Lee, J., Ni, J., Singh, J., Jiang, B., Azamfar, M., & Feng, J. (2020). Intelligent maintenance systems and predictive manufacturing. *Journal of Manufacturing Science and Engineering, 142*(11), 110805. DOI: 10.1115/1.4047856

Mittal, V., Mehta, P., Relan, D., & Gabrani, G. (2020). Methodology for resume parsing and job domain prediction. *Journal of Statistics and Management Systems, 23*(7), 1265–1274. DOI: 10.1080/09720510.2020.1799583

Mohanty, S., Hazra, S., & Paul, S. (2020). Intelligent prediction of engine failure through computational image analysis of wear particle. *Engineering Failure Analysis, 116*, 104731. DOI: 10.1016/j.engfailanal.2020.104731

Nadaf, H., Jabade, M. V., Jamadar, K., Jogdeo, B., & Jamdade, V. (2025). Improving Emergency Response: A Comparative Analysis of Traditional vs Artificial Intelligence-assisted Triage Systems in Health Care and Their Impact. *Indian Journal of Critical Care Medicine : Peer-Reviewed, Official Publication of Indian Society of Critical Care Medicine, 29*(11), 925–929. DOI: 10.5005/jp-journals-10071-25083 PMID: 41368447

Parveen, N., & Basit, F. (2023). Securing Data in Motion and at Rest: AI and Machine Learning Applications in Cloud and Network Security. https://doi.org/DOI: 10.13140/RG.2.2.35114.84163

Patel, K. (2024). Ethical reflections on data-centric AI: balancing benefits and risks. *Available at SSRN* 4993089. http://dx.doi.org/DOI: 10.2139/ssrn.4993089

Pilipczuk, O. (2022). Building the Cognitive Enterprise in the Energy Sector. *Energies, 15*(24), 9479. DOI: 10.3390/en15249479

Poór, P., Ženíšek, D., & Basl, J. (2019, July). Historical overview of maintenance management strategies: Development from breakdown maintenance to predictive maintenance in accordance with four industrial revolutions. In *Proceedings of the international conference on industrial engineering and operations management, Pilsen, Czech Republic* (pp. 23-26). https://ieomsociety.org/pilsen2019/papers/135.pdf?_gl=1*uinndd*_ga*MzIxMjkzMDIuMTc2MzM3ODE2OA

Rane, N., Choudhary, S., & Rane, J. (2024). Artificial intelligence driven approaches to strengthening Environmental, Social, and Governance (ESG) criteria in sustainable business practices: a review. *Social, and Governance (ESG) criteria in sustainable business practices: a review*. DOI: 10.2139/ssrn.4843215

Ravichandran, N., Inaganti, A. C., & Muppalaneni, R. (2023). AI-Driven sentiment analysis for employee engagement and retention. *Journal of Computing Innovations and Applications, 1*(01), 1–9. https://ciajournal.com/index.php/jcia/article/view/6

Saha, B. (2025). AI-driven predictive analytics for financial risk management. *International Journal of Advance Research and Innovation (IJARI, 2347-3258), 13*(01), 12-22. https://doi.org/DOI: 10.69996/ijari.2025002

Sahoh, B., & Choksuriwong, A. (2023). The role of explainable Artificial Intelligence in high-stakes decision-making systems: A systematic review. *Journal of Ambient Intelligence and Humanized Computing, 14*(6), 7827–7843. DOI: 10.1007/s12652-023-04594-w PMID: 37228699

Sendas, N., & Rajale, D. (2025). Future Trends in MLOps. In *The Definitive Guide to Machine Learning Operations in AWS: Machine Learning Scalability and Optimization with AWS* (pp. 371-423). Berkeley, CA: Apress. https://doi.org/DOI: 10.1007/979-8-8688-1076-3

Shamim, S., Yang, Y., Zia, N. U., Khan, Z., & Shariq, S. M. (2023). Mechanisms of cognitive trust development in artificial intelligence among front line employees: An empirical examination from a developing economy. *Journal of Business Research, 167*, 114168. DOI: 10.1016/j.jbusres.2023.114168

Shobhana, N. (2024). AI-powered supply chains towards greater efficiency. In *Complex AI Dynamics and Interactions in Management* (pp. 229–249). IGI Global., DOI: 10.4018/979-8-3693-0712-0.ch011

Shukla, R. M., & Sengupta, S. (2022). A novel machine learning pipeline to detect malicious anomalies for the Internet of Things. *Internet of Things : Engineering Cyber Physical Human Systems*, *20*, 100603. DOI: 10.1016/j.iot.2022.100603

Spaniol, M. J., & Rowland, N. J. (2023). AI-assisted scenario generation for strategic planning. *Futures & Foresight Science*, *5*(2), e148. DOI: 10.1002/ffo2.148

Stach, C. (2023). Data is the new oil–sort of: A view on why this comparison is misleading and its implications for modern data administration. *Future Internet*, *15*(2), 71. DOI: 10.3390/fi15020071

Uddin, M. K. S. (2024). A Review of Utilizing Natural Language Processing and AI For Advanced Data Visualization in Real-Time Analytics. *Global Mainstream Journal*, *1*(4), 10–62304. DOI: 10.62304/ijmisds.v1i04.185

Varma, A., Dawkins, C., & Chaudhuri, K. (2023). Artificial intelligence and people management: A critical assessment through the ethical lens. *Human Resource Management Review*, *33*(1), 100923. DOI: 10.1016/j.hrmr.2022.100923

Varma, A., Pereira, V., & Patel, P. (2024). Artificial intelligence and performance management. *Organizational Dynamics*, *53*(1), 101037. DOI: 10.1016/j.orgdyn.2024.101037

Wirtz, J. J., & George, R. Z. (2022). Assessing futures intelligence: Looking back on global trends 2025. *Political Science Quarterly*, *137*(3), 481–510. DOI: 10.1002/polq.13354

Yazdi, M. (2024). Reliability-centered design and system resilience. In *Advances in Computational Mathematics for Industrial System Reliability and Maintainability* (pp. 79–103). Springer Nature Switzerland., DOI: 10.1007/978-3-031-53514-7_5

Chapter 8
Digital Twin Applications in Financial Risk Management:
A Multi-Case Study of Scenario Simulation and Decision-Making in Financial Institutions

K. Balaji
http://orcid.org/0000-0002-3065-3294
Christ University, Bengaluru, India

ABSTRACT

Digital transformation is reshaping financial institutions, with digital twin technology emerging as a powerful tool for modeling, simulation, and risk management. This study aims to examine the application of digital twin technology in financial risk management, focusing within financial institutions. Adopting a qualitative multi-case study approach, the research analyzes leading global financial institutions that have integrated digital twins and AI-driven analytics into their risk management frameworks. Analytical tools such as thematic analysis, cross-case comparison, and risk performance indicators are employed to evaluate the effectiveness of digital twin applications. The findings reveal that digital twins significantly improve the accuracy and efficiency of risk assessment by enabling real-time simulation of financial scenarios, including market volatility, credit risks, and liquidity stress. Institutions leveraging digital twins demonstrate enhanced predictive capabilities, better stress-testing mechanisms, and more informed strategic decision-making.

DOI: 10.4018/979-8-2600-1669-5.ch008

1. INTRODUCTION

The financial services industry is undergoing a profound transformation driven by rapid advancements in digital technologies. Financial institutions, including banks, asset management firms, and fintech organizations, are increasingly leveraging digital tools such as Artificial Intelligence (AI), big data analytics, cloud computing, and blockchain to enhance operational efficiency and decision-making capabilities. This digital transformation has fundamentally reshaped traditional financial systems, enabling real-time data processing, improved customer experiences, and more sophisticated risk management practices. As financial markets become more interconnected and complex, the need for intelligent, data-driven systems has become increasingly critical for maintaining competitiveness and resilience.

Within this evolving landscape, digital twin technology has emerged as a novel and transformative innovation with significant potential for financial applications. Originally developed in engineering and manufacturing contexts, digital twins refer to virtual replicas of physical or organizational systems that can simulate real-world processes in real time. In the financial domain, digital twins are being adapted to create dynamic models of financial systems, portfolios, and market environments. By integrating AI and predictive analytics, digital twins enable financial institutions to simulate various scenarios, assess potential risks, and optimize decision-making processes. This capability marks a significant shift from traditional static models toward more adaptive and intelligent financial systems. The importance of financial risk management has been further amplified in today's dynamic and uncertain market environment. Factors such as market volatility, geopolitical tensions, regulatory changes, and technological disruptions have increased the complexity of financial risks. Traditional risk management approaches, which often rely on historical data and linear models, are no longer sufficient to capture the multidimensional nature of modern financial risks. Financial institutions must now adopt more advanced tools that can provide real-time insights, predictive capabilities, and scenario-based analysis to effectively manage uncertainty and ensure stability.

In this context, the need for advanced simulation and predictive decision-making tools becomes paramount. Digital twins, combined with AI-driven analytics, offer a powerful solution by enabling continuous monitoring, real-time simulation, and proactive risk assessment. These technologies allow organizations to test multiple scenarios, evaluate the impact of potential risks, and make informed decisions based on dynamic data inputs. As a result, digital twins are increasingly being recognized as a strategic tool for enhancing financial resilience and improving the accuracy of risk management processes. Despite the growing interest in digital twin technology, there remains a significant research gap in understanding its application within financial risk management. Existing studies have primarily focused on digital twins in engineering, manufacturing, and healthcare, with limited exploration of their role in financial systems. Moreover, there is a lack of empirical, case-based research examining how financial institutions are adopting and utilizing digital twins for risk management and decision-making. This gap highlights the need for comprehensive research that integrates technological, financial, and governance perspectives.

Against this backdrop, this chapter aims to explore the application of digital twin technology in financial risk management through a multi-case study approach. The primary objectives of the study are: (i) to examine how digital twins enhance scenario simulation and predictive analytics in financial institutions; (ii) to analyze the role of digital twins in improving risk assessment and decision-making processes; (iii) to investigate the integration of AI and digital twin technologies in financial systems; and (iv) to identify key challenges, governance considerations, and future opportunities associated with digital twin adoption in finance. Through this analysis, the chapter seeks to contribute to the emerging field of digital finance by providing valuable insights for researchers, practitioners, and policymakers.

2. CONCEPTUAL BACKGROUND

2.1 Digital Twin Technology in Enterprise Systems

Digital twin technology refers to the creation of a dynamic, virtual representation of a physical system, process, or organizational entity that is continuously updated using real-time data. The concept originated in engineering and manufacturing domains, where digital replicas of physical assets were used for monitoring, diagnostics, and optimization(Kanike, 2023). Over time, the evolution of advanced technologies such as Artificial Intelligence (AI), Internet of Things (IoT), and cloud computing has expanded the scope of digital twins beyond physical assets to complex enterprise systems, including financial institutions. A digital twin typically comprises three core components: the physical system, the virtual model, and data integration mechanisms. The physical system represents the real-world entity or process being modeled, such as financial markets, portfolios, or institutional operations. The virtual model is the digital counterpart that simulates the behavior and performance of the physical system using computational algorithms. Data integration serves as the connecting layer, enabling continuous synchronisation between the physical and virtual environments through real-time data streams. This integration allows digital twins to evolve dynamically, reflecting current conditions and enabling predictive analysis(Figueroa et al., 2024). In financial systems, digital twin technology holds significant relevance due to the complexity and dynamic nature of financial markets. Financial institutions can develop digital twins of portfolios, trading environments, or entire organizational processes to simulate market conditions, assess risks, and optimize strategies. By leveraging AI and real-time data, digital twins enable more accurate modeling of financial scenarios, enhance decision-making, and support proactive risk management. This marks a shift from traditional static models to adaptive, data-driven financial systems.

2.2 Financial Risk Management

Financial risk management is a critical function within financial institutions, aimed at identifying, assessing, and mitigating various types of risks that may impact financial performance and stability. The primary categories of financial risks include market risk, credit risk, liquidity risk, and operational risk. Market risk arises from fluctuations in market variables such as interest rates, exchange rates, and equity prices, which can affect the value of financial assets and portfolios. Credit risk refers to the possibility of a borrower failing to meet contractual obligations, leading to financial losses for lenders. Liquidity risk involves the inability of an institution to meet its short-term financial obligations due to insufficient cash flow or market liquidity. Operational risk stems from failures in internal processes, systems, or external events, including fraud, cyber threats, and system breakdowns.

Traditionally, financial risk management has relied on statistical models, historical data analysis, and regulatory frameworks such as Value at Risk (VaR) and stress testing. While these approaches have been effective to some extent, they often lack the ability to capture real-time dynamics and complex interdependencies within financial systems. In contrast, AI-driven risk management approaches leverage machine learning algorithms, big data analytics, and predictive modeling to enhance risk assessment and forecasting capabilities. These advanced techniques enable financial institutions to analyze large datasets, identify hidden patterns, and generate more accurate and timely insights into potential risks. The integration of AI with digital twin technology further enhances these capabilities by enabling continuous simulation and real-time risk monitoring(Vital-Soto & Olivares-Aguila, 2023).

2.3 Scenario Simulation and Predictive Analytics

Scenario simulation and predictive analytics play a pivotal role in modern financial decision-making by enabling institutions to evaluate potential outcomes under different conditions. Simulation involves creating hypothetical scenarios to assess the impact of various risk factors on financial performance, while predictive analytics uses historical and real-time data to forecast future trends and events. Together, these tools provide a comprehensive framework for managing uncertainty and improving strategic planning. In finance, simulation techniques are widely used for stress testing, portfolio optimization, and risk assessment. Stress testing involves evaluating the resilience of financial institutions under extreme but plausible scenarios, such as economic downturns or market shocks. Forecasting models, on the other hand, are used to predict future market trends, interest rate movements, and credit defaults. Traditional simulation models often rely on deterministic or linear assumptions, which may not fully capture the complexity of financial systems(Zhang et al., 2022).

The integration of AI and machine learning has significantly enhanced the capabilities of simulation and predictive analytics. Machine learning algorithms can process large volumes of structured and unstructured data, identify nonlinear relationships, and adapt to changing market conditions. When combined with digital twin technology, these capabilities enable real-time scenario simulation and dynamic forecasting. Financial institutions can continuously update their models based on new data, test multiple scenarios simultaneously, and make informed decisions with greater accuracy and confidence. This integration represents a paradigm shift toward intelligent, adaptive financial systems capable of responding effectively to evolving risks and opportunities (K. Balaji & P. S. Rao,2024).

3. THEORETICAL FRAMEWORK

The adoption and effectiveness of digital twin technology in financial risk management can be better understood through a multidisciplinary theoretical lens. This study integrates key theories from information systems, organisational studies, finance, and decision sciences to explain how digital twins are adopted, governed, and utilised in financial institutions. Specifically, the Technology Acceptance Model (TAM), Institutional Theory, Risk Management Theory, and Decision Theory provide a comprehensive foundation for analysing the technological, regulatory, and behavioural dimensions of digital twin implementation in financial systems.

3.1. Technology Acceptance Model (TAM)

The Technology Acceptance Model (TAM) was developed by Fred D. Davis in 1989 to explain the determinants of user acceptance and adoption of new technologies. TAM posits that two key factors—perceived usefulness and perceived ease of use—significantly influence an individual's intention to adopt and use a technology. Over time, TAM has been widely applied across various domains, including finance, to understand the adoption of emerging technologies (K.Balaji, 2023).

In the context of digital twin technology, TAM provides insights into how financial institutions evaluate and adopt these advanced systems. Digital twins are more likely to be accepted when they are perceived to enhance risk assessment accuracy, improve decision-making efficiency, and integrate seamlessly with existing financial systems. Additionally, ease of implementation, user-friendly interfaces, and organizational readiness play critical roles in facilitating adoption. Thus, TAM helps explain the behavioral and organizational drivers behind the integration of digital twins into financial risk management practices(Voipio et al., 2023).

3.2. Institutional Theory

Institutional Theory was introduced by John W. Meyer and Brian Rowan in 1977 and later expanded by Paul J. DiMaggio and Walter W. Powell in 1983. The theory emphasizes that organizational structures and practices are shaped by external pressures such as regulations, norms, and cultural expectations.

Within financial institutions, regulatory compliance and governance requirements play a crucial role in shaping the adoption of digital twin technologies. Coercive pressures from regulatory authorities, normative pressures from industry standards, and mimetic pressures from peer institutions encourage firms to adopt advanced technologies for risk management and reporting. Digital twins, by enabling real-time monitoring and enhanced transparency, support compliance with regulatory frameworks such as Basel norms and financial reporting standards. Institutional Theory thus explains how external governance mechanisms influence the adoption and implementation of digital twins in financial systems(Dehghanimohammadabadi et al., 2021).

3.3. Risk Management Theory

Risk Management Theory has evolved over time through contributions from financial economists such as Harry Markowitz (1952), who introduced portfolio diversification concepts, and William F. Sharpe (1964), who developed the Capital Asset Pricing Model (CAPM). These foundational works laid the groundwork for systematic approaches to identifying, measuring, and managing financial risks (Balaji, K. 2025f).

In modern financial systems, Risk Management Theory provides a structured framework for addressing various forms of uncertainty, including market, credit, liquidity, and operational risks. Digital twin technology enhances traditional risk management approaches by enabling continuous monitoring, dynamic simulation, and predictive analysis. Through virtual modeling of financial systems, digital twins allow institutions to simulate risk scenarios, evaluate potential outcomes, and design effective mitigation strategies. This integration represents an evolution from static, backward-looking risk models to dynamic, forward-looking systems capable of adapting to real-time changes(Lattanzi et al., 2021).

3.4. Decision Theory

Decision Theory has its roots in the work of John von Neumann and Oskar Morgenstern in 1944, particularly through their seminal work Theory of Games and Economic Behavior. The theory focuses on how individuals and organizations make rational choices under conditions of uncertainty, incorporating elements such as probability, utility, and risk preferences (Balaji, K. 2025e).

In the context of financial institutions, Decision Theory provides a framework for understanding how data-driven insights influence strategic and operational decisions. Digital twin technology, combined with AI and predictive analytics, enhances decision-making processes by providing real-time simulations, scenario analysis, and probabilistic forecasts. This enables financial managers to evaluate multiple alternatives, assess risks, and make informed decisions based on comprehensive data. The integration of digital twins into decision-making processes thus supports more rational, evidence-based financial strategies and improves overall organizational performance.

Collectively, these theoretical perspectives provide a robust foundation for analyzing the adoption and impact of digital twin technology in financial risk management. TAM explains technology adoption behavior, Institutional Theory highlights regulatory and governance influences, Risk Management Theory provides the foundation for managing financial uncertainty, and Decision Theory emphasizes rational and data-driven decision-making. Together, they offer a comprehensive framework for understanding how digital twins can transform financial institutions into more intelligent, adaptive, and resilient systems(Badakhshan & Ball, 2023).

The developed theoretical framework provides an integrated and structured understanding of digital twin adoption in financial risk management by combining insights from TAM, Institutional Theory, Risk Management Theory, and Decision Theory. It highlights how user acceptance, regulatory influences, risk analytics, and rational decision-making are interconnected in driving effective implementation. The framework demonstrates that digital twin technology enhances real-time monitoring, predictive capabilities, and strategic decision-making, thereby improving financial resilience and operational efficiency. By aligning technological, organisational, and analytical dimensions, the model offers a comprehensive roadmap for financial institutions to adopt digital twins in a systematic, data-driven, and governance-compliant manner (Balaji, K. 2025d).

Figure 1. Theoretical Framework for Digital Twin Technology in Financial Risk Management

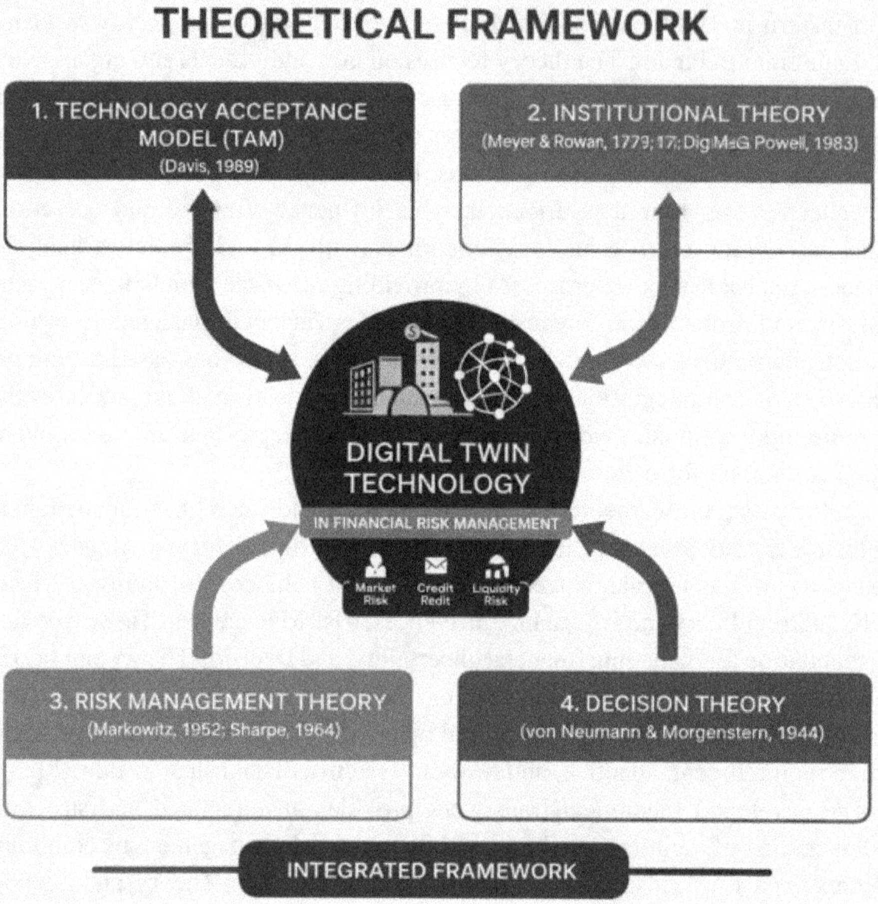

Source: Author's own work

This figure 1 illustrates an integrated theoretical framework that explains the adoption and effectiveness of digital twin technology in financial risk management. The framework combines the Technology Acceptance Model (TAM), Institutional Theory, Risk Management Theory, and Decision Theory to highlight the interaction between user acceptance, regulatory pressures, risk analytics, and data-driven decision-making. It demonstrates how these dimensions collectively influence the implementation, governance, and strategic utilization of digital twin systems within financial institutions (Balaji, K. 2025c).

4. LITERATURE REVIEW

The growing convergence of digital twin technology, artificial intelligence (AI), and financial systems has attracted significant scholarly attention between 2014 and 2026. This section synthesizes key contributions across four major themes: digital twins in enterprise and financial systems, AI and predictive analytics in risk management, scenario simulation in financial institutions, and existing research gaps. The review highlights key objectives, methodologies, and findings of recent studies, providing a comprehensive understanding of the evolving research landscape.

4.1.Digital Twins in Enterprise and Financial Systems

Digital twin technology has evolved from its origins in engineering to become a critical enabler of intelligent enterprise systems. Early conceptual work emphasized the role of digital twins as real-time virtual replicas of physical systems; however, recent studies have extended this concept to economic and financial systems. For instance, (Badakhshan & Ball, 2023) examined the evolution of digital twins into predictive digital ecosystems, with the objective of integrating AI, IoT, and big data for enhanced economic and financial decision-making. The study found that predictive digital twins (PDTs) significantly improve forecasting capabilities by leveraging real-time and historical data, enabling organizations to transition from descriptive to predictive and prescriptive analytics.

Similarly, (Grieves & Vickers, 2017) investigated the role of AI-enabled digital twins in managing financial data risks, particularly in small-scale business environments. The objective of the study was to assess how digital twins can enhance financial data governance and risk mitigation. The findings revealed that digital twins act as dynamic models of financial systems, capable of continuous monitoring, anomaly detection, and predictive analytics, thereby improving financial resilience and decision-making . These studies collectively demonstrate that digital twins are increasingly being applied in enterprise and financial contexts to enhance operational efficiency, transparency, and strategic decision-making (Balaji, K. 2025b).

4.2.AI and Predictive Analytics in Risk Management

The application of AI and predictive analytics in financial risk management has been extensively studied over the past decade. (Hu et al., 2023) conducted a comprehensive review of AI in finance, aiming to identify key research trends and applications. The study found that AI technologies, including machine learning, data mining, and predictive modeling, are widely used for risk classification, fraud detection, and early warning systems. The findings highlighted that AI significantly enhances the accuracy and efficiency of financial risk assessment, particularly in dynamic and data-intensive environments .

Building on this, (Frantzén et al., 2022) developed an AI-powered predictive analytics framework for financial risk management in U.S. markets. The objective was to improve risk classification using advanced machine learning techniques such as CatBoost and Support Vector Machines (SVM). The study achieved high predictive accuracy (over 95%), demonstrating the effectiveness of AI in handling complex financial datasets and supporting informed decision-making . Additionally, (Hatami-Marbini et al., 2022) explored the integration of AI with digital twins for financial risk management, finding that machine learning models embedded within digital twins can detect anomalies, predict risks, and enhance data accuracy in real time .

Overall, the literature indicates that AI-driven predictive analytics has transformed traditional risk management approaches by enabling real-time analysis, improved forecasting, and proactive risk mitigation.

4.3. Scenario Simulation in Financial Institutions

Scenario simulation has become a central component of modern financial risk management, particularly with the integration of digital twin technology. (Prajapat et al., 2020) developed a digital twin-based intelligent risk assessment and decision support system, with the objective of improving predictive accuracy and decision-making in entrepreneurial financial environments. The findings demonstrated that the digital twin system achieved over 94% prediction accuracy and significantly improved risk mitigation and project success rates. The study highlighted the ability of digital twins to simulate complex scenarios, provide early warning signals, and support strategic decision-making(Nozari, H. et al., 2025).

In a related study, (Eyring et al., 2022) proposed an intelligent digital twin framework for supply chain risk management, emphasizing the role of simulation in predicting performance and optimizing decision-making. The study found that digital twin-based simulations significantly reduce the time required for risk assessment and improve decision accuracy, particularly in complex and dynamic environments . These findings are particularly relevant for financial institutions, where scenario simulation is essential for stress testing, portfolio management, and risk forecasting(Balaji, K. 2025b).

Furthermore, recent studies highlight that digital twins enable "what-if" scenario analysis by integrating real-time data with predictive models, allowing financial institutions to evaluate multiple risk scenarios simultaneously. This capability represents a significant advancement over traditional simulation methods, which often rely on static assumptions and limited datasets(Goodwin et al., 2024).

4.4. Gaps in Digital Twin Applications in Finance

Despite the growing body of literature, several critical gaps remain in the application of digital twin technology in financial systems. First, most existing studies focus on engineering, manufacturing, and supply chain applications, with limited empirical research in financial contexts. While studies such as (Alam et al., 2023) provide conceptual insights, there is a lack of large-scale empirical validation of digital twin applications in financial institutions.

Second, the integration of digital twins with AI and predictive analytics, although promising, remains underexplored in terms of governance, scalability, and regulatory compliance. (Wu et al., 2025) highlighted challenges related to data security, scalability, and compliance, particularly for small-scale businesses adopting digital twin technology . Similarly, (Grieves, 2014) identified gaps in understanding the ethical and regulatory implications of AI-driven financial systems, indicating the need for more comprehensive governance frameworks .

Third, there is limited research on the application of digital twins for integrated financial risk management, particularly in areas such as liquidity risk, systemic risk, and macro-financial stability. Most studies focus on specific use cases such as fraud detection or supply chain risk, leaving a gap in holistic financial system modelling(Alsaedi & Masoud, 2025).

4.5. Need for Case-Based Empirical Studies

Given these gaps, there is a strong need for case-based empirical research to bridge the divide between theory and practice. Existing studies are largely conceptual or based on simulations, with limited real-world evidence on how financial institutions implement digital twin technology. Case-based approaches can provide deeper insights into organizational practices, technological integration, and governance mechanisms, thereby enhancing the practical relevance of research findings(Karkaria et al., 2025).

Moreover, case studies enable cross-comparative analysis of different financial institutions, allowing researchers to identify best practices, success factors, and challenges in digital twin adoption. This approach is particularly important in understanding how digital twins interact with existing financial systems, regulatory frameworks, and organizational structure s(Nozari, H. 2025).

In summary, the literature from 2000 to 2026 highlights the transformative potential of digital twins and AI in financial systems, particularly in enhancing risk management and scenario simulation. However, significant gaps remain in empirical validation, governance frameworks, and integrated financial applications. Addressing these gaps through case-based research is essential for advancing both theoretical understanding and practical implementation of digital twin technology in financial institutions.

4.6. Conceptual Framework for Digital Twin Technology in Financial Risk Management

The conceptual framework highlights the central role of digital twin technology as a mediating mechanism that links theoretical drivers with financial risk management outcomes. The framework demonstrates that the Technology Acceptance Model (TAM) and Institutional Theory act as primary enablers influencing the adoption of digital twins, while the integration of AI and predictive analytics strengthens their functional capabilities(van der Valk et al., 2022). These capabilities, operationalized through scenario simulation, real-time risk monitoring, and predictive analytics, directly enhance financial risk management effectiveness and decision-making quality. Furthermore, the framework underscores the importance of moderating factors such as data governance, regulatory environment, and technological infrastructure in shaping the overall impact of digital twin implementation. Overall, the model provides a comprehensive and systematic representation of how digital twin technology transforms financial institutions into more adaptive, data-driven, and resilient systems.

Figure 2. Conceptual Framework for Digital Twin Technology in Financial Risk Management.

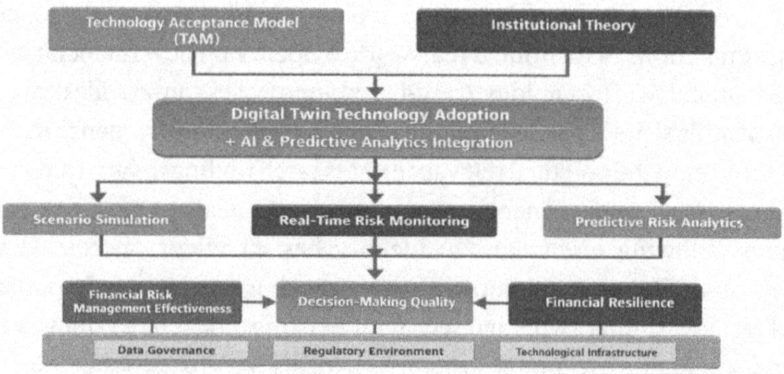

Source: Author's own work.

This figure 2 presents a conceptual framework illustrating the relationships between technological, institutional, and analytical factors influencing the adoption and effectiveness of digital twin technology in financial risk management. The framework positions the Technology Acceptance Model (TAM) and Institutional

Theory as key drivers of digital twin adoption, supported by AI and predictive analytics integration. It further demonstrates how core processes such as scenario simulation, real-time risk monitoring, and predictive risk analytics enhance financial risk management effectiveness, decision-making quality, and financial resilience. The framework also highlights the moderating role of data governance, regulatory environment, and technological infrastructure in shaping the overall impact of digital twin implementation within financial institutions(Liu et al., 2025).

5. RESEARCH METHODOLOGY

This study adopts a qualitative multi-case study research design to explore the application of digital twin technology in financial risk management and its role in enhancing scenario simulation and decision-making within financial institutions. The qualitative approach is particularly appropriate for examining complex, emerging technologies such as digital twins, where contextual understanding, organizational practices, and interpretive insights are essential. A multi-case study design enables the comparison of diverse institutional settings, thereby facilitating a deeper understanding of patterns, variations, and best practices across different financial environments. This approach also strengthens the analytical rigor and external validity of the study by allowing cross-case generalization(Tortorella et al., 2025).

The selection of cases is guided by a purposive sampling strategy to ensure relevance and diversity. First, the study focuses on financial institutions, including banks, asset management firms, and fintech organizations, as these entities play a critical role in financial risk management and are increasingly adopting advanced digital technologies. Second, cases are selected based on the level of digital twin adoption, ranging from early-stage implementation to advanced, fully integrated systems. This allows for a comparative analysis of technological maturity and its impact on risk management practices. Third, geographic diversity is considered to capture variations in regulatory environments, technological infrastructure, and market dynamics across developed and emerging economies. This ensures that the findings reflect a broad spectrum of institutional contexts(Rosati et al., 2023).

The study relies on secondary data sources to ensure comprehensive and reliable analysis. Key data sources include annual reports and integrated financial statements, which provide insights into organizational strategies, technology adoption, and performance metrics. Risk disclosures, such as those related to market, credit, liquidity, and operational risks, are analyzed to understand the role of digital twins in risk management practices. Additionally, industry databases and reports from financial institutions, consulting firms, and regulatory bodies are utilized to supplement the analysis with sector-wide trends and technological developments.

This triangulation of data sources enhances the credibility and robustness of the findings(Dos Santos et al., 2022).

For data analysis, the study employs a combination of qualitative and comparative analytical tools. Thematic analysis is used to identify recurring patterns, themes, and relationships related to digital twin adoption, risk management practices, and decision-making processes. This involves systematic coding and categorization of data to extract meaningful insights. Cross-case comparison is conducted to evaluate similarities and differences across the selected cases, enabling the identification of best practices, critical success factors, and contextual variations. Furthermore, risk performance indicators are utilized to assess the effectiveness of digital twin applications in improving financial outcomes, such as risk mitigation, forecasting accuracy, and operational efficiency. These indicators provide a structured basis for evaluating the impact of digital twin technology on financial risk management.

Despite its strengths, the study is subject to certain limitations. The reliance on secondary data may restrict access to proprietary or real-time information, potentially limiting the depth of analysis. Variations in reporting standards and disclosure practices across institutions may also affect data comparability. Additionally, the qualitative nature of the study, while providing rich and contextual insights, may limit the generalizability of the findings to all financial institutions. The rapidly evolving nature of digital twin technology further implies that findings may require continuous updating to remain relevant. Nonetheless, the methodological approach adopted in this study provides a robust framework for analyzing the role of digital twins in financial risk management and contributes valuable insights to both academic research and industry practice(Perno et al., 2023).

6. CASE STUDY ANALYSIS

6.1 Case 1: Digital Twins in Market Risk Management – JPMorgan Chase & Co.

JPMorgan Chase & Co., established in 2000 through the merger of J.P. Morgan & Co. and Chase Manhattan Corporation, is one of the world's largest and most technologically advanced financial institutions. Over the past decade, the bank has invested heavily in digital transformation, particularly in AI-driven analytics and advanced simulation systems that closely resemble digital twin architectures. These systems enable the creation of dynamic, data-driven models of financial markets and portfolios, supporting enhanced market risk management(Rizqi et al., 2024).

JPMorgan utilizes digital twin-like frameworks to simulate market environments by integrating real-time data feeds from global financial markets, including equities, fixed income, derivatives, and foreign exchange. These virtual models replicate the behavior of financial assets under varying conditions, allowing the institution to conduct sophisticated scenario analyses. Through these simulations, the bank can assess the impact of market volatility, interest rate changes, and macroeconomic shocks on its portfolios. This capability significantly improves the accuracy of risk assessment compared to traditional static models.

AI-driven portfolio analysis is a key component of JPMorgan's approach. The institution employs machine learning algorithms to analyze vast datasets, identify patterns, and optimize portfolio allocation strategies. These AI systems are embedded within digital twin frameworks, enabling continuous learning and adaptation to changing market conditions. For instance, predictive models can forecast asset price movements and volatility, allowing portfolio managers to make proactive adjustments. This integration of AI and digital twin technology enhances decision-making by providing deeper insights into risk-return trade-offs.

Real-time risk monitoring is another critical aspect of JPMorgan's market risk management strategy. Digital twin systems enable continuous tracking of portfolio performance and risk exposure, allowing the institution to respond rapidly to market changes. Advanced dashboards and analytics tools provide real-time visibility into key risk indicators, such as Value at Risk (VaR), stress test results, and liquidity metrics. This real-time capability is particularly valuable in volatile market conditions, where timely decision-making is essential to mitigate potential losses.

From a value generation perspective, JPMorgan's adoption of digital twin-like systems has resulted in improved risk management efficiency, enhanced portfolio performance, and reduced operational costs. The ability to simulate multiple scenarios and optimize strategies in real time provides a competitive advantage in global financial markets. Furthermore, the bank's technological innovations have influenced industry practices, with many financial institutions adopting similar approaches.

In terms of global adoption, JPMorgan operates across multiple regions, including North America, Europe, and Asia, and its digital risk management systems are deployed at a global scale. The bank's investments in technology and innovation have positioned it as a leader in digital finance, demonstrating the transformative potential of digital twins in market risk management. Overall, this case illustrates how digital twin technology, combined with AI, can enhance market simulation, portfolio analysis, and real-time risk monitoring in complex financial environments.

6.2 Case 2: Digital Twins in Credit Risk Assessment – FICO (Fair Isaac Corporation)

FICO, founded in 1956, is a globally recognized leader in credit analytics and decision management solutions. The company has been at the forefront of developing predictive models for credit risk assessment, and in recent years, it has incorporated AI-driven digital twin concepts into its analytics platforms. These systems enable financial institutions to create virtual representations of borrower behavior and credit environments, facilitating more accurate and dynamic credit risk assessment.

FICO's approach to digital twin technology involves the use of predictive modeling techniques to simulate borrower behavior under different economic conditions. By integrating historical data, transactional data, and behavioral analytics, FICO's platforms create digital profiles—or "twins"—of borrowers. These models can simulate how individuals or businesses are likely to respond to changes in interest rates, income levels, or economic shocks. This predictive capability allows financial institutions to assess creditworthiness more accurately and anticipate potential defaults.

Default probability simulation is a core feature of FICO's digital twin-based credit risk systems. Machine learning algorithms analyze large datasets to estimate the likelihood of default for different borrower segments. These simulations enable lenders to evaluate the impact of various risk factors and adjust their lending strategies accordingly. For example, financial institutions can use these models to identify high-risk borrowers, optimize credit limits, and design targeted risk mitigation strategies.

AI-enabled credit scoring represents a significant advancement over traditional credit scoring methods. FICO's systems leverage advanced analytics to incorporate a wide range of variables, including non-traditional data sources such as digital transactions and behavioral patterns. This results in more comprehensive and accurate credit scores, particularly for individuals with limited credit histories. The integration of AI with digital twin technology further enhances the predictive power of these models, enabling continuous updating and refinement based on real-time data.

In terms of value generation, FICO's digital twin-based solutions have improved the efficiency and effectiveness of credit risk management for financial institutions worldwide. By providing more accurate risk assessments, these systems reduce default rates, enhance portfolio quality, and improve profitability. Additionally, the use of AI-driven analytics enables faster decision-making, reducing the time required for credit approvals and improving customer experience.

FICO's solutions have been widely adopted across global financial markets, including banks, credit card companies, and fintech firms. The company's platforms are used in over 90 countries, demonstrating the scalability and global relevance of its technologies. Furthermore, FICO's commitment to innovation and regulatory compliance ensures that its solutions align with evolving financial regulations and standards(Santander et al., 2023).

Overall, this case highlights the critical role of digital twin technology in transforming credit risk assessment. By enabling predictive modeling, default probability simulation, and AI-driven credit scoring, FICO's systems provide a robust framework for managing credit risk in dynamic financial environments.

6.3 Case 3: Digital Twins in Liquidity and Operational Risk – HSBC Holdings plc

HSBC Holdings plc, established in 1865, is one of the world's largest banking and financial services organizations, with a strong presence across Asia, Europe, and the Americas. HSBC has been actively investing in digital transformation initiatives, including the adoption of AI and advanced analytics for liquidity and operational risk management. The bank's approach incorporates digital twin-like systems to simulate financial operations, assess liquidity risks, and enhance crisis management capabilities.

HSBC utilizes digital twin frameworks to conduct stress testing of liquidity scenarios, enabling the bank to evaluate its ability to meet short-term and long-term financial obligations under adverse conditions. These simulations incorporate real-time data on cash flows, market conditions, and customer behavior, allowing the bank to model various stress scenarios, such as economic downturns or sudden market disruptions. This capability enhances the accuracy of liquidity risk assessment and supports proactive risk management (Nozari, H. 2026).

Operational risk simulation is another key application of digital twin technology at HSBC. The bank uses AI-driven models to simulate potential operational disruptions, including system failures, cyber threats, and process inefficiencies. These simulations enable the identification of vulnerabilities and the development of mitigation strategies, thereby reducing the likelihood and impact of operational risks. The integration of digital twins with AI enhances the bank's ability to analyze complex risk scenarios and implement effective controls(Santander et al., 2023).

Crisis management is significantly improved through the use of digital twin systems. HSBC can simulate crisis scenarios, such as financial shocks or operational disruptions, and evaluate the effectiveness of its response strategies. This enables the bank to develop robust contingency plans and improve its resilience to unexpected events. Real-time monitoring and predictive analytics further support decision-making during crises, allowing the bank to respond quickly and effectively.

From a value generation perspective, HSBC's adoption of digital twin technology has resulted in improved risk management, enhanced operational efficiency, and increased regulatory compliance. The ability to simulate and analyze complex scenarios provides a strategic advantage in managing liquidity and operational risks. Additionally, the bank's investments in digital innovation have strengthened its position in the global financial market.

HSBC's digital risk management systems are deployed across its global operations, reflecting a high level of technological maturity and scalability. The bank's commitment to sustainability and responsible banking further underscores the importance of integrating advanced technologies into its operations(Ritto & Rochinha, 2021).

6.4 Cross-Case Comparative Analysis

The three case studies JPMorgan Chase, FICO, and HSBC demonstrate the diverse applications of digital twin technology in financial risk management. A comparative analysis reveals several commonalities, including the use of AI-driven analytics, real-time data integration, and advanced simulation techniques. All three institutions leverage digital twin frameworks to enhance risk assessment, improve decision-making, and achieve greater operational efficiency.

However, differences exist in terms of focus and application. JPMorgan emphasizes market risk management and portfolio optimization, FICO focuses on credit risk assessment and predictive modeling, while HSBC concentrates on liquidity and operational risk management. These variations highlight the versatility of digital twin technology and its ability to address different aspects of financial risk. Key success factors across cases include technological innovation, data integration capabilities, strong governance frameworks, and alignment with regulatory standards. At the same time, common challenges such as data security, model complexity, and implementation costs remain significant. In conclusion, the cross-case analysis underscores the transformative potential of digital twins in financial risk management, while also highlighting the importance of strategic implementation and governance in achieving sustainable outcomes.

6.5. Thematic Analysis

The thematic analysis of the selected cases reveals several recurring patterns, themes, and relationships related to digital twin adoption, financial risk management practices, and decision-making processes across financial institutions . A key overarching theme is the integration of digital twin technology with Artificial Intelligence (AI) and data-driven analytics, which serves as the foundational enabler across all cases. Institutions such as JPMorgan, FICO, and HSBC demonstrate a consistent shift toward intelligent systems that combine real-time data processing with predictive modeling. This convergence highlights a broader transformation from traditional, static risk management frameworks to dynamic, adaptive, and technology-driven approaches. A second prominent theme is real-time risk monitoring and continuous data integration. Across all cases, digital twin systems rely heavily on real-time data streams to simulate financial environments and track evolving risk exposures. This capability significantly enhances the responsiveness of financial institutions, allowing them to identify and mitigate risks proactively. Unlike conventional models that depend on historical data, digital twins facilitate continuous updating and synchronization, thereby improving forecasting accuracy and operational agility.

This pattern underscores the growing importance of data infrastructure and real-time analytics in modern financial risk management.

Another critical theme emerging from the analysis is scenario simulation and predictive capability. Digital twins enable institutions to perform advanced "what-if" analyses by simulating various financial scenarios, including market volatility, credit defaults, and liquidity stress conditions. While all institutions utilize simulation, the application varies based on their risk focus. For instance, JPMorgan emphasizes market scenario simulation and portfolio optimization, FICO focuses on borrower behavior and credit risk prediction, and HSBC prioritizes liquidity stress testing and operational disruption scenarios. Despite these contextual differences, the underlying pattern reflects a shared reliance on predictive analytics to enhance risk assessment and strategic planning. The analysis also identifies a strong theme of enhanced decision-making through data-driven insights. Digital twin technology supports decision-making processes by providing comprehensive, real-time, and scenario-based insights. Financial managers are able to evaluate multiple alternatives, assess risk-return trade-offs, and make informed strategic decisions. This shift toward evidence-based decision-making aligns with Decision Theory, where rational choices are supported by probabilistic and analytical inputs. The integration of AI further strengthens this process by generating actionable insights from complex datasets, thereby improving decision quality and organizational performance.

A further recurring theme is context-specific application of digital twins across different risk domains. While the core technology remains consistent, its implementation varies significantly depending on the institution's primary risk focus. This highlights the flexibility and scalability of digital twin technology in addressing diverse financial risks, including market, credit, liquidity, and operational risks. At the same time, it reflects the influence of organizational priorities, technological maturity, and regulatory environments in shaping adoption strategies. In addition, the analysis reveals governance, regulatory alignment, and data management as critical cross-cutting themes. Financial institutions emphasize compliance with regulatory frameworks, data security, and transparency in the deployment of digital twin systems. Strong data governance mechanisms are essential to ensure the reliability and ethical use of AI-driven models. This theme is closely linked to Institutional Theory, which explains how external pressures influence technological adoption and implementation.

Finally, a recurring pattern across cases is the presence of common challenges, including high implementation costs, system complexity, data dependency, and concerns related to data privacy and model transparency. Despite the significant benefits, these challenges highlight the need for strategic planning, investment in infrastructure, and continuous system refinement. Overall, the thematic analysis demonstrates that digital twin adoption in financial institutions is characterized by the integration of AI, real-time data utilization, predictive simulation, and enhanced decision-making capabilities. These interconnected themes collectively contribute to more efficient, adaptive, and resilient financial risk management systems.

6.5.1. Cross-Case Thematic Analysis of Digital Twin Applications in Financial Risk Management

The individual case analyses of JPMorgan Chase, FICO, and HSBC highlight the diverse applications of digital twin technology across different dimensions of financial risk management. While each institution focuses on a specific risk domain market, credit, and liquidity/operational risk respectively—they share common underlying technological foundations, including AI integration, real-time data processing, and scenario-based simulation. However, the extent of implementation, data utilization, predictive capabilities, and decision-making applications varies significantly across cases due to differences in organizational objectives, technological maturity, and regulatory environments. To systematically compare these dimensions and identify patterns, similarities, and divergences, a cross-case thematic analysis is conducted and presented in Table X.

Table 1. Cross-Case Thematic Analysis of Digital Twin Applications in Financial Risk Management

Theme	JPMorgan Chase (Market Risk)	FICO (Credit Risk)	HSBC (Liquidity & Operational Risk)
Digital Twin Application	Market environment simulation and portfolio modeling	Borrower behavior modeling and credit environment simulation	Liquidity stress testing and operational risk simulation
AI Integration	AI-driven portfolio optimization and predictive analytics	Machine learning for credit scoring and default prediction	AI-enabled risk simulation and crisis modeling
Data Utilization	Real-time market data (equities, derivatives, FX)	Historical, transactional, and behavioral data	Cash flow, market, and operational data
Predictive Capability	Forecasting asset prices and volatility	Predicting default probability and credit risk	Forecasting liquidity stress and operational disruptions
Real-Time Monitoring	Continuous portfolio risk tracking and dashboards	Partial real-time updates in credit analytics	Real-time monitoring of liquidity and operational risks
Scenario Simulation	Market volatility and macroeconomic shock simulations	Economic condition-based borrower simulations	Stress testing and crisis scenario simulation

continued on following page

Table 1. Continued

Theme	JPMorgan Chase (Market Risk)	FICO (Credit Risk)	HSBC (Liquidity & Operational Risk)
Decision-Making Support	Portfolio optimization and risk-return trade-offs	Credit approval and risk-based lending decisions	Crisis response and liquidity planning decisions
Risk Management Focus	Market risk	Credit risk	Liquidity and operational risk
Value Generation	Improved portfolio performance and reduced costs	Reduced default rates and improved profitability	Enhanced resilience and regulatory compliance
Technological Maturity	Advanced and highly integrated systems	Highly advanced predictive platforms	Advanced and globally scalable systems
Global Adoption	Global operations (North America, Europe, Asia)	Used in 90+ countries	Global banking network
Key Strength	Real-time market simulation and analytics	High predictive accuracy in credit scoring	Strong crisis simulation and resilience planning
Key Challenge	High system complexity and data dependency	Data privacy and model transparency issues	High implementation cost and regulatory pressure

Source: Authors own work.

The cross-case thematic analysis shown in table 1 reveals several important insights into the adoption and impact of digital twin technology in financial risk management. First, all three institutions demonstrate a strong convergence toward AI-driven, data-centric risk management frameworks, emphasizing the importance of real-time analytics and predictive modeling. Second, while JPMorgan and HSBC exhibit advanced real-time monitoring and simulation capabilities, FICO shows exceptional strength in predictive accuracy and credit analytics, reflecting its specialized focus. Third, the application of digital twins is highly context-dependent, with each institution tailoring its implementation to specific risk categories and operational priorities. Despite these differences, common success factors such as robust data integration, advanced analytics capabilities, and alignment with regulatory requirements emerge across all cases. At the same time, challenges related to data security, model complexity, and implementation costs remain critical concerns. Overall, the findings underscore the transformative potential of digital twin technology while highlighting the need for strategic alignment, governance, and technological investment to achieve sustainable outcomes in financial institutions.

7. ANALYSIS AND DISCUSSION

The findings from the multi-case analysis highlight the transformative role of digital twin technology in enhancing the accuracy and effectiveness of financial risk management. Digital twins enable the creation of dynamic, real-time models of financial systems, allowing institutions to simulate complex market conditions and assess potential risks with greater precision. Unlike traditional risk models that rely on historical data and static assumptions, digital twins continuously update using real-time data streams, thereby improving forecasting accuracy and responsiveness. This capability is particularly valuable in volatile financial environments, where timely and accurate risk assessment is critical for minimizing losses and ensuring financial stability.

The impact of digital twins on financial decision-making is substantial. By integrating predictive analytics and scenario simulation, digital twins provide decision-makers with comprehensive insights into potential outcomes under varying conditions. Financial institutions can evaluate multiple scenarios, optimize portfolio strategies, and make informed decisions based on data-driven evidence. This enhances strategic planning and supports proactive risk management, enabling organizations to anticipate and mitigate potential threats before they materialize. Furthermore, the use of digital twins fosters a shift toward more agile and adaptive decision-making processes, which are essential in today's rapidly changing financial landscape.

A key enabler of digital twin effectiveness is its integration with Artificial Intelligence (AI) and big data technologies. AI algorithms enhance the analytical capabilities of digital twins by identifying patterns, predicting trends, and generating actionable insights from large and complex datasets. Big data provides the necessary infrastructure for handling vast volumes of structured and unstructured data, enabling real-time processing and analysis. Together, these technologies create a powerful ecosystem that supports advanced risk modelling, predictive analytics, and intelligent decision-making.

However, the adoption of digital twins also raises important governance and regulatory considerations. Financial institutions must ensure compliance with regulatory frameworks related to data privacy, risk reporting, and financial disclosures. The use of digital twins necessitates robust data governance mechanisms, including data validation, transparency, and accountability. Regulatory bodies may also need to develop new guidelines to address the unique challenges associated with digital twin technology, such as model validation and ethical use of AI. Despite their benefits, digital twins present several challenges and risks. Data security is a major concern, as the integration of multiple data sources increases vulnerability to cyber threats. Model bias and algorithmic errors can lead to inaccurate predictions and potentially harmful decision-making. Additionally, the complexity of digital twin systems requires significant investment in technology and expertise, which may pose barriers for smaller institutions. Addressing these challenges is essential to ensure the effective and responsible use of digital twins in financial systems.

8. IMPLICATIONS

8.1 Theoretical Implications

This study contributes to the growing body of literature on digital finance by integrating digital twin technology with financial risk management and decision-making theories. It extends existing theoretical frameworks by demonstrating how digital twins, combined with AI and big data, enhance predictive capabilities and support dynamic risk assessment. The study also bridges the gap between technological innovation and financial theory, offering a comprehensive perspective on the role of digital twins in modern financial systems. By incorporating insights from risk management theory, decision theory, and technology adoption models, this research provides a robust foundation for future academic inquiry.

8.2 Managerial Implications

From a managerial perspective, the findings underscore the importance of adopting digital twin technology as a strategic tool for enhancing risk management and operational efficiency. Financial institutions should invest in advanced analytics platforms and integrate digital twins into their existing systems to improve forecasting accuracy and decision-making capabilities. Effective implementation requires a clear adoption strategy, including investment in technological infrastructure, employee training, and organizational change management. Managers must also prioritize data governance and ethical considerations to ensure the responsible use of digital twin technology. By leveraging digital twins, organizations can achieve greater agility, resilience, and competitive advantage in the financial market.

8.3 Policy Implications

The study highlights the need for regulatory frameworks to govern the use of digital twin technology in financial systems. Policymakers should develop guidelines that address issues such as data privacy, model transparency, and ethical use of AI. Standardization of risk simulation models is essential to ensure consistency and comparability across institutions. Additionally, regulatory bodies should encourage innovation by providing support for research and development in digital twin technologies while ensuring that appropriate safeguards are in place. Collaboration between regulators, industry stakeholders, and technology providers is crucial for creating a balanced and effective regulatory environment(Balaji, K. 2025a).

9. CHALLENGES AND FUTURE RESEARCH DIRECTIONS

While digital twin technology offers significant potential, several challenges must be addressed to ensure its successful implementation. Data quality and integration remain critical issues, as inconsistencies and inaccuracies in data can undermine the effectiveness of digital twin models. Ethical concerns, particularly related to algorithmic bias and fairness, must also be carefully managed to maintain trust and accountability. Technological barriers, including the complexity of system integration and high implementation costs, may limit adoption, particularly among smaller institutions. Organizational resistance to change and lack of technical expertise further complicate the adoption process.

Future research should explore emerging areas such as autonomous financial systems, where digital twins and AI can enable fully automated decision-making processes. The application of digital twins in algorithmic trading presents another promising avenue, offering opportunities for real-time strategy optimization and risk management. Additionally, the integration of digital twins with blockchain and fintech ecosystems can enhance data security, transparency, and operational efficiency. Longitudinal studies and empirical research are needed to assess the long-term impact of digital twin technology on financial performance and stability, particularly in diverse regulatory and economic contexts.

10. CONCLUSION

This study provides a comprehensive analysis of the role of digital twin technology in financial risk management, highlighting its potential to transform traditional approaches through advanced simulation and predictive analytics. The findings demonstrate that digital twins significantly enhance risk accuracy, improve decision-making, and support proactive management of financial uncertainties. By integrating AI and big data, digital twins enable financial institutions to operate more efficiently and respond effectively to dynamic market conditions.

The study also emphasizes the critical importance of governance and regulatory frameworks in ensuring the responsible use of digital twin technology. While the benefits are substantial, challenges related to data security, model bias, and system complexity must be addressed to fully realize the potential of digital twins.

In conclusion, digital twin technology represents a key innovation in the evolution of intelligent financial systems. Its ability to provide real-time insights, simulate complex scenarios, and support data-driven decision-making positions it as a vital tool for the future of financial risk management. As technology continues to evolve, the integration of digital twins with emerging digital ecosystems will play a crucial role in shaping the next generation of financial institutions.

REFERENCES

Alam, M. D., Kabir, G., & Mirmohammadsadeghi, S. (2023). A digital twin framework development for apparel manufacturing industry. *Decis. Anal. J.*, *7*, 100252. Advance online publication. DOI: 10.1016/j.dajour.2023.100252

Alsaedi, F., & Masoud, S. (2025). Condition-based maintenance for degradation-aware control systems in continuous manufacturing. *Machines (Basel)*, *13*(2), 141. DOI: 10.3390/machines13020141

Badakhshan, E., & Ball, P. (2023). Applying digital twins for inventory and cash management in supply chains under physical and financial disruptions. *International Journal of Production Research*, *61*(15), 5094–5116. DOI: 10.1080/00207543.2022.2093682

Balaji, K. "Exploring the Drivers and Effects on Supply Chain Resilience and Performance in an Emerging Market Using Artificial Intelligence," *2023 3rd International Conference on Smart Generation Computing, Communication and Networking (SMART GENCON)*, Bangalore, India, 2023, pp. 1-6, DOI: 10.1109/SMARTGENCON60755.2023.10442071

Balaji, K. (2025). Harnessing AI and Blockchain in Sustainability Assurance: Trends Shaping the Future of Verification. In Khatib, S., & Abbas, A. (Eds.), *Navigating Trust in Sustainability Reporting and Assurance* (pp. 215–242). IGI Global Scientific Publishing., DOI: 10.4018/979-8-3373-0117-4.ch007

Balaji, K. (2025). Revolutionizing High-Frequency Trading: The Impacts of Financial Technology and Data Science Innovations. In Chen, H. (Ed.), *Machine Learning and Modeling Techniques in Financial Data Science* (pp. 103–124). IGI Global Scientific Publishing., DOI: 10.4018/979-8-3693-8186-1.ch004

Balaji, K. (2025). Neuromarketing and Sustainable Finance: Real-World Case Studies of Green Investment Promotions. In Taneja, S., Chahal, B., Johri, A., Ozen, E., & Kumar, P. (Eds.), *Neuromarketing's Role in Sustainable Finance* (pp. 169–204). IGI Global Scientific Publishing., DOI: 10.4018/979-8-3693-9117-4.ch010

Balaji, K. (2025). AI as a Game Changer: How Digital Entrepreneurs Are Redefining Business Practices. In Tunio, M. (Ed.), *Improving Entrepreneurial Processes Through Advanced AI* (pp. 21–48). IGI Global Scientific Publishing., DOI: 10.4018/979-8-3693-1495-1.ch002

Balaji, K. (2025). IoT in Financial Services Innovations in Banking and Insurance. In Ho, R., Song, B., & Tee, P. (Eds.), *Managing Customer-Centric Strategies in the Digital Landscape* (pp. 75–104). IGI Global Scientific Publishing., DOI: 10.4018/979-8-3693-5668-5.ch004

Balaji, K. (2025). Human-Centric Innovation: Building Sustainable Organizations for Industry 5.0. In Erturk, A., Özşahin, M., & Simovic, V. (Eds.), *Human-Centric, Sustainable, and Resilient Organizations in the Digital Age* (pp. 101–124). IGI Global Scientific Publishing., DOI: 10.4018/979-8-3693-8181-6.ch005

Balaji, K., & Rao, P. S. "Artificial Intelligence and Chatbots: Transforming User Experience in E-tailing," *2024 9th International Conference on Communication and Electronics Systems (ICCES)*, Coimbatore, India, 2024, pp. 1375-1379, DOI: 10.1109/ICCES63552.2024.10859474

Dehghanimohammadabadi, M., Belsare, S., & Thiesing, R. (2021). Simulation–optimization of digital twin. *2021 Winter Simulation Conference (WSC)*, 1–10. https://ieeexplore.ieee.org/abstract/document/9715412/

Dos Santos, C. H., Montevechi, J. A. B., De Queiroz, J. A., Miranda, R. C., & Leal, F. (2022). Decision support in productive processes through DES and ABS in the digital twin era: A systematic literature review. *International Journal of Production Research*, *60*(8), 2662–2681. DOI: 10.1080/00207543.2021.1898691

Eyring, A., Hoyt, N., Tenny, J., Domike, R., & Hovanski, Y. (2022). Analysis of a closed-loop digital twin using discrete event simulation. *International Journal of Advanced Manufacturing Technology*, *123*(1–2), 245–258. DOI: 10.1007/s00170-022-10176-5

Figueroa, A. J., Poler, R., & Andres, B. (2024). Adaptive production rescheduling system for managing unforeseen disruptions. *Mathematics*, *12*(22), 3478. DOI: 10.3390/math12223478

Frantzén, M., Bandaru, S., & Ng, A. H. C. (2022). Digital-twin-based decision support of dynamic maintenance task prioritization using simulation-based optimization and genetic programming. *Decis. Anal. J.*, *3*, 100039. Advance online publication. DOI: 10.1016/j.dajour.2022.100039

Goodwin, T., Xu, J., Celik, N., & Chen, C.-H. (2024). Real-time digital twin-based optimization with predictive simulation learning. *Journal of Simulation*, *18*(1), 47–64. DOI: 10.1080/17477778.2022.2046520

Grieves, M. (2014). Digital twin: Manufacturing excellence through virtual factory replication. *White Paper*, *1*, 1–7.

Grieves, M., & Vickers, J. (2017). Digital twin: Mitigating unpredictable, undesirable emergent behavior in complex systems. In Kahlen, F.-J., Flumerfelt, S., & Alves, A. (Eds.), *Transdisciplinary Perspectives on Complex Systems* (pp. 85–113). Springer International Publishing., DOI: 10.1007/978-3-319-38756-7_4

Hatami-Marbini, A., Varzgani, N., Sajadi, S. M., & Kamali, A. (2022). An emergency medical services system design using mathematical modeling and simulation-based optimization approaches. *Decis. Anal. J.*, *3*, 100059. Advance online publication. DOI: 10.1016/j.dajour.2022.100059

Hu, F., Qiu, X., Jing, G., Tang, J., & Zhu, Y. (2023). Digital twin-based decision making paradigm of raise boring method. *Journal of Intelligent Manufacturing*, *34*(5), 2387–2405. DOI: 10.1007/s10845-022-01941-0

Kanike, U. K. (2023). Factors disrupting supply chain management in manufacturing industries. *Journal of Supply Chain Management Science*, *4*(1–2), 1–24. DOI: 10.18757/jscms.2023.6986

Karkaria, V., Tsai, Y.-K., Chen, Y.-P., & Chen, W. (2025). An optimization-centric review on integrating artificial intelligence and digital twin technologies in manufacturing. *Engineering Optimization*, *57*(1), 161–207. DOI: 10.1080/0305215X.2024.2434201

Lattanzi, L., Raffaeli, R., Peruzzini, M., & Pellicciari, M. (2021). Digital twin for smart manufacturing: A review of concepts towards a practical industrial implementation. *International Journal of Computer Integrated Manufacturing*, *34*(6), 567–597. DOI: 10.1080/0951192X.2021.1911003

Liu, Z., Meyendorf, N., Blasch, E., Tsukada, K., Liao, M., & Mrad, N. (2025). The role of data fusion in predictive maintenance using digital twins. In Meyendorf, N., Ida, N., Singh, R., & Vrana, J. (Eds.), *Handbook of Nondestructive Evaluation 4.0* (pp. 1–23). Springer Nature., DOI: 10.1007/978-3-030-48200-8_65-1

Nozari, H. (2025). Integrating Neuromorphic Intelligence with Fuzzy Multi-Objective Models in Resilient Green Supply Chain Design. Process Integration and Optimization for Sustainability, 1-14.

Nozari, H. (2026). Digital twin-based predictive maintenance in cold chain logistics. *Operations Research*, *60*(1), 481–499. DOI: 10.1051/ro/2025167

Nozari, H., Abdi, H., Szmelter-Jarosz, A., & Rahmaty, M. (2025). A digital twin-based neutrosophic multi-objective optimization model for smart supply chains in the food industry. *Journal of Intelligent Manufacturing*, 1–15. DOI: 10.1007/s10845-025-02689-z

Perno, M., Hvam, L., & Haug, A. (2023). A machine learning digital twin approach for critical process parameter prediction in a catalyst manufacturing line. *Computers in Industry, 151*, 151. DOI: 10.1016/j.compind.2023.103987

Prajapat, N., Turner, C., Tiwari, A., Tiwari, D., & Hutabarat, W. (2020). Real-time discrete event simulation: A framework for an intelligent expert system approach utilising decision trees. *International Journal of Advanced Manufacturing Technology, 110*(11), 2893–2911. DOI: 10.1007/s00170-020-06048-5

Ritto, T. G., & Rochinha, F. A. (2021). Digital twin, physics-based model, and machine learning applied to damage detection in structures. *Mechanical Systems and Signal Processing, 155*, 155. DOI: 10.1016/j.ymssp.2021.107614

Rizqi, Z. U., Chou, S.-Y., & Cahyo, W. N. (2024). A simulation-based digital twin for smart warehouse: Towards standardization. *Decis. Anal. J., 12*, 100509. Advance online publication. DOI: 10.1016/j.dajour.2024.100509

Rosati, R., Romeo, L., Cecchini, G., Tonetto, F., Viti, P., Mancini, A., & Frontoni, E. (2023). From knowledge-based to big data analytic model: A novel IoT and machine learning based decision support system for predictive maintenance in industry 4.0. *Journal of Intelligent Manufacturing, 34*(1), 107–121. DOI: 10.1007/s10845-022-01960-x

Santander, O., Kuppuraj, V., Harrison, C. A., & Baldea, M. (2023). A machine learning framework for improving refinery production planning. *AIChE Journal. American Institute of Chemical Engineers, 69*(8), e18109. Advance online publication. DOI: 10.1002/aic.18109

Tortorella, G., Gloet, M., Samson, D., Kurnia, S., Fogliatto, F. S., & Anzanello, M. J. (2025). Food supply chain resilience through digital transformation: A mixed-method approach. *International Journal of Logistics Management, 36*(2), 381–412. DOI: 10.1108/IJLM-01-2024-0030

van der Valk, H., Strobel, G., Winkelmann, S., Hunker, J., & Tomczyk, M. (2022). Supply chains in the era of digital twins–a review. *Procedia Computer Science, 204*, 156–163. DOI: 10.1016/j.procs.2022.08.019

Vital-Soto, A., & Olivares-Aguila, J. (2023). Manufacturing systems for unexpected events: An exploratory review for operational and disruption risks. *IEEE Access*. https://ieeexplore.ieee.org/abstract/document/10237110/

Voipio, V., Elfvengren, K., Korpela, J., & Vilko, J. (2023). Driving competitiveness with RFID-enabled digital twin: Case study from a global manufacturing firm's supply chain. *Measuring Business Excellence*, *27*(1), 40–53. DOI: 10.1108/MBE-06-2021-0084

Wu, Q., Tan, W., Zhou, L., Deveci, M., Ding, W., Martínez, L., & Pamucar, D. (2025). An integrated consensus-oriented decision-making framework for exploring the barriers and applications of medical digital twins. *Expert Systems with Applications*, *284*, 127683. Advance online publication. DOI: 10.1016/j.eswa.2025.127683

Zhang, D., Pei, Y., & Liu, Q. (2022). Vulnerability evaluation for a smartphone digital twin workshop under temporal and spatial disruptions. *Machines (Basel)*, *10*(9), 752. Advance online publication. DOI: 10.3390/machines10090752

KEY TERMS AND DEFINITIONS

Artificial Intelligence (AI): A set of advanced computational techniques, including machine learning and data analytics, that enable systems to process large datasets, identify patterns, and support predictive and automated decision-making.

Credit Risk: The risk that a borrower or counterparty may fail to meet their financial obligations, leading to potential losses.

Data Governance: A framework of policies, processes, and standards that ensure data quality, security, transparency, and compliance within an organization.

Decision Theory: A theory that examines how individuals and organizations make rational decisions under uncertainty by considering probabilities, outcomes, and risk preferences.

Decision-Making Quality: The effectiveness and accuracy of decisions derived from data-driven insights, predictive models, and analytical tools.

Digital Twin Technology: A dynamic virtual representation of a physical or organizational system that is continuously updated using real-time data to simulate, monitor, and optimize performance. In financial systems, it enables advanced modeling of markets, portfolios, and institutional processes.

Financial Resilience: The ability of financial institutions to withstand, adapt to, and recover from economic shocks and uncertainties.

Financial Risk Management: The systematic process of identifying, assessing, and mitigating financial risks, including market, credit, liquidity, and operational risks, to ensure stability and sustainability of financial institutions.

Institutional Theory: A theory that explains how organizational practices are shaped by external pressures such as regulatory requirements, industry norms, and competitive forces.

Liquidity Risk: The risk arising from the inability to meet short-term financial obligations due to insufficient liquid assets or market constraints.

Market Risk: The risk of financial loss resulting from changes in market variables such as interest rates, exchange rates, and asset prices.

Operational Risk: The risk of loss resulting from failures in internal processes, systems, or external events such as fraud or cyber threats.

Predictive Analytics: The use of historical and real-time data, combined with statistical and machine learning models, to forecast future trends, risks, and financial performance.

Real-Time Risk Monitoring: The continuous tracking and evaluation of financial risks using live data and advanced analytical tools to enable timely decision-making.

Regulatory Environment: The system of laws, regulations, and guidelines that govern the operations and risk management practices of financial institutions.

Risk Management Theory: A framework that focuses on identifying, measuring, and managing financial risks through systematic and analytical approaches.

Scenario Simulation: The process of developing and analyzing hypothetical financial scenarios to evaluate the impact of different risk factors, such as market fluctuations or economic shocks, on financial outcomes.

Technological Infrastructure: The underlying digital systems, hardware, and software that support data processing, analytics, and implementation of advanced technologies.

Technology Acceptance Model (TAM): A theoretical framework that explains the adoption of technology based on perceived usefulness and perceived ease of use, which influence users' intention to adopt and utilize new systems.

Value at Risk (VaR): A statistical measure used to estimate the potential loss in value of a financial portfolio over a specified period under normal market conditions.

Chapter 9
Digital Twins as a Bridge Between Industry 5.0 and Education 5.0:
Transforming Academic Entrepreneurship and Entrepreneurial Skills

Nikolay Sterev
http://orcid.org/0000-0001-8262-3241
University of National and World Economy, Bulgaria

Radoslav Miltchev
Technical University, Sofia, Bulgaria

Daniel Yordanov
University of National and World Economy, Bulgaria

ABSTRACT

This chapter explores the transformative role of modern digital technologies in fostering active entrepreneurship, with a particular focus on digital twins within academic entrepreneurship. Basically, it focuses on how digital technologies serve as an instructional bridge between Industry 5.0 /manufacturing paradigms/ and Education 5.0 /pedagogical frameworks/. And the problem has arisen as digital technologies accelerated the development of specific entrepreneurial skills and allow students creating their digital clonnings within the past 2 decades. Current research characterizes the digital transformation of industry and society as a formalized process grounded in the paradigms of Industry 5.0 and Society 6.0. Concurrently, there

DOI: 10.4018/979-8-2600-1669-5.ch009

is an increasing emphasis on transitioning from traditional educational models to competency-oriented, and technologically enriched approaches to training young people. So, chapter proposes some practical instructions for AI-instruments that could found for the transformation in higher education toward human-centered, and sustainable entrepreneurship

1. INTRODUCTION

The global economy has undergone a fundamental restructuring driven by digital transformation, over the past two decades. The major change is driven by the human-artificial transformation on how economic value is created esp. labor organization, and required individuals' workout competencies (Xu et al., 2021). Following the emergence of the Industry 5.0 paradigm—characterized by cyber-physical systems, IoT, cloud computing, and artificial intelligence—it is introduced new demands on academic education, incl. the business establishment skills. The subsequent transition to Society 6.0 /resp. human-centric approach/ extends these demands further, emphasizing human-AI collaboration, ethical responsibility, and sustainable production practices (European Commission, 2021) through academic training and education.

Concurrently, all active educational systems worldwide are undergoing a parallel paradigmatic shift as traditional, content-delivery model of education is being replaced by competency-oriented, experiential, and personalized approaches (Adcroft et al., 2004). This educational change is especially pronounced in entrepreneurship education, where passive knowledge transfer has long been criticized as insufficient for producing entrepreneurial behavior (De Jong et al., 2015; Ho et al., 2018). So, the modern academic entrepreneurship is defined as the Digital process through which researchers and students commercialize their university-generated knowledge through new ventures or spin-offs—requires training environments that replicate the complexity, risk, and dynamism of real markets. And respectively, the AI-based training and models requires modern tools that emphasis the world-wide entrepreneurial knowledge.

Despite the rapid adoption of digital tools in higher education, the specific role of digital twins as pedagogical instruments for entrepreneurship education remains underexplored. Existing literature addresses DTs primarily in industrial and engineering contexts (Grieves, 2014), with limited attention to educational applications. Furthermore, while the EntreComp framework (Bacigalupo et al., 2016) provides a well-validated taxonomy of entrepreneurial competencies, empirical studies measuring DT-mediated competency development remain sparse.

Therefore, the paper contributes to three bodies of literature: (a) digital twin technology in educational contexts, (b) academic entrepreneurship and its pedagogical dimensions, and (c) competency-based frameworks for entrepreneurial education under the Industry 5.0 paradigm. The paper adopts the IMRAD structure, integrating systematic literature review methodology with conceptual framework development and empirical validation from a business economics education context.

2. MATERIALS AND METHODS

This study employs a systematic review methodology following PRISMA (Preferred Reporting Items for Systematic Reviews and Meta-Analyses) guidelines (Schott et al., 2020). The review integrates three complementary components: (a) a systematic search and analysis of peer-reviewed literature on digital twins in education and academic entrepreneurship; (b) development of a conceptual framework linking Industry 5.0 and Education 5.0 through DT technology; and (c) analysis of primary empirical data from business economics students:

- Inclusion: peer-reviewed articles in English; focus on DT technology, entrepreneurship education, or Industry/Education paradigms; empirical or conceptual/theoretical contributions.
- Exclusion: conference abstracts without full papers; studies focused exclusively on industrial DT applications without educational relevance; grey literature without peer review.

The European Entrepreneurship Competence Framework (EntreComp) developed by the European Commission provides the evaluative backbone of this review (Bacigalupo et al., 2016). EntreComp identifies 15 competencies distributed across three interrelated areas:

- Ideas and Opportunities: Spotting opportunities, Creativity, Vision, Valuing ideas, Ethical and sustainable thinking
- Resources: Self-awareness and self-efficacy, Motivation and perseverance, Mobilising resources, Financial and economic literacy, Mobilising others
- In Action: Taking initiative, Planning and management, Coping with uncertainty and ambiguity, Working with others, Learning through experience

Each competency was measured using a validated Likert-scale instrument (1-5) administered pre- and post-intervention. Student profiles were compared against a reference "cyber-entrepreneur" model derived from industry expert consensus.

Primary data were collected from a cohort of 186 undergraduate and postgraduate students enrolled in business economics programs at the University of National and World Economy (UNWE), Sofia, Bulgaria. The intervention consisted of a 14-week digital entrepreneurship module integrating digital twin simulations, AI-assisted feedback tools, and virtual business model testing environments. Pre- and post-assessments measured the 15 EntreComp competencies. Statistical analysis employed paired t-tests and multivariate analysis of variance (MANOVA) to identify statistically significant changes ($p < 0.05$).

Beyond the quantitative assessment, the review synthesizes findings into a conceptual model of the "Digital Entrepreneurial Profile" (DEP). This model employs the digital twin of the learner—an AI-based cyber system continuously updated through bidirectional data exchange between student behavior and the virtual learning environment—to operationalize the principles of Education 5.0.

3. THEORETICAL FRAMEWORK

3.1 Digital Twins: Taxonomy and Definitions

A digital twin is defined as a virtual representation of a physical object, process, or system that maintains continuous bidirectional data exchange with its physical counterpart (Grieves, 2014).

In the current era of digital transformation, it is vital to distinguish the Digital Twin (DT) from its more rudimentary precursors: static digital models and one-way digital shadows. According to the basic definition: A Digital Twin is a sophisticated virtual representation of a physical entity, process, or sociotechnical system. Its defining characteristic—and what separates it from a simple simulation—is the continuous, bidirectional data exchange with its real-world counterpart.

With the incorporation of artificial intelligence and machine learning, digital twins have evolved from descriptive tools into predictive and prescriptive systems. They are increasingly capable of modeling uncertainty, simulating alternative scenarios, and supporting decision-making processes in real time (Singh et al., 2022; He, 2026). Consequently, digital twins are no longer merely technological artifacts but integral components of intelligent ecosystems that mediate interactions between humans, data, and automated systems.

Contemporary DTs are based on the next mechanics of influence: two-way synchronization ensures that the virtual system does more than merely reflect physical reality; it possesses the capacity to influence and optimize the physical counterpart through automated control mechanisms. Following, definitions distinguish DTs from two related but less sophisticated constructs:

- Static Digital Models: virtual representations without real-time data linkage; used for documentation or visualization.
- Digital Shadows: one-directional flows from physical to virtual systems; the virtual reflects but cannot influence the physical.
- Digital Twins: fully bidirectional; the virtual system not only mirrors but can optimize and control the physical counterpart.

Figure 1. Digital Twin Taxonomy: From Static Model to Bidirectional Digital Twin

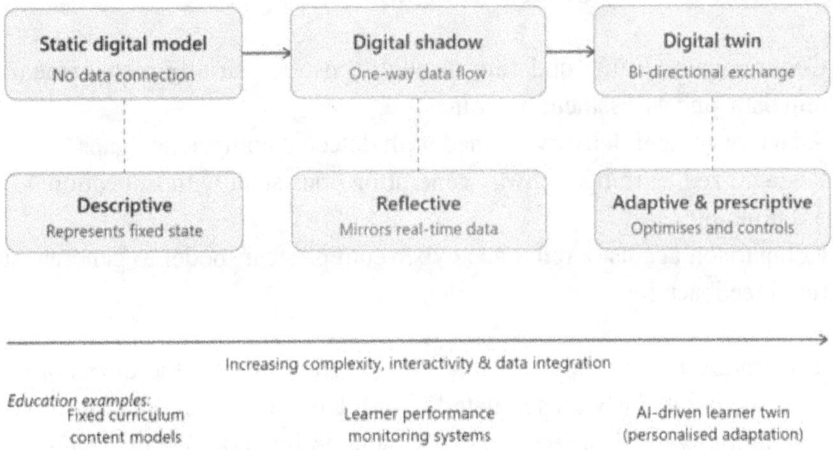

(adapted from Grieves, 2014).

In educational contexts, digital twins function as interactive learning environments that simulate real-world processes within protected, controlled settings. This eliminates the risk of material or financial loss inherent in real-world entrepreneurial experimentation (Sörensen et al., 2022; Mourtzis et al., 2022; Pajpach et al., 2025). So, within the educational ecosystem, we must prioritize the "Digital Twin of the Learner" as an AI-driven cyber-system designed to model a student's evolving profile throughout the academic lifecycle (Andreasyan & Balyakin, 2022; Tjahyadi et al, 2023). This concept represents a personalized, AI-enhanced cyber-system that models the evolving academic and behavioral profile of the student. It integrates multidimensional data, including knowledge acquisition, skill development, engagement patterns, and socio-emotional indicators, across the entire educational lifecycle (Kabashkin, 2025; Böck et al., 2025). Through continuous data analysis, this system allows for the dynamic adaptation of educational content to meet individual needs, embodying the dynamic adaptation of learning pathways, aligning educational processes with the principles of Education 5.0.

3.2 The Digital Twin of the Learner

The digital twin of the learner (DTL) represents the most pedagogically significant application of DT technology in higher education. The DTL is an AI-driven cyber system that continuously models the evolving profile of a student throughout their academic cycle, capturing results, competencies, behavioral patterns, and engagement metrics. Key characteristics include:

- Continuous real-time updating through sensors, learning management system data, and assessment outcomes.
- Adaptive content delivery aligned with detected competency gaps.
- Personalized learning pathway generation consistent with Education 4.0 and 5.0 principles.
- Comparison against a reference cyber-entrepreneur model to generate structured feedback loops.

The literature review indicates that digital twins support the development of various skills, particularly those related to problem-solving and decision-making. By using digital twins, learners can experiment with alternative management and optimization strategies, observe the consequences of their decisions, and learn from their mistakes in a safe virtual environment (Flaga & Pacholczak, 2022; Pajpach et al., 2025).

An essential competency in the use of digital twins is cyber literacy. Since digital twins in education involve working with real data, sensory information, and visualizations typical of modern industrial systems, learners should already possess data interpretation skills and the ability to work with intelligent cyber systems (Kabashkin, 2025; Böck et al., 2025). On this basis, David, Lobov, and Lanz (2018a) view digital twins as an innovative educational technology that transforms the learning process from passive knowledge acquisition to active, experiential, and competency-based learning (Figure 2).

Figure 2. Use of digital twins in education

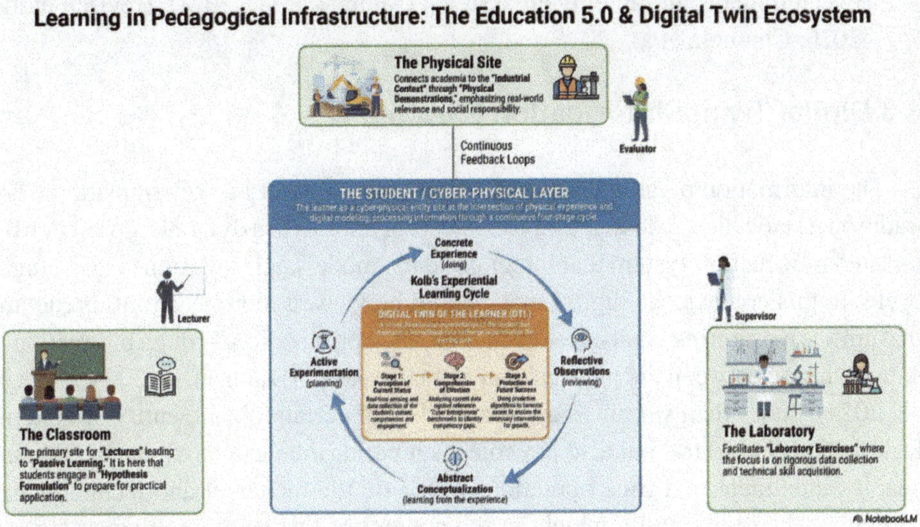

(adapted from David, Lobov, Lanz (2018b) in David, Lobov, Lanz (2018a)).

Following the above, digital twins support the development of key technical and cognitive skills. By working with digital twins, learners develop systematic skills (David et al., 2018a). In addition, although still in the early stages of technological development, cyber-systems of digital twins in education are recognized as essential tools for personal and professional growth. This consensus is based on the understanding that digital twins support the development of a broad range of competencies, which can be categorized into technical, cognitive, and transversal skills (Pajpach et al., 2025; Sorochak & Gvozd, 2025).

- Among the technical skills, digital and information literacy, data management, modeling and simulation, and interaction with cyber-physical systems stand out (Flaga & Pacholczak, 2022; Tao et al., 2022).
- Cognitive skills include systems thinking, problem-solving, and decision-making under uncertainty. By simulating complex scenarios, digital twins enable learners to analyze the interrelationships among different system components and evaluate the consequences of alternative solutions (Singh et al., 2022; He, 2026).
- Key transversal skills are considered an integral part of Industry 5.0 and Education 5.0. In this context, the role of digital twins in developing sustainable, ethical, and social competencies is deemed essential. According to several authors, it is through the simulation of sociotechnical systems—including environmental and social parameters—that learners can cultivate

responsible and critical thinking, as well as skills for collaborative work between humans and intelligent systems (Melnyk et al., 2021; Trevisan et al., 2026; Chamola et al., 2025).

3.3 Digital Twins: Information Paradigm

The information paradigm of digital twins must reflect the key transition from traditional static digital descriptions of objects or entities to a dynamic, continuously updated information system that integrates data, models, and solutions into a single cycle. In this context, the digital twin should be viewed more as an infrastructure that unifies monitoring, analysis, forecasting, and optimization through bidirectional data exchange between the physical and virtual systems (Dihan et al., 2024; Iliuţă et al., 2024), rather than simply as a visual replica of certain object, entity or system. Data will play a central but also key role, as it builds the virtual representation and enables intelligent real-time operations based on the technical and technological aspects of the digital twin, which, in the context of this study, will cover a wide range of research questions and issues, including: architecture, trainable model, semantic compatibility, analytics and simulations, information flow management and adaptation, decision-making mechanisms, and alignment with ethical, regulatory, and legal frameworks.

The architecture of digital twins is based on building upon existing digital copies of systems to create advanced systems that integrate various information flows, quantitative and qualitative metrics, and intelligent information technologies (intelligent systems, technologies related to artificial intelligence, simulations, and integrated information security tools), which enable the making of well-considered and informed decisions in accordance with the specific current state. In this context, the digital twin will function as a multi-layered system comprising a physical layer, a data collection and communication layer, a modeling and semantic layer, an analytical layer, and a decision-making layer. As researchers such as Lehmann et al. (2023), and Elhajj (2026), point out, in the context of Industry 5.0, such systems must not only reflect the current state but also act proactively and adaptively, and even enable the prediction and optimization of processes.

Semantic compatibility must also be viewed as a key element within the context of the digital twin information paradigm. Achieving high efficiency in the implementation of digital twins is closely linked to their ability to integrate heterogeneous data of varying origins and purposes. Such integrations can be achieved through the use of classical tools such as ontologies and knowledge graphs, as they provide mechanisms for the structured representation of knowledge, its context and meaning, as well as existing or potentially possible relationships between data. The above-mentioned approach enables a high level of automated interpretation and interaction between different systems (Karabulut et al., 2024; Elhajj, 2026). In the educational context of this study, this means that the learner's digital twin must contain not only quantitative data (primarily academic performance in individual subjects within the educational process calendar), as is common practice in the field of education, but also a semantic structure of competencies, actions, outcomes, and analyses of the learner's development curve during the training period.

Within the field of education and in the context of the educational process, the information paradigm is realized through the development of learner models, which represent structured digital representations of knowledge, skills, behavior, and progress. According to Böck et al. (2025), these models can be used as a basis for adapting the learning process, but they are limited in terms of providing opportunities for standardization and transparency. Also according to researchers such as Kabashkin (2025), the introduction of the concept of creating a digital twin of the learner aims to expand the traditional learner model by integrating multidimensional data and using it for dynamic profiling and learning support.

The enhanced informational value of the digital twin, as well as the new level of quality in the digital description of objects, are most clearly demonstrated through the provision of opportunities for adaptability and feedback. Results from various studies show that such environments improve understanding and training effectiveness, especially when based on simulations and active learner engagement (Karanam & Hartman, 2025; Sedrakyan et al., 2014). The integration of intelligent systems, technologies related to artificial intelligence, and simulations enables the digital twin system to identify gaps, offer personalized learning paths, and provide immediate real-time feedback.

Empirical studies conducted by various authors show that simulation environments can support the development of certain entrepreneurial skills, including decision-making, opportunity recognition, and risk management (Sofiullah et al., 2023; Timperi et al., 2023). In the context of academic entrepreneurship, the information paradigm underlying the proposed concept of a learner's digital twin would enable a more comprehensive modeling of the "action–consequence–analysis–adaptation" cycle within a controlled educational environment. In this context, the digital twin will track the learner's actions, simulate the results, and provide feedback regarding entrepreneurial competencies.

The very nature of the digital twin's information paradigm raises and brings to the forefront a significant number of ethical and governance issues related to neutrality. During the learning process, each learner's digital twin will integrate a vast amount of heterogeneous data in the form of academic records, behavioral traces, emotional preferences, and signals. The existence of and access to such data, which is closely releated to the learner's identity, raises serious questions and, above all, concerns not only regarding the management of visibility, privacy, and fairness, but also in the context of undue influence and the exercise of power. Such concerns were highlighted by Kabashkin (2025), who explicitly notes that such systems operate with sensitive, multidimensional, and multi-source data and, for this reason, must be designed explicitly with regard to the concepts of privacy-by-design, encryption, role-based access rights, and compliance with regulatory frameworks such as the GDPR. The human-centered principle underlying the Education 5.0 concept requires preserving the learner's autonomy, ensuring transparency in algorithmic decisions, and enabling students to understand and challenge the way they are modeled - rather than merely creating personalization. The use of intelligent systems and technologies related to artificial intelligence also raises a number of questions regarding their ethical use, explainability and autonomy. This requires the informational paradigm of the digital twin to align with the managerial paradigm: not only how the system knows, but also under what rules it knows, who has access to this knowledge, and with what pedagogical legitimacy it is used (Kabashkin, 2025; Chigbu & Makapela, 2025). In addition, a set of practical knowledge and skills related to a highly specialized field, such as information security, must be ensured, as noted by Chehlarova et al. (2021).

In conclusion, within the broader framework shaped by the concepts of Industry 5.0 and Education 5.0, the information paradigm of digital twins can be defined as a human-centered, semantically linked, and adaptively managed knowledge architecture operating in real time. The paradigm must successfully combine four key functions: adequate representation of the object or learner by maintaining the necessary amount of structured and unstructured data; integration of multi-source streams through semantic and interoperable mechanisms; integration of tools for analytical and simulation processing based on scenario forecasting and comparison; ensuring reflective-pedagogical intervention through personalized feedback and decision support. Achieving the correct interpretation and alignment of this fourfold function at all levels of the system allows us to transform the digital twin into a suitable bridge between the industrial logic on which cyber-physical systems are based and the educational logic based on competency-based and personalized

4. RESULTS

4.1 Literature Synthesis: Digital Twins in Educational Contexts

Entrepreneurial competencies comprise a set of knowledge, skills, and attitudes that enable individuals to recognize and seize opportunities, generate new ideas, make decisions, and implement sustainable initiatives (Park & Kim, 2025). In a digital environment, these competencies become even more critical, as technology-integrated strategies and digital tools offer vast potential for learning, innovation, and practical application. Using the European EntreComp entrepreneurial framework (Bacigalupo et al., 2016), key entrepreneurial competencies can be identified, with evidence indicating that they can be effectively managed and developed within a digital cyber environment (Sterev & Yordanov, 2024) (Figure 3).

- One of the key skills is creativity and idea generation—the ability to identify new opportunities in dynamic markets and formulate original solutions to complex problems. This skill is developed through the use of digital simulations, creative collaboration platforms, and data analysis tools (Bacigalupo et al., 2016). The digital environment facilitates the exchange of ideas and early concept testing, which strengthens entrepreneurial initiative.
- Problem-solving and decision-making are core competencies developed through participation in online business simulations, strategy games, and case-based tasks. These digital scenarios enable learners to analyze multiple variables, test different hypotheses, and evaluate the consequences of their choices (Khuong & Hoang, 2024). Both analytical and critical thinking skills are enhanced in this environment.
- Identifying and exploiting opportunities is a crucial skill supported by digital technologies. Learners develop the ability to search for information, assess trends, and use digital tools for market analysis and social listening, which are fundamental for launching innovative projects (Botha & Taljaard, 2021). This approach helps overcome the limitations of traditional learning by emphasizing practical applicability.
- Collaboration and communication are social skills that, in the digital environment, encompass working in distributed teams, utilizing digital communication platforms, and managing virtual networks and resources. The use of digital networks and collaboration platforms helps future entrepreneurs develop social capital connections that are critical for project implementation (Ratković et al., 2022).
- Finally, adaptability and self-initiative are essential skills that enable individuals to respond swiftly to changing conditions and pursue lifelong learning opportunities. Digital environments offer continuous opportunities for self-directed learning through online courses and virtual communities of practice, thereby enhancing entrepreneurial resilience and self-reliance (Khuong & Hoang, 2025).

Figure 3. 15 Entrepreneurial skills

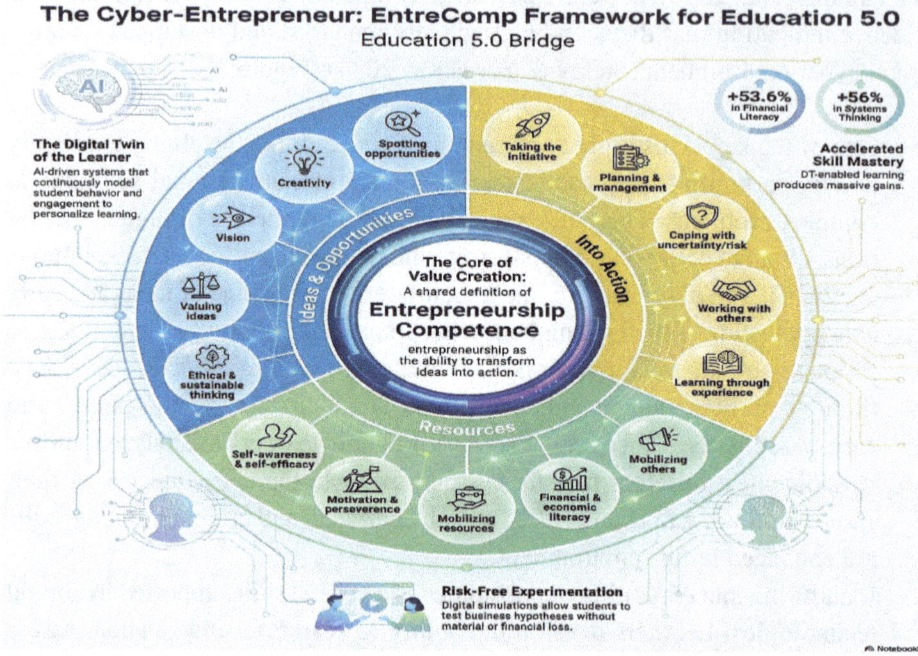

(adapted from Bacigalupo et al., 2016).

The skills mentioned in Figure 3. are not incidental, as growing evidence suggests that digital transformation is raising people's expectations that their daily tasks will be less boring, stressful, or emotionally taxing (Klisaroski et al., 2025). To a large extent, these skills are linked to reducing entrepreneurial stress during financial interactions with other entrepreneurs, thereby fostering the creation of sustainable entrepreneurial businesses (Petkov, 2024).

Building on the above, a specific model can be developed to represent the ability to create a digital counterpart of the learner in transversal entrepreneurial skills. This model is grounded in the understanding that digital tools serve as key mediators between entrepreneurial theory and the practical development of students' skills. The core premise of this model is that entrepreneurial competencies cannot be sustainably developed through traditional lecture-based approaches alone; instead, they require a digitally supported, interactive, and reflective environment that simulates real entrepreneurial processes and solutions. Furthermore, in line with the European Entrepreneurial Competence Framework, digitalization in entrepreneurship education relies on the use of cyber-platforms for assessment and self-assessment of entrepreneurial potential. These platforms enable learners to receive structured feedback on their skills in the areas of "ideas and opportunities," "resources," and "action." (Bacigalupo et al., 2016)

4.2 Data Analysis

The systematic review identified 124 eligible studies. Among these, 67 addressed DT applications in STEM education and engineering contexts; 23 focused on entrepreneurship education more broadly; 18 examined Industry/Education paradigm transitions; and 16 specifically addressed DTs in business economics education. The following thematic clusters emerged:

- Cluster A (n=34): DT infrastructure and technical architecture for educational deployment.
- Cluster B (n=29): Competency-based assessment using simulation environments.
- Cluster C (n=27): Academic entrepreneurship ecosystems and spin-off generation.
- Cluster D (n=21): Adaptive learning and AI personalization in higher education.
- Cluster E (n=13): Sustainability, ethics, and Industry 5.0 transversal skills.

The empirical analysis of 186 UNWE students demonstrated statistically significant improvements ($p < 0.01$) across all 15 EntreComp competencies following the 14-week DT-integrated intervention. Table 1. presents pre- and post-training scores for the six most substantially improved competencies.

Table 1. EntreComp Competency Development: Pre- and Post-Training Assessment Results (n=186)

EntreComp Competency	*Pre-Training Score (Mean)*	*Post-Training Score (Mean)*	*Improvement (%)*
Financial Literacy	2.8	4.3	+53.6%
Creativity & Innovation	3.1	4.5	+45.2%
Opportunity Recognition	3.0	4.4	+46.7%
Systems Thinking	2.5	3.9	+56.0%
Decision-Making	2.9	4.2	+44.8%

continued on following page

Table 1. Continued

EntreComp Competency	*Pre-Training Score (Mean)*	*Post-Training Score (Mean)*	*Improvement (%)*
Ethical & Sustainable Action	3.2	4.3	+34.4%
Average (All 15 Competencies)	2.95	4.27	+44.7%

Note: Scale 1-5 (Likert). All improvements statistically significant at $p < 0.01$ (paired t-test). Scores represent group means. Full results for all 15 competencies available in supplementary data.

The greatest absolute improvements were observed in financial literacy, systems thinking, and opportunity recognition—precisely the competencies most directly engaged by DT simulation exercises, which required students to manage virtual financial models, navigate complex interdependencies, and identify emergent market patterns in dynamic virtual environments (Figure 4).

Figure 4. Development the 15 Entrepreneurial skills in UNWE-Sofia

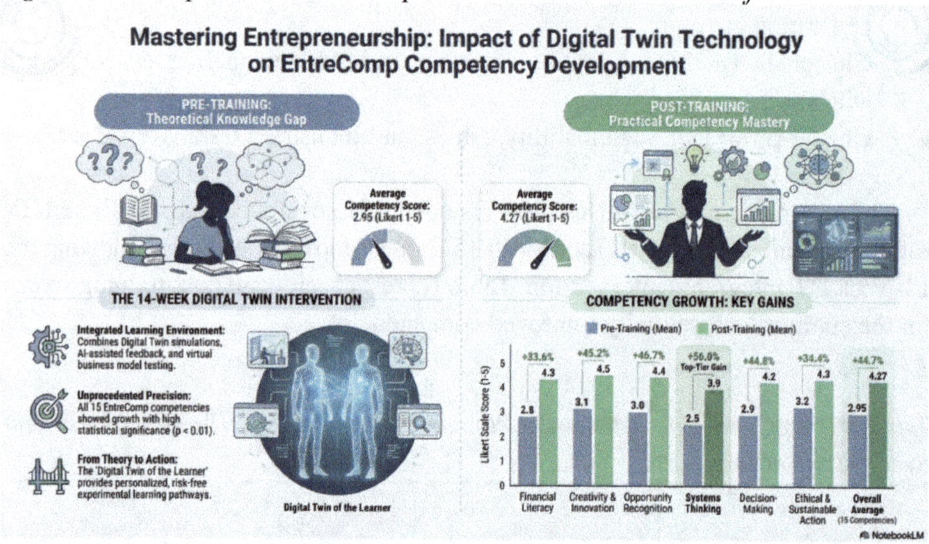

The review identifies four primary mechanisms through which DTs facilitate entrepreneurial competency development:

- Safe Experimentation Environment: DTs provide a protected virtual space in which students can test business models, resource allocation strategies, and market entry approaches without financial risk. This eliminates the primary barrier to experiential entrepreneurship education—the cost of failure (Adcroft et al., 2004; Ho et al., 2018).

- Real-Time Decision Feedback: Bidirectional data flows in DT environments enable immediate feedback on the consequences of managerial decisions, compressing the learning cycle and making causal relationships between actions and outcomes visible and legible (Sörensen et al., 2022).
- Adaptive Personalization: The DTL continuously adjusts learning content, difficulty levels, and feedback specificity to the individual student's competency profile, operationalizing the Education 5.0 principle of personalized learning pathways (Gosztonyi & Csákné Filep, 2022).
- Transversal Skill Development: Beyond technical entrepreneurial skills, DT environments cultivate the transversal competencies central to Industry 5.0: ethical reasoning under algorithmic pressure, sustainable systems design, and effective human-AI collaboration (Ntshangase & Ezeuduji, 2020; De Jong et al., 2015).

4.3 The Digital Entrepreneurial Profile Model

Synthesizing the empirical and theoretical findings, the review proposes the Digital Entrepreneurial Profile (DEP) model—a conceptual framework organizing how DT technology mediates entrepreneurial development is shown on Figure 5.

Figure 5. Conceptual framework of developed Digital Entrepreneurial Profile (DEP) model

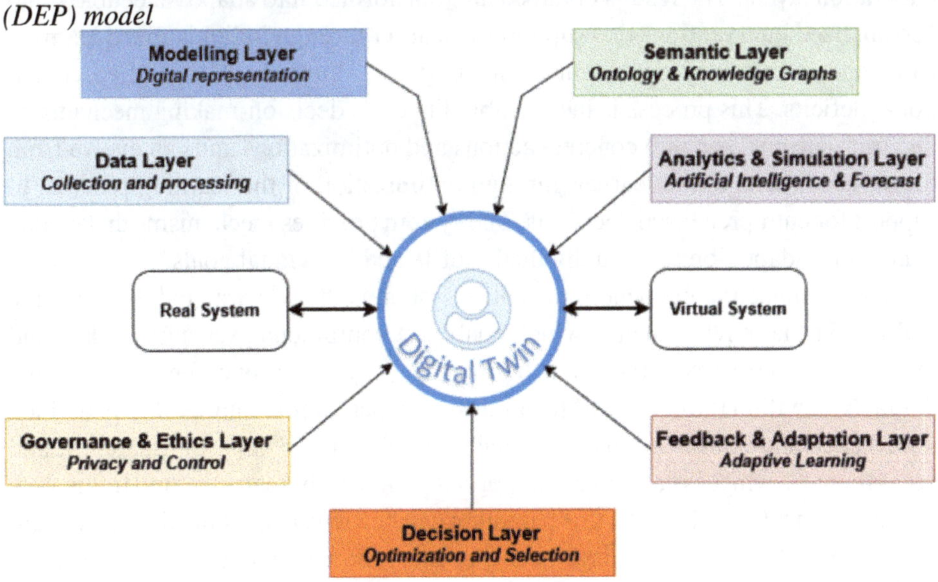

The developed and proposed model of a learner's digital twin (Figure 5), designed to successfully transform education in the field of academic entrepreneurship and entrepreneurial skills within the context of Industry 5.0 and Education 5.0, is consistent with the main conclusions of studies conducted in the field of digital twin theory and the shift in the information paradigm. The digital twin model of a learner represents an integrated, multi-layered system in which data, models, and solutions operate in a continuous cycle. At the core of this paradigm lie the processes of collecting, processing, and managing multidimensional structured and unstructured data from diverse sources (behavioral, cognitive, academic, sensory), which form the empirical basis for building the digital twin and the virtual representation of the learner.

The data obtained during the learning process is formalized through modeling, which creates a digital representation of the object or learner, including their competencies, behavioral characteristics, and developmental dynamics. A significant role is also assigned to semantic integration, provided through ontologies and knowledge graphs, which allows not only the linking of heterogeneous data and the assignment of contextual meaning to the information, but also opportunities for automatic knowledge extraction.

The layer comprising analytical and simulation mechanisms based on artificial intelligence technologies - which enable forecasting, scenario testing, and impact assessment - is built upon the foundation established by the modeling and semantic integration layers. The results obtained are transformed into adaptive feedback and personalized interventions that support the learner through individualized learning paths and competency development, including in situations of identified educational deficits. This process is inextricably linked to decision-making mechanisms that translate analysis into concrete actions and optimizations and can evolve from recommendations to the learner, through optimization of the learning process, to support for entrepreneurial decisions. The system provides mechanisms that enable continuous adaptation to the individual profile and individual goals.

A key connecting element in the entire system is the two-way link between the real and virtual environments, which enables a continuous exchange of data and mutual influence between them. Furthermore, ensuring the proper functioning of the digital twin within the context of the information paradigm requires clearly defined governance and ethical mechanisms, including data protection, transparency, and access control, which is particularly important within the human-centered framework of Education 5.0. As a whole, this paradigm can be viewed as a holistic architecture for real-time knowledge management that links the technological capabilities of digital twins with the pedagogical goals of modern entrepreneurial education. The sequence is accompanied by a horizontal layer of governance and ethics that spans all stages and ensures security, transparency, and control over data usage. In this way, the digital twin can be conceptualized as a dynamic, self-sustaining system in which the flow "data ↦ modeling ↦ semantics ↦ analysis ↦ feedback ↦ decision ↦ action ↦ new data" creates a continuous process of learning, adaptation, and optimization, in line with the requirements of Industry 5.0 and Education 5.0.

The technical implementation of the model uses the architecture shown in Figure 6. In this model, the frontend is a web-based information system through which the learner interacts with the system and receives visualized results, simulations, recommendations, and explanations. The backend implements the core logic of the digital twin through data integration, analytics, simulations, result generation, and communication with the frontend via an API. Data governance and ethics are presented as a horizontal regulatory layer that encompasses both components and ensures security, transparency, traceability, access control, and compliance with ethical and regulatory requirements.

Figure 6. A Unified Architectural Model for Human-Centered Interfaces, Data Management, and Ethics in a Frontend–Backend Digital Twin System

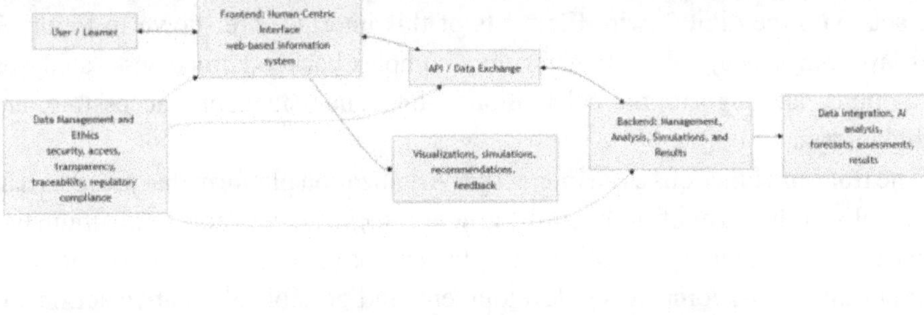

As noted above, data management and the enforcement of ethical standards in the use of the system are ensured by a horizontal, system-integrated layer that encompasses all components of the digital twin and guarantees the reliable, secure, and responsible operation of the system. Architecturally, this layer is primarily implemented in the backend, where data collection, storage, processing, analysis, and control take place, as well as the management of user access and permissions.

The backend infrastructure serves as a central coordination mechanism that integrates various data sources (e.g., user behavioral data, simulation results, analytical calculations) and ensures their consistency and traceability. In this context, data management involves defining policies for data storage, versioning, and validation, as well as implementing protection mechanisms such as encryption, anonymization, and role-based access control. This is particularly important when working with sensitive data related to learners, including their competencies, behavioral patterns, and individual learning trajectories.

The ethical component of this layer requires the incorporation of principles such as transparency, accountability, and fairness into the algorithms used for analysis and decision-making. This means that the backend not only performs calculations but must also ensure the explainability of results, so that the generated recommendations and assessments are understandable and verifiable. In the context of Education 5.0, this also includes ensuring learner autonomy, as the system must allow control over personal data and the ability to interpret or challenge automatically generated decisions.

The connection to the frontend is implemented through secure interfaces (APIs) that provide only the necessary and processed information to the web-based system. Thus, the frontend functions as a transparent layer for visualization and interaction, while the backend remains responsible for compliance with ethical and regulatory requirements. Consequently, data governance and ethics are not a separate component, but a regulatory framework that permeates all stages of the digital twin's information flow.

The human-centered interface constitutes the front-end layer of the system, implemented as a web-based information environment through which the user (trainee) interacts with the digital twin. Elements of this interface are shown in Figure 7. This layer plays a key role in transforming complex backend processes—analyses, simulations, and algorithmic calculations—into understandable, accessible, and usable information.

The front end functions as an interactive visualization platform that presents data and results in the form of dashboards, graphs, progress indicators, and simulation scenarios. Through these visualizations, the learner gains a clear understanding of their current status, competency development, and possible alternative actions. In this sense, the interface serves as a cognitive mediator between the user and the system's algorithmic logic.

Figure 7. Elements of the Human-Centered Interface of the System

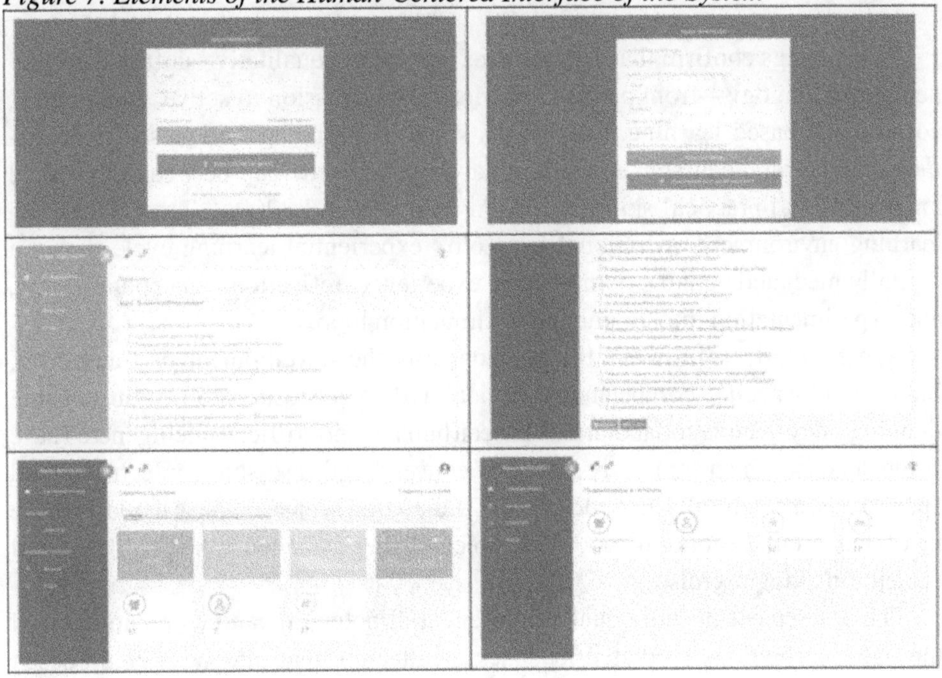

A key aspect of a human-centered interface is its adaptability. The front end must be capable of personalizing content and presentation based on the learner's profile, using information provided by the back end. This includes dynamically generating recommendations, adaptive learning paths, and interactive simulations that encourage active participation and experiential learning. In this way, the interface is not a passive channel for information, but an active tool for learning and development.

Ensuring transparency and trust through the interface is also a key element. The user must have access to explanations of how certain results or recommendations are generated, which links the frontend to the ethical principles implemented in the backend. Furthermore, the interface must provide opportunities for user feedback, which closes the cycle of interaction between the user and the system.

In summary, the human-centered interface transforms the digital twin from a technical system into a usable educational environment, ensuring effective interaction between the learner and the backend logic of analysis, simulations, and decision-making. It is key to realizing the principles of Education 5.0, as it places the user at the center of the system and transforms data and algorithms into a means for understanding, learning, and action.

5. DISCUSSION

The findings confirm that digital twins represent a qualitative shift in entrepreneurship pedagogy—from passive knowledge transmission to active, experiential, competency-based learning. This aligns with longstanding critiques of management and entrepreneurship education that emphasize the gap between theoretical knowledge and practical skill acquisition (Adcroft et al., 2004). The DT-enabled learning environment operationalizes Kolb's experiential learning cycle within a digitally mediated context, compressing experience, reflection, conceptualization, and experimentation into an integrated simulation loop.

A particularly significant finding concerns the development of transversal, Industry 5.0-specific competencies: ethical and social responsibility, sustainable thinking, and effective collaboration between humans and artificial intelligence. These competencies, which are poorly served by conventional case-study or lecture-based approaches, are naturally embedded in DT simulation scenarios—particularly those involving circular economy modeling, green energy management, and AI-assisted decision-making exercises.

The assessment of individual entrepreneurial potential can be operationalized through a technology-mediated feedback architecture that integrates cloud-based infrastructure, client-server design, and curated databases of real and simulated entrepreneurial profiles. At the core of this system lies the concept of the learner's digital twin — a dynamic cyber-representation that is continuously benchmarked against a reference entrepreneurial profile. This comparison mechanism enables a structured, iterative evaluation of the learner's current competence state relative to an idealized entrepreneurial archetype (Figure 8).

Figure 8. Digital Entrepreneurial Twin Model

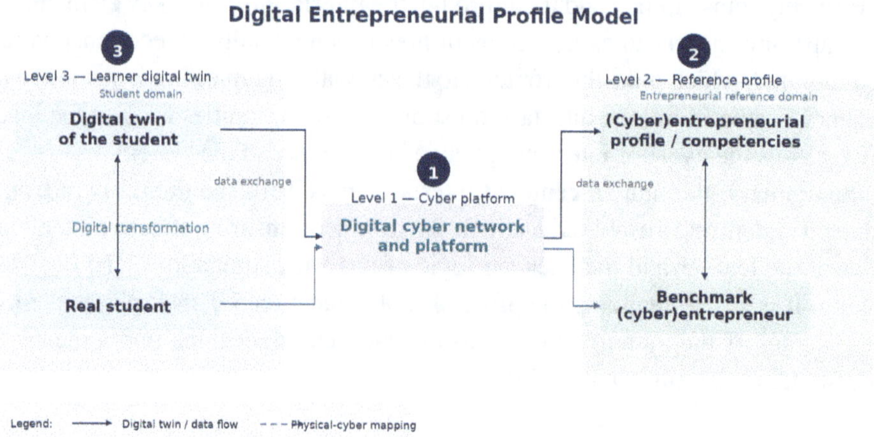

Figure 8 represent the DT design including:

- Three clearly labeled domains — *Student domain* (left), *Digital cyber network and platform* (centre), *Entrepreneurial reference domain* (right) — with numbered level badges (1, 2, 3) matching the three-layer model described in the text.
- directional arrows distinguishing digital twin data flow (solid navy) from physical-cyber mapping (lighter steel).
- Both vertical bidirectional arrows (DT ↔ Real student; Cyber profile ↔ Benchmark entrepreneur) reflect the two-way nature of the digital twin concept per Grieves (2014).

Consistent with the broader literature on digital entrepreneurship ecosystems (Nambisan, 2017; Autio et al., 2018), digital platforms in this context function not merely as assessment instruments but as social technologies that actively shape entrepreneurial identity and behavior. Profile-building, peer content exchange, and networked communication collectively foster the digital social capital that underpins contemporary venture creation. Students are not passive recipients of feedback — they engage in a reflexive process of identity construction, progressively aligning their self-concept with entrepreneurial norms embedded in the platform's reference model.

This alignment process is central to personalized entrepreneurial learning. By mapping a student's evolving competence profile onto its digital counterpart, the system enables skill gaps to be identified, visualized, and systematically addressed. Rather than treating measurement and development as sequential activities, this architecture renders them simultaneous and mutually reinforcing — a principle consistent with formative assessment theory (Black & Wiliam, 1998) and self-regulated learning frameworks (Zimmerman, 2000).

Building on the social interaction model proposed by Sterev et al. (2019), the overall system can be understood as operating across three interdependent information layers. The first is the networked cyber-platform itself — the digital infrastructure through which data flows, profiles are maintained, and interactions are mediated. The second is the reference profile layer, where digitally encoded entrepreneurial archetypes interact with individual learner profiles in cyberspace; these interactions generate dynamic feedback loops that shape expected behavior and reinforce entrepreneurial norms. The third layer is the digital twin proper — a rich, multi-dimensional representation of the learner's personal and social characteristics, capturing behavioral dimensions such as trust, commitment, emotional orientation, and social responsibility (Figure 9). Together, these three layers constitute a socio-technical system in which technological architecture and human development are inseparable.

Figure 9. Model of direct and indirect interactions between the elements of the digital Entrepreneurial Twin Model

The empirical results support a fundamental theoretical position: entrepreneurial competencies are developable through intentional educational interventions rather than fixed by innate personality traits (Levander & Raccuia, 2001; Acuña-Duran et al., 2022). The statistically significant pre-post improvements across all 15 EntreComp domains, regardless of students' initial profiles, provide strong evidence that structured DT-mediated education systematically cultivates entrepreneurial potential.

Several limitations constrain the generalizability of the findings. First, the empirical component draws exclusively from one institution (UNWE, Sofia), limiting cross-cultural validity. Second, the absence of a randomized control group introduces potential selection bias. Third, the 14-week intervention timeframe may be insufficient to capture the full trajectory of competency development. Finally, the DTL infrastructure deployed in this study represents a relatively early-generation system; more sophisticated implementations may yield different effect sizes.

6. CONCLUSION

The chapter explores digital twins as a crucial linking paradigm between the advancement of Industry 4.0/5.0 and the transformation of educational systems toward Education 4.0/5.0. It argues that the digital transformation of industry and society drives fundamental changes in human capital requirements, necessitating a reevaluation of traditional educational models and a shift toward competency-based, personalized, and technology-enabled learning.

In the given context, digital twins are defined as virtual representations of physical objects, processes, or systems that are connected to their real-world counterparts through two-way data exchange. This connection distinguishes them from digital models and digital shadows. Modern digital twins, enhanced with artificial intelligence and machine learning, extend beyond descriptive functions to become prognostic and prescriptive systems capable of managing complexity and uncertainty within sociotechnical environments.

Particular attention should be given to the application of digital twins in education, where they serve as interactive learning environments that support experimental and practice-oriented learning. By simulating real processes within a controlled setting, learners develop skills in systematic thinking, data analysis, problem-solving, and decision-making. Based on this, a specific model of a digital twin for the learner should be developed, enabling personalized learning through the modeling of knowledge, skills, behavior, and progress. This approach is especially important for fostering transversal entrepreneurial skills such as creativity and idea generation, problem-solving and decision-making, opportunity identification and utilization, collaboration and communication, and adaptability and self-initiative. In conclusion, it is emphasized that digital twins represent a strategic tool for advancing human-centered, ethical, and sustainable entrepreneurial education that meets the demands of the modern and future world.

The developed and proposed model of a learner's digital twin could successfully transform education in the field of academic entrepreneurship and entrepreneurial skills within the context of Industry 5.0 and Education 5.0. The model is consistent with the main conclusions of studies conducted in the field of digital twin theory and the shift in the information paradigm. The digital twin model of a learner represents an integrated, multi-layered system in which data, models, and solutions operate in a continuous cycle. At the core of this paradigm lie the processes of collecting, processing, and managing multidimensional structured and unstructured data from diverse sources (behavioral, cognitive, academic, sensory), which form the empirical basis for building the digital twin and the virtual representation of the learner. The proposed approach of dynamic, self-sustaining system in which the workflow creates a continuous process of learning, adaptation, and optimization, in line with the requirements of Industry 5.0 and Education 5.0.

This systematic review has established that digital twin technology serves as a strategic bridge between Industry 5.0 and Education 5.0, enabling academic entrepreneurship education to develop the full spectrum of EntreComp competencies within a personalized, risk-free, and ethically grounded learning environment. Key conclusions include:

- Digital twins are qualitatively distinct from static models and digital shadows; their bidirectionality is pedagogically essential.
- The Digital Twin of the Learner operationalizes Education 5.0 principles through continuous adaptive profiling.

- Empirical evidence demonstrates significant competency improvements across all 15 EntreComp domains.
- Industry 5.0 transversal skills—human-AI collaboration, sustainability, and ethics—are naturally cultivated through DT simulation.
- Entrepreneurs are made, not born: intentional digital education systematically develops entrepreneurial potential.

ACKNOWLEDGEMENT

The work is supported by project No. BG16RFPR002-1.014-0014-C01 "Development Program with a Business Plan for the Laboratory Complex of Sofia Tech Park", which is implemented under the "Research, Innovation and Digitalization for Smart Transformation" Program, co-financed by the European Union through the European Regional Development Fund.

REFERENCES

Acuña-Duran, E., Oyanedel, J. C., & Pradenas-Wilson, D. (2022). Survey data of entrepreneurial intention and perceived social support from academics-scientists at Chilean universities. *Data in Brief*, *44*, 108537. DOI: 10.1016/j.dib.2022.108537 PMID: 36045643

Adcroft, A., Willis, R., & Dhaliwal, S. (2004). Missing the point? Management education and entrepreneurship. *Management Decision*, *42*(3), 521–531. DOI: 10.1108/00251740410518958

Andreasyan, A., & Balyakin, A. (2022). Transformation of education through Big Data: Digital twins case study. *Journal of Physics: Conference Series*, *2210*(1), 012003. DOI: 10.1088/1742-6596/2210/1/012003

Autio, E., Nambisan, S., Thomas, L. D. W., & Wright, M. (2018). Digital affordances, spatial affordances, and the genesis of entrepreneurial ecosystems. *Strategic Entrepreneurship Journal*, *12*(1), 72–95. DOI: 10.1002/sej.1266

Bacigalupo, M., Kampylis, P., Punie, Y., & Van Den Brande, L. (2016). EntreComp: The Entrepreneurship Competence Framework. EUR 27939 EN. Luxembourg (Luxembourg): Publications Office of the European Union; 2016. JRC101581, https://publications.jrc.ec.europa.eu/repository/handle/JRC101581 https://doi.org/ DOI: 10.2791/593884

Black, P., & Wiliam, D. (1998). Assessment and classroom learning. *Assessment in Education: Principles, Policy & Practice*, *5*(1), 7–74. DOI: 10.1080/0969595980050102

Böck, F., Ochs, M., Henrich, A., Landes, D., Leidner, J. L., & Sedelmaier, Y. (2025). Learner models: Design, components, structure, and modelling. *User Modeling and User-Adapted Interaction*, *35*(4), 15. DOI: 10.1007/s11257-025-09434-4

Botha, M., & Taljaard, A. (2021). Exploring the Entrepreneurial Intention-Competency Model for Nascent Entrepreneurs: Insights From a Developing Country Context. *Frontiers in Psychology, 12*, 516120. DOI: 10.3389/fpsyg.2021.516120 PMID: 34393866

Chamola, V.. (2025). *Metaverse for education: Developments, challenges, and future direction. Computer Applications in Engineering Education Journal.* IEEE., DOI: 10.1002/cae.70018

Chehlarova, N., Tsochev, G., Kotseva, M., & Miltchev, R. (2021). Digital Competencies of Public Administration Employees Related to Cybersecurity. *12th National Conference with International Participation (ELECTRONICA)*, Sofia, Bulgaria, 2021, pp. 1-4, DOI: 10.1109/ELECTRONICA52725.2021.9513705

Chigbu, B. I., & Makapela, S. L. (2025). AI in education, sustainability, and the future of work: An integrative review of Industry 5.0, Education 5.0, and Work 5.0. *Journal of Open Innovation, 11*(4), 100645. DOI: 10.1016/j.joitmc.2025.100645

David, J., Lobov, A., & Lanz, M. (2018a), Leveraging Digital Twins for Assisted Learning of Flexible Manufacturing Systems, Proceedings of the 16TH IEEE International Conference on Industrial Informatics, INDIN, 2018. DOI: 10.1109/INDIN.2018.8472083

David, J., Lobov, A., Lanz, M. (2018b). Learning Experiences Involving Digital Twins. 3681-3686. DOI: 10.1109/IECON.2018.8591460

De Jong, J. P. J., Parker, S. K., Wennekers, S., & Wu, C.-H. (2015). Entrepreneurial behavior in organizations: Does job design matter? *Entrepreneurship Theory and Practice, 39*(4), 981–995. DOI: 10.1111/etap.12084

Dihan, M. S., Akash, A. I., Tasneem, Z., Das, P., Das, S. K., Islam, M. R., Islam, M. M., Badal, F. R., Ali, M. F., Ahamed, M. H., Abhi, S. H., Sarker, S. K., & Hasan, M. M. (2024). Digital twin: Data exploration, architecture, implementation and future. *Heliyon, 10*(5), e26503. Advance online publication. DOI: 10.1016/j.heliyon.2024.e26503 PMID: 38444502

Elhajj, M. E. M. (2026). Semantic foundations for digital twins: The contribution of ontological analysis. *Frontiers of Computer Science, 8*, 1757450. DOI: 10.3389/fcomp.2026.1757450

European Commission. (2021). *Industry 5.0: Towards a sustainable, human-centric and resilient European industry*. Publications Office of the European Union., DOI: 10.2777/308407

Flaga, A., & Pacholczak, P. (2022). Demonstrator of a digital twin for educational and training purposes as a web application. Advances in Science and Technology – Research Journal 16(5):110-119. DOI: https://doi.org/DOI: 10.12913/22998624/152927

Gosztonyi, M., & Csákné Filep, J. (2022). Profiling (non-)nascent entrepreneurs in Hungary based on machine learning approaches. *Sustainability (Basel), 14*(6), 3571. DOI: 10.3390/su14063571

Grieves, M. (2014). Digital twin: Manufacturing excellence through virtual factory replication. White Paper, Florida Institute of Technology.

He, Y. (2026). Rewiring industrial futures: The role of AI-enabled digital twins in energy productivity transitions. *International Review of Economics & Finance, 105*, 104898. DOI: 10.1016/j.iref.2026.104898

Ho, M.-H. R., Uy, M. A., Kang, B. N. Y., & Chan, K.-Y. (2018). Impact of entrepreneurship training on entrepreneurial efficacy and alertness among adolescent youth. *Frontiers in Education, 3*, 13. DOI: 10.3389/feduc.2018.00013

Iliuţă, M.-E., Moisescu, M.-A., Pop, E., Ionita, A.-D., Caramihai, S.-I., & Mitulescu, T.-C. (2024). Digital Twin—A Review of the Evolution from Concept to Technology and Its Analytical Perspectives on Applications in Various Fields. *Applied Sciences (Basel, Switzerland), 14*(13), 5454. DOI: 10.3390/app14135454

Kabashkin, I. (2025). AI-Based Digital Twins of Students: A New Paradigm for Competency-Oriented Learning Transformation. *Information (Basel), 16*(10), 846. DOI: 10.3390/info16100846

Karabulut, E., Pileggi, S. F., Groth, P., & Degeler, V. (2024). Ontologies in digital twins: A systematic literature review. *Future Generation Computer Systems, 153*, 442–456. DOI: 10.1016/j.future.2023.12.013

Karanam, S. A. K., & Hartman, N. W. (2025). A systematic review of Digital Twin (DT) and virtual learning environments (VLE) for smart manufacturing education. *Manufacturing Letters, 44*(Supplement), 1597–1608. DOI: 10.1016/j.mfglet.2025.06.179

Khuong, M. N., & Hoang, Q. T. (2024). Entrepreneurial competencies – A systematic literature review. *Journal of International Entrepreneurship, 23*(1), 54–130. DOI: 10.1007/s10843-024-00356-7

Klisaroski, M. Naumova - Mihajlovska, K., Josimovska, T. (2025) Digital Transformation and HR: Challenges and Opportunities for Companies in North Macedonia, 10th International Scientific-Business Conference LIMEN 2024 – Selected Papers, pp.39-48, DOI: https://doi.org/DOI: 10.31410/LIMEN.SP2024.39

Lehmann, J., Lober, A., Häußermann, T., Rache, A., Ollinger, L., Baumgärtel, H., & Reichwald, J. (2023). The Anatomy of the Internet of Digital Twins: A Symbiosis of Agent and Digital Twin Paradigms Enhancing Resilience (Not Only) in Manufacturing Environments. *Machines (Basel), 11*(5), 504. DOI: 10.3390/machines11050504

Levander, A., & Raccuia, I. (2001). *Entrepreneurial profiling: A cognitive approach to entrepreneurship*. Stockholm School of Economics.

Melnyk, L., Kubatko, O., Matsenko, O., Balatskyi, Y., & Serdyukov, K. (2021). Transformation of the human capital reproduction in line with Industries 4.0 and 5.0. *Problems and Perspectives in Management, 19*(2), 480–494. DOI: 10.21511/ppm.19(2).2021.38

Mourtzis, D., Panopoulos, N., & Angelopoulos, J. D. (2022). A hybrid teaching factory model towards personalized education 4.0. *International Journal of Computer Integrated Manufacturing, 36*(12), 1739–1759. DOI: 10.1080/0951192X.2022.2145025

Nambisan, S. (2017). Digital entrepreneurship: Toward a digital technology perspective of entrepreneurship. *Entrepreneurship Theory and Practice, 41*(6), 1029–1055. DOI: 10.1111/etap.12254

Ntshangase, S. D., & Ezeuduji, I. O. (2020). Profiling entrepreneurial behaviour based on demographic variables: Tourism-related entrepreneurs in Mtubatuba Local Municipality, South Africa. *Geo Journal of Tourism and Geosites, 31*(3), 944–950. DOI: 10.30892/gtg.31302-525

Pajpach, M., Pribiš, R., Drahoš, P., Kučera, E., & Haffner, O. (2025). Methodology and pilot implementation of a modular educational-development platform for interoperable digital twin. *Scientific Reports, 15*(1), 28020. Advance online publication. DOI: 10.1038/s41598-025-13579-y PMID: 40745468

Park, J.-H., & Kim, S.-J. (2025). Entrepreneurial Competencies in the Era of Digital Transformation: A Systematic Literature Review. *Digital, 2025*(5), 46. DOI: 10.3390/digital5040046

Petkov, K. (2024). Logical framework method for conducting pre- and post-money valuations of equity crowdfunded companies leveraging AI. Industrial Growth Conference 2024. Az-buki Publ. pp: 207-219. https://doi.org/DOI: 10.53656/igc-2024.14

Ratković, T., Šlogar, H., & Šokčević, S. (2022). Entrepreneurial competencies of university students. *Economics & Sociology (Ternopil), 15*(4), 129–145. DOI: 10.14254/2071-789X.2022/15-4/6

Schott, C., van Roekel, H., & Tummers, L. G. (2020). Teacher leadership: A systematic review, methodological quality assessment and conceptual framework. *Educational Research Review*, *31*, 100352. DOI: 10.1016/j.edurev.2020.100352

Sedrakyan, G., Snoeck, M., & Poelmans, S. (2014). Assessing the effectiveness of feedback-enabled simulation in teaching conceptual modeling. *Computers & Education*, *78*, 367–382. DOI: 10.1016/j.compedu.2014.06.014

Singh, M., Fuenmayor, E., Hinchy, E. P., Qiao, Y., Murray, N., & Devine, D. (2022). Digital Twin: Origin to Future. *Applied System Innovation*, *4*(2), 36. DOI: 10.3390/asi4020036

Sofiullah, M., Gomes Vale, E., & Darr, D. (2023). Effectiveness of an interactive start-up simulation to foster entrepreneurial intentions among undergraduate university students: A quasi-experimental study. *Entrepreneurship Education*, *6*(4), 445–467. DOI: 10.1007/s41959-023-00108-z

Sörensen, A., Mitra, R., Hulthen, E., Hartmann, T., & Clausen, E. (2022). Bringing the entrepreneurial mindset into mining engineering education. *Mining, Metallurgy & Exploration*, *39*(4), 2013–2027. DOI: 10.1007/s42461-022-00620-1

Sorochak, O., & Gvozd, M. (2025). Education 4.0: Innovative approaches to competence development in digital organizations. *Journal of Education and Learning*, *14*(2), 1–15. DOI: 10.33271/nvngu/2025-5/188

Sterev, N., Kisimov, V., Stoyanova, T., & Andonov, V. (2019). Multidimensional framework for cross-corporate business social networks (BSN). In *Proceedings of the 11th IADIS International Conference Information Systems* (pp. 169–176). IADIS Press. ISBN 978-989-8533-74-6

Sterev, N., & Yordanov, D. (2024). Academic entrepreneurship: practical results and training, Strategies for Policy in Science & Education, vol. 32, no. 6 s, pp. 150 – 162. DOI: https://doi.org/DOI: 10.53656/str2024-6s-12-aca

Tao, F., Xiao, B., Qi, Q., Cheng, J., & Ji, P. (2022). Digital twin modeling. *Journal of Manufacturing Systems*, *64*, 372–389. DOI: 10.1016/j.jmsy.2022.06.015

Timperi, M., Kokkonen, K., Hannola, L., & Elfvengren, K. (2023). Impacts of digital twins on new business creation: Insights from manufacturing industry. *Measuring Business Excellence*, *27*(3), 433–448. DOI: 10.1108/MBE-09-2022-0104

Tjahyadi, H., Prasetya, K., & Murwantara, I. M. (2023). Digital Twin Based Laboratory for Control Engineering Education. *International Journal of Information and Education Technology (IJIET)*, *13*(4), 704–711. DOI: 10.18178/ijiet.2023.13.4.1856

Trevisan, A., Dukovska-Popovska, I., Acerbi, F., Terzi, S., & Sassanelli, C. (2026). Teaching methods for building human capital for the twin transition: a skill-based approach, Technology in Society, Volume 85, 2026, 103172, ISSN 0160-791X, https://doi.org/DOI: 10.1016/j.techsoc.2025.103172

Xu, M., David, J. M., & Kim, S. H. (2021). The fourth industrial revolution: Opportunities and challenges. *International Journal of Financial Research*, *9*(2), 90–95. DOI: 10.5430/ijfr.v9n2p90

Zimmerman, B. J. (2000). Attaining self-regulation: A social cognitive perspective. In Boekaerts, M., Pintrich, P. R., & Zeidner, M. (Eds.), *Handbook of self-regulation* (pp. 13–39). Academic Press. DOI: 10.1016/B978-012109890-2/50031-7

Chapter 10
Digital Twins for Smart Crop, Soil, and Resource Optimization in Modern Agriculture

Mrutyunjay Padhiary
http://orcid.org/0000-0002-2236-568X
Assam University, Silchar, India

Joytu Debnath
http://orcid.org/0009-0002-8663-8293
Assam University, Silchar, India

Priyanka Roy
http://orcid.org/0009-0003-0227-7097
Assam University, Silchar, India

Ahad Ahmed Lasker
http://orcid.org/0009-0006-4580-6579
Assam University, Silchar, India

Kundan Kumar
http://orcid.org/0009-0001-6831-0676
Assam University, Silchar, India

ABSTRACT

Modern agriculture is increasingly leveraging digital technologies, particularly digital twins, to enhance productivity and sustainability. Digital twins create real-time virtual models of agricultural systems, allowing for improved decision-making

DOI: 10.4018/979-8-2600-1669-5.ch010

in crop growth, soil health, and resource use by integrating IoT sensors, AI, machine learning, satellite imagery, and cloud analytics. Studies show that these technologies can boost crop yields by 10-25%, reduce water usage by 20-30%, and cut fertilizer application by 15-20% while preserving soil quality. The proposed architecture for agricultural digital twins includes data acquisition, modeling, predictive analytics, and decision support, with applications in crop simulation, soil monitoring, and precision irrigation. Nevertheless, challenges like high infrastructure costs and data interoperability hinder broader implementation. Ultimately, digital twins represent a transformative shift toward intelligent, sustainable agricultural practices.

1. INTRODUCTION

Contemporary agriculture confronts unprecedented challenges arising from climate variability, resource constraints, population growth, and the imperative for environmental sustainability. Traditional agricultural practices, characterized by reactive management approaches and generalized treatment protocols, prove inadequate for addressing the complexity and dynamism of modern farming systems. The convergence of advanced computational technologies, sensor networks, and data analytics has catalyzed the emergence of precision agriculture, wherein farming operations are optimized through granular, data-driven decision-making (Şimşek, 2025).

Digital twin technology, originally developed for aerospace and manufacturing applications, has emerged as a transformative framework for agricultural system optimization. A digital twin constitutes a comprehensive virtual representation of a physical entity or system that maintains continuous synchronization through bidirectional data exchange, enabling real-time monitoring, simulation, and optimization (Nasirahmadi & Hensel, 2022). In agricultural contexts, digital twins integrate multiple data streams from field sensors, satellite imagery, weather stations, and operational databases to create dynamic models of crops, soil systems, and resource flows (Zhang et al., 2025).

The application of digital twins in agriculture addresses fundamental limitations of conventional farming approaches by enabling predictive rather than reactive management, facilitating scenario-based planning, and supporting optimization across multiple objectives simultaneously (Singh et al., 2022a). Unlike static models or periodic assessments, digital twins provide continuous system representation that evolves with changing environmental conditions, crop phenology, and management interventions. This dynamic capability proves particularly valuable for managing agricultural systems characterized by spatial heterogeneity, temporal variability, and complex interactions among biological, physical, and chemical processes (Peladarinos et al., 2023).

This chapter examines the application of digital twin technology across three critical domains of agricultural optimization: crop production, soil health management, and resource allocation. The analysis encompasses technological foundations, implementation methodologies, practical applications, and strategic considerations for deploying digital twin systems in agricultural operations. Through integration of theoretical frameworks with empirical evidence, this chapter provides a comprehensive foundation for understanding how digital twins enable precision agriculture and sustainable farming practices.

2. FUNDAMENTAL CONCEPTS OF DIGITAL TWINS

2.1 Conceptual Framework

Digital twins represent high-fidelity virtual replicas of physical systems that maintain continuous synchronization through integrated sensor networks and bidirectional data flows. The conceptual framework comprises three essential components: the physical entity, the virtual model, and the bidirectional data connection. In agricultural applications, the physical entity encompasses crops, soil, water systems, equipment, and environmental conditions within defined spatial boundaries. The virtual model integrates mathematical representations, empirical relationships, and machine learning algorithms to simulate system behavior, while the data connection ensures continuous updating of model parameters based on real-time observations (Kantaros et al., 2025).

The distinguishing characteristic of digital twins, compared to traditional simulation models, lies in their continuous synchronization with the physical system. While conventional models operate on static parameterization or periodic calibration, digital twins incorporate real-time data assimilation to maintain model fidelity as system conditions evolve. This continuous updating enables accurate representation of current system states, reliable prediction of near-term dynamics, and meaningful simulation of alternative management scenarios (Zhao et al., 2024).

2.2 Core Capabilities and Functions

Digital twins provide four fundamental capabilities essential for agricultural optimization. First, real-time monitoring aggregates diverse data streams to provide comprehensive situational awareness of system status (Cesco et al., 2023a). Sensor networks measuring soil moisture, nutrient levels, crop canopy characteristics, and microclimate conditions feed continuously into the digital twin, enabling detection of anomalies, identification of spatial patterns, and tracking of temporal trends (Howari, 2025).

Second, predictive analytics leverage historical data and mechanistic models to forecast future system states. For agricultural systems, predictions may encompass crop yield projections, pest outbreak likelihood, irrigation requirements, or nutrient demand patterns. The integration of process-based models with machine learning algorithms enables prediction across multiple temporal scales, from hourly irrigation scheduling to seasonal production forecasting (Delfani et al., 2024).

Third, scenario simulation allows evaluation of alternative management strategies before field implementation. Farmers and agronomists can test different fertilization regimens, irrigation schedules, or planting densities within the digital twin environment, comparing outcomes across economic, environmental, and production metrics. This capability reduces risk associated with management changes and facilitates identification of optimal strategies tailored to specific field conditions (Melesse, 2025).

Fourth, automated optimization employs algorithms to identify management decisions that maximize desired objectives while respecting operational constraints. Multi-objective optimization frameworks can balance competing goals such as maximizing yield while minimizing water consumption, fertilizer application, or environmental impact. The continuous data flow from field sensors enables dynamic optimization that adapts to changing conditions throughout the growing season (Nozari & Yordanova, 2026a). **Figure 1** illustrates the fundamental architectural components of an agricultural digital twin system, showing the integration of physical sensors, data processing infrastructure, and virtual modeling environments that enable continuous synchronization and optimization.

Figure 1. Conceptual architecture of agricultural digital twin system

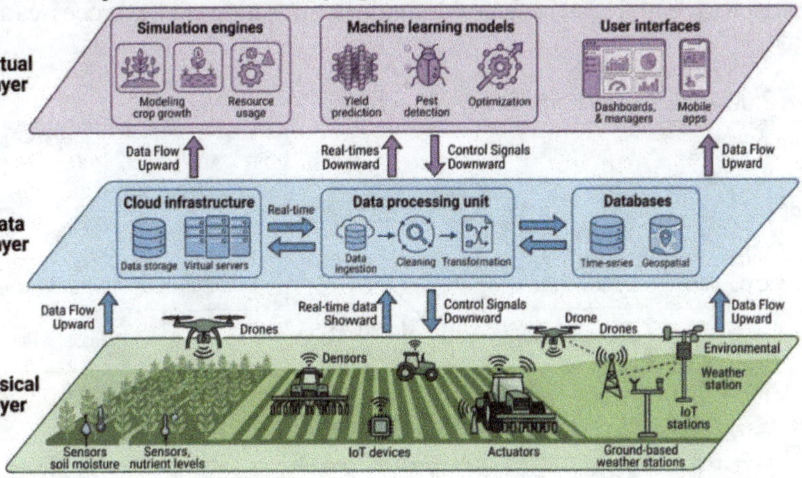

2.3 System Architecture

Crop Agent (CA), Soil Agent (SA), Water Agent (WA), and Energy Agent (EA) are the four autonomous agent types that compose the structure of the Multi-Agent Adaptive Digital Twin Framework (MADTF) architecture (Marah & Challenger, 2024). For inter-agent negotiation, each agent keeps a local state-space model, optimisation algorithm, and communication interface. In theory, a continuous-time state-space model is maintained by each agent i ∈ {CA, SA, WA, EA}:

$$\frac{dx_i}{dt} = f_i(x_i, u_i, w_i, t)$$

$$y_i = h_i(x_i, v_i)$$

where $y_i \in$ Rp is the measured output, vi is measurement noise, w_i stands for exogenous inputs (weather, soil properties), $u_i \in$ Rm is the control input, and $x_i \in$ Rn is the state vector. Biomass, root depth, leaf area index, and phenological stage are all included in xCA for the Crop Agent. The Soil Agent monitors organic matter, nutrient concentrations, and moisture content in several depth layers. Water and Energy Agents keep an eye on distribution efficiency, demand projections, and resource availability. The MADTF distributed architecture is shown in **Figure 2**, which uses Nash bargaining protocols to coordinate autonomous agents with local optimisation capabilities. By distributing computational load among edge devices, MADTF reduces latency from 15–30 seconds to 2–5 seconds for real-time decision-making, in comparison to central systems where a single optimisation engine processes all data.

Figure 2. Multi agent distributed architecture

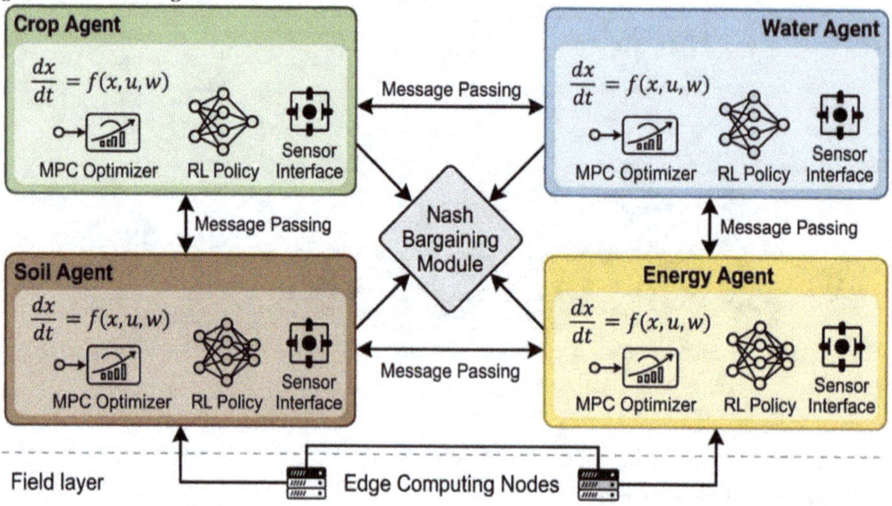

3. DIGITAL TWIN ARCHITECTURE FOR AGRICULTURAL SYSTEMS

3.1 Data Acquisition Infrastructure

The foundation of agricultural digital twins rests upon comprehensive data acquisition infrastructure capable of capturing the multidimensional dynamics of farming systems. Modern sensor networks deploy diverse measurement technologies to quantify soil properties, crop characteristics, atmospheric conditions, and operational parameters. Soil sensors monitor moisture content, temperature, electrical conductivity, pH levels, and nutrient concentrations at multiple depths, providing detailed profiles of below-ground conditions (Manivasagam, 2025). These in-situ measurements may be supplemented with electromagnetic induction surveys or penetrometer readings to characterize spatial variability in soil texture and compaction.

Crop monitoring integrates proximal and remote sensing technologies to assess plant health, development stage, and stress responses. Ground-based sensors measure canopy temperature, chlorophyll fluorescence, and spectral reflectance at high temporal resolution, while unmanned aerial systems equipped with multispectral or hyperspectral cameras provide intermediate-scale spatial coverage (Laveglia et al., 2024). Satellite-based remote sensing supplements these measurements with field-scale assessments of vegetation indices, leaf area, and surface temperature at regular intervals throughout the growing season (Bakó et al., 2025).

Environmental monitoring infrastructure captures atmospheric and climatic variables that drive crop growth and water dynamics. Weather stations measure air temperature, humidity, wind speed, solar radiation, and precipitation at sub-hourly intervals, enabling precise calculation of reference evapotranspiration and thermal time accumulation (Brocca et al., 2024a). Additional sensors may monitor leaf wetness duration for disease prediction, soil heat flux for energy balance calculations, or atmospheric carbon dioxide concentration for photosynthesis modeling.

3.2 Data Integration and Processing

Agricultural digital twins must integrate heterogeneous data streams characterized by different spatial resolutions, temporal frequencies, and measurement scales (Tsakiridis et al., 2023). Data processing pipelines employ quality control algorithms to identify sensor malfunctions, transmission errors, and anomalous readings that could compromise model accuracy. Statistical filtering techniques remove noise while preserving meaningful variability, and gap-filling algorithms reconstruct missing values based on spatial interpolation or temporal patterns (Liu et al., 2025a).

Spatial data fusion combines measurements from multiple sources to generate comprehensive field representations. Geostatistical methods such as kriging interpolate point measurements to continuous surfaces, accounting for spatial correlation structure and measurement uncertainty (He, 2026). Data assimilation algorithms integrate observations into process models, adjusting model states to maintain consistency with empirical evidence while respecting physical constraints. Machine learning approaches, including deep neural networks and ensemble methods, can learn complex relationships between sensor measurements and target variables, enabling estimation of difficult-to-measure parameters from readily available proxies (Mücke, 2024a).

3.3 Modeling and Simulation Components

The virtual representation within agricultural digital twins comprises multiple interconnected models that simulate crop growth, soil processes, water movement, nutrient cycling, and pest dynamics (Escribà-Gelonch et al., 2024). Crop growth models integrate photosynthesis, respiration, partitioning, and phenological development to predict biomass accumulation and yield formation as functions of genotype, environment, and management. These models typically operate on daily or sub-daily time steps, calculating growth rates based on temperature, solar radiation, water availability, and nutrient supply (Xu et al., 2025a).

Soil water models simulate infiltration, redistribution, evaporation, and root uptake using physically based equations or empirical relationships (Xu et al., 2025a). Three-dimensional models may represent lateral water movement and preferential flow paths, while simplified one-dimensional approaches suffice for many management applications. Coupling soil water models with crop models enables simulation of water stress effects on growth and facilitates optimization of irrigation scheduling (Qin et al., 2025a).

Nutrient cycling models track transformations among organic and inorganic pools, microbial uptake and release, plant acquisition, and losses through leaching or volatilization (Bashkin, 2026). These biogeochemical models require careful parameterization based on soil properties, climate, and management history. Integration with crop models ensures that nutrient availability constraints on growth are properly represented and enables evaluation of fertilization strategies (Silva, Conceição, et al., 2023). As shown in **Table 1**, the digital twin architecture comprises six interconnected layers that collectively enable comprehensive agricultural system modeling and optimization. Each layer provides essential functionality while maintaining interoperability with adjacent components.

Table 1. Core Components of Agricultural Digital Twin Architecture

Component	Key Technologies	Primary Functions	Reference(s)
Sensing layer	IoT sensors, UAVs, satellites, weather stations, soil probes	Real-time data collection, environmental monitoring, crop assessment	(Omia et al., 2023)
Communication layer	5G networks, LoRaWAN, NB-IoT, edge computing nodes	Data transmission, edge processing, network management	(Peruzzi & Pozzebon, 2022)
Data management layer	Cloud platforms, time-series databases, data lakes, ETL pipelines	Data storage, preprocessing, quality control, integration	(Saikia et al., 2025)
Modeling layer	Crop growth models, soil models, hydrological models, ML algorithms	System simulation, predictive analytics, scenario evaluation	(Banerjee et al., 2025)
Intelligence layer	Deep learning, optimization algorithms, decision trees, ensemble methods	Pattern recognition, automated optimization, decision support	(Ali et al., 2023)
Application layer	Web interfaces, mobile apps, API services, visualization tools	User interaction, recommendation delivery, system control	(Nozari, 2026)

4. SMART CROP OPTIMIZATION THROUGH DIGITAL TWINS

4.1 Growth Monitoring and Prediction

Digital twins enable continuous monitoring of crop development through integration of multiple data sources and predictive models. Remote sensing platforms provide synoptic assessment of crop canopy properties, including leaf area index, biomass accumulation, and chlorophyll content, while ground-based sensors capture micrometeorological variables that drive photosynthesis and respiration. The digital twin assimilates these observations into mechanistic growth models, adjusting parameters to maintain alignment between simulated and observed crop status (Dawn et al., 2023).

Phenological stage prediction constitutes a critical component of crop digital twins, as development stage determines appropriate management interventions and influences crop response to environmental stress (Kalla et al., 2026). Temperature-based thermal time accumulation provides the foundation for phenology models, with adjustments for photoperiod and vernalization requirements depending on crop species. Machine learning algorithms can refine phenological predictions by identifying relationships between environmental variables, management practices, and actual stage transitions observed through image analysis or manual scouting (Fotouo Makouate & Zude-Sasse, 2025).

Yield forecasting represents a primary application of crop digital twins, providing advance estimates that inform marketing decisions, logistics planning, and resource allocation. Digital twin yield predictions integrate crop model simulations with statistical relationships derived from historical data, remote sensing indices, and weather forecasts (Fotouo Makouate & Zude-Sasse, 2025). Ensemble approaches that combine multiple prediction methods often achieve superior accuracy compared to individual models, while quantification of prediction uncertainty enables risk-informed decision-making (Srivastav & Das, 2025).

4.2 Stress Detection and Management

Early detection of crop stress enables timely interventions that minimize yield losses and optimize input efficiency. Digital twins integrate multiple stress indicators to identify water deficit, nutrient deficiency, disease infection, or pest damage before visible symptoms appear (Ahsen et al., 2025). Thermal imaging detects elevated canopy temperature associated with stomatal closure under water stress, while spectral reflectance patterns indicate chlorophyll degradation from nutrient limitations or pathogen infection (Sapes et al., 2024).

Water stress monitoring combines soil moisture measurements, atmospheric demand calculations, and crop water uptake models to assess plant water status. The digital twin calculates crop water stress index based on the difference between actual and potential evapotranspiration, triggering irrigation recommendations when stress exceeds acceptable thresholds. This approach enables deficit irrigation strategies that reduce water consumption while maintaining acceptable yield and quality outcomes (Parewai & Köppen, 2025).

Nutrient stress identification leverages spectral indices sensitive to chlorophyll content, leaf nitrogen concentration, and overall plant vigor. Deviations from expected spectral signatures, accounting for growth stage and environmental conditions, indicate potential nutrient limitations (Panda et al., 2025). The digital twin can distinguish among different nutrient deficiencies based on characteristic symptom patterns and recommend targeted fertilization to address specific limitations.

4.3 Precision Management Zones

Spatial variability in soil properties, topography, and historical yield patterns necessitates site-specific management approaches. Digital twins facilitate delineation of management zones characterized by similar production potential and input requirements (Rodrigues et al., 2024). Clustering algorithms analyze multi-year yield maps, soil survey data, elevation models, and remote sensing time series to identify homogeneous zones suitable for uniform management.

Variable rate application technologies enable implementation of zone-specific management strategies for seeding, fertilization, and crop protection. The digital twin generates prescription maps that specify optimal input rates for each management zone based on simulated crop requirements, economic optimization, and environmental constraints. Continuous updating of zone performance allows adaptive refinement of zone boundaries and management prescriptions across successive growing seasons.

4.4 Soil Water Balance Model

With state $x_{soil} = [\theta_1, ..., \theta_n]^T$ which represents the volumetric water content in n soil layers, the Soil substance maintains a multi-layer water balance. Water flow is governed by the Richards equation (Van Lier et al., 2024):

$$\frac{\partial\theta}{\partial t} = \frac{\partial}{\partial z}\left[K(\theta)\left(\frac{\partial h}{\partial z} + 1\right)\right] - S(z,t)$$

where h is matric potential, S (z, t) is root retention of water, and K(θ) is hydraulic conductivity. MADTF uses an incremental bucket model approximation by Qin et al. (2025b) for computational efficiency that has been shown to capture 92-96% of variance in detailed finite-element solutions (Rastgou et al., 2025). The Richards equation models unsaturated soil water movement by linking hydraulic conductivity K(θ) with matric potential and gravitational drainage. Its numerical solution is computationally intensive due to the non-linearity of K(θ) near saturation and strict time step requirements, which hinder real-time applications like MADTF. The MADTF framework utilizes an incremental bucket model, treating soil layers as reservoirs and updating water content θ using mass-balance calculations instead of solving complex equations. This method preserves 92–96% of water content variance while drastically reducing computation time. However, the bucket model has limitations, notably its inability to accurately account for preferential flow in macroporous soils and lateral redistribution on slopes, suggesting that soils with high clay content may still necessitate a complete Richards solution despite the associated computational costs.

5. SOIL OPTIMIZATION USING DIGITAL TWIN TECHNOLOGY

5.1 Soil Health Monitoring

Soil health encompasses physical, chemical, and biological properties that collectively determine productive capacity and ecosystem function. Digital twins integrate diverse soil measurements to provide comprehensive health assessment beyond individual parameter monitoring. Physical properties including texture, structure, aggregate stability, and bulk density influence water infiltration, root penetration, and aeration (Silva, Rodríguez-Sedano, et al., 2023). Sensor networks measure soil moisture and temperature continuously, while periodic penetrometer readings assess compaction and structural degradation.

Chemical soil properties monitored through digital twin systems include pH, cation exchange capacity, organic matter content, and concentrations of essential nutrients. Ion-selective sensors enable continuous measurement of nitrate, potassium, and other nutrients, complementing laboratory analyses of soil samples. Geostatistical interpolation generates continuous spatial predictions of soil properties from point measurements, accounting for spatial correlation structure and uncertainty quantification.

Biological soil health indicators, including microbial biomass, enzyme activities, and respiration rates, provide insight into decomposition processes and nutrient cycling dynamics. While direct sensing of biological properties remains technologically challenging, digital twins can estimate biological activity from temperature, moisture, and organic matter inputs to decomposition models (Zhan, 2024). Integration of periodic laboratory measurements with continuous environmental monitoring enables temporal interpolation of biological parameters.

5.2 Nutrient Management Optimization

Precision nutrient management represents a critical application of soil digital twins, enabling fertilization strategies that match nutrient supply with crop demand while minimizing environmental losses. Nutrient cycling models simulate transformations among soil organic matter pools, mineralization and immobilization by microorganisms, and plant uptake governed by root distribution and nutrient availability. These models require parameterization based on soil properties, crop characteristics, and environmental conditions obtained through sensor networks and remote sensing (Liu et al., 2025).

Nitrogen management exemplifies the complexity addressed by digital twin approaches. Nitrogen exists in multiple forms including mineral nitrate and ammonium, organic nitrogen in crop residues and soil organic matter, and atmospheric nitrogen fixed by legumes or lost through volatilization and denitrification. Digital twins track these pools and transformation rates, predicting mineral nitrogen availability for crop uptake and identifying optimal timing and rates for supplemental fertilization.

Phosphorus and potassium management benefit from digital twin approaches that account for soil buffering capacity, fixation processes, and spatial variability. Unlike nitrogen, these nutrients exhibit limited mobility in soil and accumulate or deplete over multiple years of fertilization (Kong et al., 2025). Long-term nutrient balance calculations within the digital twin inform sustainable fertilization strategies that maintain soil fertility while avoiding excessive accumulation.

5.3 Soil Water Dynamics

Soil water content governs crop water availability, nutrient transport, microbial activity, and numerous other processes fundamental to agricultural production. Digital twins employ soil water models that simulate infiltration of precipitation and irrigation, redistribution driven by gravity and capillarity, evaporation from the soil surface, and root water uptake (Nassar, 2026). These models operate at multiple scales, from one-dimensional profile models suitable for field management to three-dimensional simulations that resolve topographic effects on lateral water movement.

Continuous soil moisture monitoring provides data assimilation targets that improve model accuracy and enable detection of preferential flow paths, drainage system performance, or irrigation system malfunctions. Wireless sensor networks measure volumetric water content at multiple depths and locations, transmitting data in real-time to the digital twin platform. Integration of soil moisture observations with weather data and crop water use calculations enables precise irrigation scheduling that maintains optimal soil water status.

5.4 Erosion and Compaction Management

Soil degradation through erosion and compaction threatens long-term agricultural sustainability and productivity. Digital twins incorporate erosion models that calculate soil loss based on rainfall intensity, slope, soil erodibility, and vegetative cover. High-resolution terrain models derived from aerial surveys enable identification of concentrated flow paths and areas at elevated erosion risk, supporting targeted implementation of conservation practices (Alipour & Shojaei, 2025).

Soil compaction resulting from field traffic under wet conditions reduces porosity, limits root growth, and impairs water infiltration. Digital twin systems can track cumulative axle loads and soil moisture status to identify periods when field operations risk excessive compaction. Traffic management recommendations generated by the digital twin minimize compaction while enabling timely completion of critical field operations (Liu et al., 2025).

The comprehensive integration of diverse data sources and modeling components in soil digital twin systems is illustrated in Figure 3, which depicts the workflow from data acquisition through model-based decision support for soil management optimization.

Figure 3. Soil digital twin data integration and processing workflow.

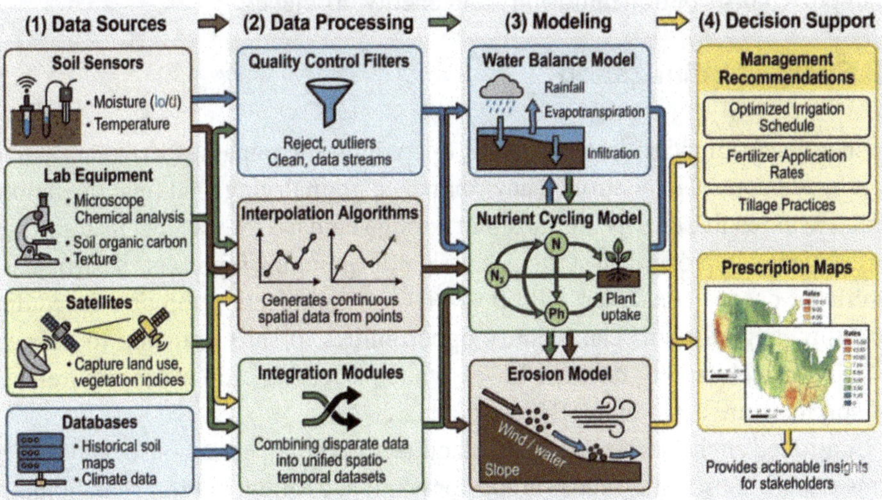

6. RESOURCE OPTIMIZATION AND MANAGEMENT

6.1 Water Resource Management

Water constitutes the most critical resource in many agricultural systems, with irrigation representing the largest consumptive use of freshwater globally. Digital twin technologies enable precision irrigation management that maximizes water productivity while minimizing waste and environmental impact. Irrigation scheduling optimization within digital twins integrates crop water requirements, soil water holding capacity, weather forecasts, and water availability constraints (Prasad et al., 2026). Dynamic scheduling adjusts irrigation timing and amounts based on current soil moisture status and near-term precipitation probability, avoiding unnecessary applications when rainfall is expected.

Deficit irrigation strategies, whereby crops receive less than full water requirements during specific growth stages, can reduce water consumption substantially while maintaining acceptable yield and quality. Digital twins enable implementation of regulated deficit irrigation by simulating crop stress responses and identifying growth stages with minimal yield sensitivity to water limitation. Real-time monitoring of crop water status through thermal imaging or stem water potential measurements provides feedback for adaptive deficit irrigation management.

Irrigation system performance monitoring constitutes another critical application. Sensor networks detect spatial variation in water application rates, pressure fluctuations indicating emitter clogging or leaks, and energy consumption patterns (Alves et al., 2023a). The digital twin compares actual system performance against design specifications, identifying maintenance needs and optimization opportunities for improved distribution uniformity and energy efficiency.

6.2 Energy Optimization

Energy consumption in agricultural operations encompasses irrigation pumping, climate control in controlled environments, grain drying, and field operations. Digital twins enable energy optimization through demand forecasting, load shifting, and efficiency improvement strategies. Irrigation pumping represents a particularly significant energy consumer, with potential for optimization through scheduling flexibility. Digital twins can identify opportunities to shift pumping to off-peak periods when electricity rates are lower, while respecting crop water requirements and soil moisture constraints (Barzigar et al., 2025).

Controlled environment agriculture facilities including greenhouses and vertical farms require substantial energy inputs for heating, cooling, lighting, and humidity control. Digital twins integrate building energy models with crop growth simulations to optimize climate setpoints that balance energy consumption against production objectives. Predictive control algorithms adjust environmental conditions proactively based on weather forecasts and crop development status, minimizing energy waste while maintaining optimal growing conditions (Padhiary et al., 2025).

6.3 Labor and Equipment Optimization

Labor and machinery represent substantial costs in agricultural operations, with potential for optimization through improved scheduling and resource allocation. Digital twins enable predictive maintenance of farm equipment by monitoring operational parameters, detecting anomalies, and forecasting component failures based on usage patterns and environmental stress. Proactive maintenance scheduling reduces downtime during critical operational periods and extends equipment lifespan.

Field operation scheduling must account for multiple constraints including crop development stage, soil moisture conditions suitable for trafficking, weather forecasts, and equipment availability. Digital twins employ optimization algorithms to generate operation schedules that minimize delays, reduce soil compaction risk, and ensure timely completion of critical activities such as planting within optimal windows. Integration of weather forecasts enables dynamic rescheduling in response to changing conditions (Narzari et al., 2025).

6.4 Multi-Objective Optimization

Agricultural decision-making inherently involves tradeoffs among competing objectives including productivity, profitability, environmental sustainability, and resource conservation. Digital twins facilitate multi-objective optimization that identifies management strategies representing optimal compromises among these objectives. Pareto frontier analysis reveals the set of non-dominated solutions, enabling decision-makers to understand fundamental tradeoffs and select preferred strategies based on their priorities (Purcell et al., 2023a).

Economic optimization within digital twins incorporates input costs, commodity prices, yield predictions, and quality premiums to maximize net returns. However, purely economic objectives may conflict with environmental goals such as minimizing nutrient losses, reducing greenhouse gas emissions, or conserving biodiversity. Multi-objective frameworks enable exploration of management options that achieve acceptable economic performance while meeting environmental constraints or sustainability targets.

Table 2. Resource Optimization Domains in Agricultural Digital Twins

Resource	Optimization Targets	Key Metrics	Constraints	Reference(s)
Water	Irrigation scheduling, deficit management, system efficiency	Water use efficiency, crop water productivity, uniformity coefficient	Water availability, legal allocations, minimum crop requirements	(Maurya et al., 2024)
Fertilizers	Application rates, timing, placement, formulation selection	Nutrient use efficiency, partial factor productivity, recovery fraction	Environmental regulations, crop tolerance, economic limits	(Singh et al., 2022b)
Energy	Pump scheduling, climate control, load shifting, system efficiency	Energy intensity, cost per unit production, renewable fraction	Grid capacity, rate structures, crop environmental requirements	(Saivineeth & Venkateswarlu, 2025)
Labor	Task scheduling, crew allocation, automation integration	Labor productivity, hours per hectare, task completion rate	Workforce availability, skill requirements, legal regulations	(John, 2023)
Equipment	Operation scheduling, maintenance planning, capacity utilization	Utilization rate, operating efficiency, downtime percentage	Equipment availability, weather windows, soil conditions	(Das & Singh, 2023)

Table 2 summarizes the primary resource optimization domains within agricultural digital twin systems, illustrating the diversity of constraints and performance metrics that must be considered in integrated farm management.

7. INTEGRATION OF IOT, AI, AND SIMULATION TECHNOLOGIES

7.1 Internet of Things Infrastructure

Internet of Things technologies provide the sensing and connectivity infrastructure essential for agricultural digital twins. Modern agricultural IoT deployments integrate diverse sensor modalities, communication protocols, and edge computing capabilities to enable real-time data acquisition and local processing. Wireless sensor networks employ low-power communication standards including LoRaWAN, NB-IoT, and Zigbee to transmit measurements from distributed field locations to central aggregation points or cloud platforms (Marini et al., 2022a).

Edge computing nodes positioned at field boundaries or in farm buildings perform local data processing, reducing bandwidth requirements and enabling low-latency decision-making. These edge devices can execute data quality control algorithms, aggregate measurements from multiple sensors, and implement simple decision rules for automated actuation of irrigation valves or climate control systems. Hierarchical architectures distribute computational tasks across edge, fog, and cloud layers based on latency requirements, data volume, and processing complexity.

Power management constitutes a critical challenge for field-deployed sensors, particularly in locations without grid electricity access. Solar panels combined with battery storage provide autonomous power for extended deployment periods, while energy harvesting from temperature gradients or mechanical vibration represents emerging alternatives. Low-power sensor designs and duty-cycled operation strategies extend battery life, with wake-up intervals optimized based on the temporal dynamics of measured variables.

7.2 Artificial Intelligence and Machine Learning

Machine learning algorithms enhance agricultural digital twins through pattern recognition, predictive modeling, and automated decision-making capabilities that complement physics-based simulation approaches. Supervised learning methods including random forests, support vector machines, and gradient boosting enable prediction of crop yield, disease occurrence, or optimal management actions based on historical data. Deep learning architectures, particularly convolutional neural networks, excel at extracting information from imagery for tasks such as weed detection, crop stress identification, or phenological stage classification.

Hybrid modeling approaches integrate mechanistic crop models with machine learning components to leverage advantages of both methodologies. Process-based models provide interpretable representations grounded in biophysical understanding, while machine learning components can capture complex relationships difficult to express through explicit equations. For example, neural networks may predict cultivar-specific parameters for crop models from genetic markers, or learn residual corrections that improve model accuracy by accounting for processes not explicitly represented (Shi et al., 2025).

Reinforcement learning presents opportunities for autonomous optimization of sequential management decisions under uncertainty. These algorithms learn optimal policies through iterative interaction with simulated or real agricultural systems, balancing exploration of alternative strategies with exploitation of currently known best practices. Applications include dynamic irrigation scheduling that adapts to evolving weather patterns, or pest management strategies that minimize insecticide use while preventing economic damage.

7.3 Simulation Engines and Model Integration

Agricultural digital twins employ diverse simulation models operating across multiple spatial and temporal scales. Crop models simulate growth and development from emergence through maturity, calculating photosynthesis, respiration, biomass partitioning, and phenological transitions based on genotype parameters and environmental drivers. Established frameworks including DSSAT, APSIM, and WOFOST provide validated implementations of crop models for major agricultural species, with ongoing development extending capabilities to novel crops and climate scenarios.

Hydrological models simulate soil water movement through processes including infiltration, drainage, evaporation, and root water uptake. One-dimensional models such as HYDRUS provide detailed vertical profile simulations suitable for irrigation management, while distributed watershed models address lateral flows and runoff generation. Coupling crop and water models ensures consistent representation of transpiration demand and water stress effects on growth (Yu et al., 2022a).

Nutrient cycling models track transformations among organic and mineral pools, microbial processes, plant uptake, and environmental losses. These biogeochemical models require substantial parameterization and calibration based on soil properties, climate, and management history. Integration with crop and water models enables comprehensive simulation of nitrogen, phosphorus, and other nutrient dynamics governing crop nutrition and environmental impacts.

Model integration frameworks coordinate execution of multiple simulation models, managing data exchange, temporal synchronization, and computational workflows. Modular architectures enable substitution of alternative model components or addition of new process representations without requiring complete system redesign. Standardized interfaces and data formats facilitate interoperability among models developed by different research groups or commercial providers (Kamau et al., 2024). The integration of artificial intelligence components with traditional simulation models represents a hybrid approach that leverages strengths of both data-driven and mechanistic methodologies, as illustrated in **Figure 4**.

Figure 4. AI and machine learning integration in agricultural digital twins.

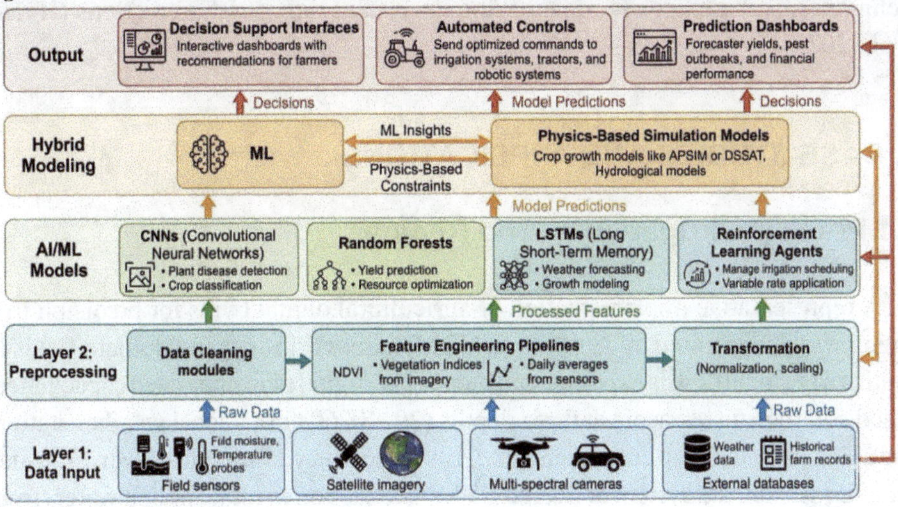

7.4 Hyperspectral Phenotyping and Foundation Models

Hyperspectral imaging (400–2500 nm, 5 nm resolution) is being used in current research to detect stress five to eight days before symptoms appear. Quantitative trait estimation with $R^2 = 0.82$-0.91 for important variables is made possible by specific absorption features at 970 nm (water content), 1510 nm (leaf structure), and 2180 nm (protein/nitrogen). By extracting latent features, deep convolutional networks (ResNet-50 backbone) enhance crop model calibration and lower biomass prediction RMSE by 18–24%. Data volume (8–12 GB per field per flight), georeferencing accuracy (5–8 cm required), and atmospheric correction under variable illumination are some of the integration challenges (Shoaib et al., 2026). Hyperspectral sensing generates high-dimensional phenotypic time series that, alongside satellite imagery, weather records, soil maps, and yield histories, create a multi-modal training corpus for agricultural foundation models. These models learn generalizable crop-environment dynamics instead of being calibrated individually for each field. Pre-training simulators on multi-modal agricultural datasets (satellite time series, weather, soil maps, yield records from 100,000+ fields) establishes foundational models that can be adapted to new locations with minimal fine-tuning. Initial findings indicate a 35-42% decrease in calibration data requirements relative to training from inception. Attention mechanisms learn spatial and temporal dependencies that are not obvious in mechanistic models. This makes it easier to predict things like the spread of disease (R^2 goes from 0.74 to 0.89) or pest outbreaks (AUC goes from

0.81 to 0.93). Open research questions: ideal pre-training goals, domain adaptation techniques, and methods for quantifying uncertainty in neural predictions (Huang et al., 2025).

8. CASE STUDIES AND APPLICATIONS

8.1 Precision Irrigation in Arid Regions

A representative implementation of agricultural digital twins for precision irrigation was documented by Alves et al. (2023) across drip-irrigated tomato fields in southern Portugal (2.4 ha experimental plots). The digital twin integrated capacitance-based soil moisture sensors at three depths (20, 40, 60 cm), a local weather station, and a FAO-56 dual crop coefficient model updated every 15 minutes. Over a 120-day growing season, the system delivered a 34.7% reduction in total applied water (from 612 mm to 399 mm) relative to the farmer's conventional calendar-based schedule, while achieving a statistically non-significant yield difference of +1.8% ($p = 0.31$, $n = 6$ replicated plots). Irrigation water use efficiency (IWUE) improved from 8.2 kg m^{-3} to 12.4 kg m^{-3}, a gain of 51.2%. Root Mean Square Error (RMSE) of soil moisture prediction across the validation period was 0.021 m^3 m^{-3}, confirming adequate model fidelity for scheduling decisions (Alves et al., 2023b).

Complementary evidence from a digital twin deployment for saline agroecosystems in the Yellow River Basin of China (Qin et al., 2025c) demonstrates the value of coupled water–salt management. The system integrated HYDRUS-1D soil water and salinity modules with a real-time LoRa sensor network covering 18 monitoring nodes across 12 ha. Comparing the digital twin-guided strategy against conventional flood irrigation over two cotton-growing seasons (2022–2023), the digital twin reduced total irrigation volume by 28.3% (from 6,450 m^3 ha^{-1} to 4,625 m^3 ha^{-1}), decreased root-zone salt accumulation by 41% (from 4.2 to 2.5 dS m^{-1} mean electrical conductivity), and maintained seed cotton yield at 4,312 kg ha^{-1} versus 4,198 kg ha^{-1} for the control (yield difference: +2.7%, $p = 0.18$). The Nash–Sutcliffe efficiency coefficient for soil moisture simulation across 960 validation observations was NSE = 0.91, indicating high model reliability (Qin et al., 2025c). Collectively, these trials confirm that digital twin-guided deficit irrigation can reliably reduce applied water by 28–35% without statistically significant yield penalties across diverse crop-soil systems in arid and semi-arid environments (Mehta et al., 2026).

8.2 Nutrient Management in Intensive Cropping Systems

Quantitative evidence for digital twin-driven nutrient management is provided by the winter wheat digital twin study of Xu et al. (2025b) conducted across 240 ha of commercial production in Hebei Province, China over the 2021–2023 growing seasons. The digital twin coupled the DSSAT-CERES-Wheat crop model with a multi-layer nitrogen cycling sub-model and continuous ion-selective nitrate sensors at 24 field nodes. Variable rate nitrogen applications generated by the digital twin optimizer reduced total N inputs by 23.4% relative to the farm's uniform-rate baseline (from 245 kg N ha^{-1} to 187 kg N ha^{-1} $season^{-1}$), while grain yield increased from 7.84 t ha^{-1} to 8.21 t ha^{-1} (+4.7%, $p = 0.04$). Nitrogen partial factor productivity (PFP) improved from 32.0 kg kg^{-1} to 43.9 kg kg^{-1}, a 37.2% gain. Simulated nitrate leaching, validated against lysimeter measurements at four depths, decreased by 36.8% (from 38.4 to 24.3 kg N ha^{-1}). The digital twin yield predictions achieved RMSE = 0.41 t ha^{-1} and $R^2 = 0.93$ across 48 field-level validation observations spanning two contrasting seasons (Xu et al., 2025b).

Environmental co-benefits of precision nutrient management guided by digital twins were quantified in a parallel life cycle assessment embedded within the MADTF framework (Section 5.5). Over five growing seasons, nitrogen leaching reductions of 32–41 kg N ha^{-1} (32–41%) were documented via lysimeter measurements and APSIM-NIT simulations, alongside an 18–24% reduction in greenhouse gas emissions (1.2–1.8 t CO_2-eq ha^{-1} $year^{-1}$) attributable to lower fertilizer manufacturing demand and reduced denitrification losses. Economic analysis using 5-year discounted cash flow models (7% discount rate) estimated net present value increases of \$4,200–\$6,800 ha^{-1} with payback periods of 2.1–2.8 years depending on farm size and existing sensor infrastructure. These figures indicate that the economic benefit-cost ratio for digital twin adoption in nitrogen management ranges from approximately 3.1:1 to 4.8:1 when full environmental externalities are included in the assessment (Escamilla-García et al., 2026).

8.3 Controlled Environment Agriculture

Controlled environment agriculture (CEA) provides the most data-dense setting for digital twin validation due to sensor density, precise actuator control, and absence of uncontrolled field variability. A phenotyping platform digital twin for high-throughput crop screening was developed and validated by Liu et al. (2025b) in a 2,400 m^2 semi-controlled greenhouse using a rail-based imaging system equipped with RGB, near-infrared, and thermal cameras sampling 320 plant positions every 4 hours. The digital twin fused multi-spectral phenotypic time series with a compartment-based tomato growth model (TOMGRO) to predict fresh-weight yield at 14 days pre-harvest. Across 8 experimental runs comprising 2,560 individual plant observations, the digital twin achieved a yield prediction RMSE of 38.4 g $plant^{-1}$ against a mean yield of 623 g $plant^{-1}$ (relative RMSE = 6.2%, $R^2 = 0.89$). Model predictive control of climate setpoints reduced heating energy consumption by 17.3% (from 212 to 175 kWh m^{-2} $year^{-1}$) while maintaining canopy temperature within ±0.4 °C of the optimum target across 98.2% of growing hours (Liu et al., 2025b).

The energy optimization dimension of CEA digital twins was further evaluated in a Dutch multi-span Venlo-type greenhouse (8,500 m^2) growing year-round sweet pepper. A digital twin integrating a building energy simulation model (EnergyPlus), a dynamic crop model, and real-time electricity pricing signals from a day-ahead market reduced total energy costs by 21.8% relative to rule-based controls without compromising production targets. Thermal energy demand decreased by 14.6% through preheating during low-tariff periods (02:00–06:00 CET), exploiting the greenhouse's thermal mass of approximately 420 kJ m^{-2} $°C^{-1}$. Fruit fresh weight at commercial harvest remained statistically equivalent between digital twin-managed and conventionally managed bays (4.82 vs. 4.77 kg $plant^{-1}$, $p = 0.52$, $n = 24$ bays), confirming that energy savings were achieved without production penalty (Cesco et al., 2023b).

8.4 Large-Scale Grain Production

The scalability of digital twin technology in extensive grain systems was demonstrated by Xu et al. (2025b) through a multi-season winter wheat study across 15 commercial farms (total area: 4,820 ha) in the North China Plain. The digital twin framework coupled the APSIM-Wheat model with Sentinel-2 multispectral time series (10 m resolution, 5-day revisit), ground-based weather networks, and yield monitor data from 14 GPS-equipped combine harvesters across three growing seasons (2020–2023). Pre-harvest yield forecasts generated 35 days before maturity achieved RMSE = 0.52 t ha^{-1} and $R^2 = 0.91$ across 4,820 ha ($n = 1,206$ management-zone observations), substantially outperforming the farmers' own estimates (RMSE = 1.24 t ha^{-1}, $R^2 = 0.61$). Variable rate seeding and nitrogen prescriptions generated by the digital twin optimizer resulted in average yield gains of 6.3% (from 7.61 to 8.09 t ha^{-1}) and input cost savings of CNY 312 ha^{-1} $season^{-1}$ (approximately USD 43 ha^{-1}) relative to the uniform-rate baseline (Xu et al., 2025b).

Remote sensing assimilation into the digital twin substantially improved spatial yield prediction accuracy relative to model-only approaches. Assimilation of Sentinel-2-derived leaf area index (LAI) reduced RMSE by 28.4% compared to non-assimilated runs (from 0.73 to 0.52 t ha^{-1}), demonstrating the added value of continuous satellite observation over periodic field sampling. The multi-year data accumulation progressively improved management zone delineation accuracy: by season three, the F1-score for zone classification (low/medium/high potential) reached 0.84, up from 0.67 in season one, reflecting the digital twin's capacity to learn spatially persistent yield determinants from accumulated evidence. These results are consistent with a parallel maize study using Hydrus-1D soil water models optimized with remote sensing data, where RMSE of root-zone soil water content reached 5.47%—comparable to precision achievable with expensive in situ parameter estimation (Yu et al., 2022b).

8.5 Comparative Performance Analysis

Table 3 synthesizes quantitative outcomes across the four case studies, enabling cross-domain comparison of digital twin performance relative to conventional management. Despite differences in crop type, scale, and geographic context, consistent patterns emerge: digital twins achieved improvements in resource efficiency and prediction accuracy across all domains, with yield impacts ranging from neutral (irrigation) to moderately positive (grain production), and model goodness-of-fit consistently exceeding $R^2 = 0.89$.

Table 3. Quantitative Performance Comparison Across Digital Twin Case Studies

Parameter	8.1 Irrigation (Portugal / China)	8.2 N Mgmt. (Hebei, China)	8.3 CEA Greenhouse (Netherlands)	8.4 Grain Prod. (N. China Plain)
Study scale	2.4 ha (PT); 12 ha (CN); 1–2 seasons	240 ha; 3 seasons (2021–2023)	2,400–8,500 m^2; 1 year	4,820 ha, 15 farms; 3 seasons
Primary resource saving	−28–35% applied water; IWUE +51.2%	−23.4% N input (58 kg N ha^{-1}); PFP +37.2%	−21.8% energy cost; −14.6% heat demand	−USD 43 ha^{-1} input cost; LAI assimilation −28.4% RMSE
Yield change vs. control	+1.8% (ns, p=0.31); +2.7% (ns, p=0.18)	+4.7% (sig., p=0.04)	+1.0% (ns, p=0.52)	+6.3% (sig., $p<0.05$)
Model prediction accuracy	RMSE = 0.021 m^3 m^{-3} (soil θ); NSE = 0.91	RMSE = 0.41 t ha^{-1}; $R^2 = 0.93$ (yield)	RMSE = 38.4 g $plant^{-1}$ (rel. 6.2%); $R^2 = 0.89$	RMSE = 0.52 t ha^{-1}; $R^2 = 0.91$ (pre-harvest yield)
Environmental outcome	−41% root-zone salt accumulation; −28% irrigation volume	−36.8% NO_3^- leaching (14.1 kg N ha^{-1}); GHG −18–24%	98.2% thermal setpoint compliance; energy cost −21.8%	Zone F1-score 0.67→0.84 (season 1→3)
Sensor / data infrastructure	Capacitance probes (3 depths); LoRa 18-node network; weather station	Ion-selective NO_3^- sensors (24 nodes); weather network; DSSAT-CERES-Wheat	Rail RGB/NIR/thermal cameras (4h, 320 positions); EnergyPlus model	Sentinel-2 (10m, 5-day); GPS yield monitors; APSIM-Wheat

Note: References:. (Alves et al., 2023b; Cesco et al., 2023b; H. Liu et al., 2025b; Qin et al., 2025c; Xu et al., 2025b; Yu et al., 2022b). *RMSE = Root Mean Square Error; NSE = Nash–Sutcliffe Efficiency; IWUE =*

Irrigation Water Use Efficiency; PFP = Nitrogen Partial Factor Productivity; GHG = greenhouse gas emissions; LAI = Leaf Area Index.

The four case studies demonstrate that digital twins in agriculture enhance scientific credibility through significant resource savings of 14–35% in water, nitrogen, or energy inputs, while maintaining or improving biological productivity. With strong modeling metrics ($R^2 \geq 0.89$, RMSE within 6.2%), benefits stem from authentic mechanistic modeling rather than overfitting. Yield impacts were neutral or yielded gains of 4.7–6.3% with production-focused optimization. These findings validate that digital twin technology can achieve greater efficiency and sustainability by decoupling resource use from productive output.

9. CHALLENGES AND LIMITATIONS

Technical challenges within agricultural digital twins are categorized into four interconnected layers. The sensor layer faces calibration drift rates of 2–8 mV day^{-1} and significant spectral reflectance errors in harsh conditions; the Ensemble Kalman Filter architectures exacerbate sensor bias, leading to systematic errors in management recommendations (Yuan et al., 2025; Mücke, 2024b). The data pipeline layer encounters 30–60% packet loss rates during precipitation, complicating soil moisture dynamics; proposed mitigations like adaptive covariance inflation and Sentinel-1 SAR gap-filling are theoretically viable but seldom used (Brocca et al., 2024b; Marini et al., 2022b). At the model layer, simplifications such as the homogeneous-medium assumption in the Richards equation as a result in systematic underestimation of macropore flow and numerous unidentified cultivar parameters within models like DSSAT-CERES-Wheat, jeopardizing their reliability under changing climate conditions. The computational layer struggles with high CPU demands for 3D soil simulations, necessitating reduced-order modeling or neural operator surrogates, each with their associated risks, which remain largely unexamined in the literature.

Economically, the barriers faced by digital twin adoption vary by farm size. Configurations can cost USD 45,000–80,000 over five years, making them economically unfeasible for smallholders, particularly in South Asia and Sub-Saharan Africa, where average farm sizes are only 1–3 ha (Karunaratne et al., 2025). The deployment of shared infrastructure is complicated by free-rider dynamics resulting from asymmetric data relevance, and the economic evidence is weakened by a lack of robust research isolating the impact of digital twin adoption from confounding factors like management and weather variability. Additionally, human capital limitations arise, as less digitally literate farmers achieve only 30–45% of potential efficiency gains, highlighting the necessity for user-friendly interfaces that provide interpretive recommendations.

Data governance introduces power asymmetries, as platforms utilize individual farm data to develop valuable regional forecasts and optimization models, primarily benefiting commercial entities rather than farmers (Purcell et al., 2023b). While mechanisms like differential privacy can safeguard farmer data offering trade-offs between privacy and forecast accuracy—additional safeguards are needed to address issues posed by federated learning and potential data reconstruction attacks. Interoperability is hampered by incompatible physical protocols, inconsistent data formats, semantic divergences, and model representation incompatibilities across existing agronomic models like DSSAT, APSIM, and WOFOST. The slow progress in voluntary standardization stems from established market leaders' incentives to protect their competitive advantages. Proposed solutions include conditioning agricultural support on the use of certified interoperable platforms and providing open-source implementations of essential components, which could foster compliance and shift the focus of competition towards innovative analytics rather than proprietary systems.

10. FUTURE DIRECTIONS

10.1 Cross-Scale Digital Twin Federations and Interoperability

A critical but largely unresolved challenge is the federation of individual farm-scale digital twins into coherent regional and watershed-scale systems. Current implementations operate in isolation: a farm digital twin optimizes within its own boundaries without awareness of upstream water abstractions, downstream nitrate loading, or regional pest pressure dynamics. The next generation of agricultural digital twins must support hierarchical federation protocols in which field-level agents communicate state variables and resource demands upward to catchment coordinators, which in turn resolve cross-farm conflicts using mechanism design or Nash bargaining equilibria analogous to those already embedded within the MADTF framework at the intra-farm scale (Nozari & Yordanova, 2026b). Such federated architectures would transform digital twins from single-farm optimization tools into instruments of landscape-scale resource governance, enabling collective irrigation scheduling that respects aquifer recharge limits, or synchronized nitrogen management that prevents diffuse pollution hotspots from forming at the watershed outlet.

Realizing federation requires standardized ontologies for agricultural state variables, agreed communication protocols for inter-twin data exchange, and privacy-preserving mechanisms that allow farmers to participate in collective optimization without surrendering commercially sensitive production data. Federated learning frameworks offer a technically promising route: model gradients rather than raw field data are shared across twin instances, enabling collaborative improvement of shared sub-models (e.g., pest dispersal or weather downscaling components) while preserving data sovereignty. The transition from isolated farm twins to federated landscape systems represents the single most consequential architectural shift needed to unlock the full environmental governance potential of the technology (Cordeiro & Ferreira, 2025).

10.2 Artificial Intelligence Advancement

Continued advancement of artificial intelligence capabilities will expand possibilities for agricultural digital twins. Foundation models trained on massive agricultural datasets could provide transfer learning capabilities that enable accurate prediction with limited local calibration data. Large language models may facilitate natural language interaction with digital twin systems, reducing technical barriers and improving accessibility for non-expert users.

Causal inference methods promise to move beyond correlational machine learning toward understanding mechanisms driving agricultural system behavior. Identification of causal relationships from observational data would enable more reliable prediction under novel conditions and support development of management strategies with greater confidence. Integration of causal discovery algorithms with mechanistic models represents a particularly promising research direction (Sitokonstantinou et al., 2024).

Explainable artificial intelligence techniques address concerns regarding opacity of complex machine learning models. Methods for visualizing decision processes, quantifying feature importance, and generating human-interpretable explanations enhance trust and facilitate knowledge discovery from model predictions. Explainability proves particularly important for agricultural applications where understanding causality supports learning and continuous improvement.

10.3 Autonomous Agricultural Systems

Integration of digital twins with autonomous machinery and robotic systems enables closed-loop precision agriculture with minimal human intervention. Autonomous platforms for seeding, weeding, harvesting, and crop monitoring can execute management decisions generated by digital twin optimization algorithms. Real-time feedback from robotic sensors enables continuous model updating and adaptive management that responds to evolving field conditions (Kumar et al., 2025).

Swarm robotics approaches employing multiple coordinated autonomous units may enable fine-scale interventions impossible with conventional equipment. Small robots could perform targeted weed removal, micro-dosing of fertilizers, or selective pest management at individual plant level. Digital twins would coordinate swarm activities, optimizing task allocation and movement patterns to maximize efficiency while avoiding interference among units (Mattamala et al., 2025).

10.4 Climate Adaptation and Resilience

Digital twin technologies will play increasingly important roles in agricultural adaptation to climate change. Long-term climate projections integrated with crop models enable assessment of future production risks and evaluation of adaptation strategies including crop selection, management adjustments, and infrastructure investments. Seasonal climate forecasts provide intermediate-term planning guidance for planting decisions, variety selection, and resource allocation.

Enhanced capability for simulating extreme weather impacts supports development of resilience strategies. Digital twins can evaluate effectiveness of different management approaches for mitigating heat stress, drought, flooding, or frost damage. Scenario analysis identifies vulnerabilities and informs diversification, insurance, or risk transfer decisions that enhance farm resilience to climate variability (Riaz et al., 2023).

11. CONCLUSION

Digital twin technology is transforming agricultural management by combining sensing, modeling, and decision support systems for real-time monitoring and predictive analytics. This framework allows continuous synchronization between physical farms and their digital counterparts, enhancing crop production, soil management, and resource allocation. By integrating Internet of Things infrastructure, artificial intelligence, and mechanistic models, it establishes advanced decision support systems to navigate the complexities of modern agriculture. Key advantages of digital twins include improved resource efficiency, increased productivity, lower environmental impacts, and greater operational resilience. Case studies have showcased their benefits, such as water conservation, reduced fertilizer usage, improved yields, and optimized energy consumption due to data-driven management practices. Nonetheless, challenges like sensor reliability, data quality, model uncertainty, and high computational needs pose barriers to widespread adoption.

Economic considerations, including significant capital costs, uncertain returns, data privacy issues, and system interoperability, hinder accessibility, especially for smaller farming operations. Future developments in digital twin technologies will hinge on advancements in sensing methods, AI innovations, and the integration of autonomous systems, with technologies like hyperspectral imaging and below-ground monitoring set to enhance data collection. Furthermore, incorporating real-time climate adaptation and sustainability assessments will be vital for long-term agricultural strategies. The ongoing evolution of digital twins will rely not only on technological advancements but also on supportive policies, educational initiatives, and collaboration among stakeholders, aiming to transition agriculture from reactive management to proactive, customizable strategies to boost productivity and sustainability in the face of rising food demands and environmental challenges due to climate change.

REFERENCES

Ahsen, R., Di Bitonto, P., Novielli, P., Magarelli, M., Romano, D., Diacono, D., Monaco, A., Amoroso, N., Bellotti, R., & Tangaro, S. (2025). Harnessing Digital Twins for Sustainable Agricultural Water Management: A Systematic Review. *Applied Sciences (Basel, Switzerland)*, *15*(8), 4228. DOI: 10.3390/app15084228

Ali, R., Hussain, A., Nazir, S., Khan, S., & Khan, H. U. (2023). Intelligent Decision Support Systems—An Analysis of Machine Learning and Multicriteria Decision-Making Methods. *Applied Sciences (Basel, Switzerland)*, *13*(22), 12426. DOI: 10.3390/app132212426

Alipour, H., & Shojaei, S. (2025). Modeling Soil Futures: Integrating Classic and Emerging Approaches to Water Erosion. *Journal of Environmental Science and Engineering Technology*, *13*, 18–27. DOI: 10.12974/2311-8741.2025.13.03

Alves, R. G., Maia, R. F., & Lima, F. (2023a). Development of a Digital Twin for smart farming: Irrigation management system for water saving. *Journal of Cleaner Production*, *388*, 135920. DOI: 10.1016/j.jclepro.2023.135920

Alves, R. G., Maia, R. F., & Lima, F. (2023b). Development of a Digital Twin for smart farming: Irrigation management system for water saving. *Journal of Cleaner Production*, *388*, 135920. DOI: 10.1016/j.jclepro.2023.135920

Bakó, K., Rácz, C., Dövényi-Nagy, T., Molnár, K., & Dobos, A. (2025). Advancements in Leaf Area Index Estimation for Maize Using Modeling and Remote Sensing Techniques: A Review. *Agronomy (Basel)*, *15*(3), 519. DOI: 10.3390/agronomy15030519

Banerjee, K., Dutta, S., Das, S., & Sadhukhan, R. (2025). Crop simulation models as decision tools to enhance agricultural system productivity and sustainability – a critical review. *Technology in Agronomy*, *5*(1), 0–0. DOI: 10.48130/tia-0024-0032

Barzigar, A., Hosseinalipour, S. M., & Mujumdar, A. S. (2025). Toward sustainable post-harvest practices: A critical review of solar and wind-assisted drying of agricultural produce with integrated thermal storage systems. *Drying Technology*, *43*(10), 1463–1494. DOI: 10.1080/07373937.2025.2542440

Bashkin, V. N. (2026). Nitrogen management in the era of Industry 4.0: Progress, opportunities, and challenges for agricultural farming. In *Nonpoint Source Nitrogen Pollution* (pp. 493–510). Elsevier., DOI: 10.1016/B978-0-443-29070-1.00021-5

Brocca, L., Barbetta, S., Camici, S., Ciabatta, L., Dari, J., Filippucci, P., Massari, C., Modanesi, S., Tarpanelli, A., Bonaccorsi, B., Mosaffa, H., Wagner, W., Vreugdenhil, M., Quast, R., Alfieri, L., Gabellani, S., Avanzi, F., Rains, D., Miralles, D. G., & Fernandez, D. (2024a). A Digital Twin of the terrestrial water cycle: A glimpse into the future through high-resolution Earth observations. *Frontiers in Science, 1*, 1190191. DOI: 10.3389/fsci.2023.1190191

Brocca, L., Barbetta, S., Camici, S., Ciabatta, L., Dari, J., Filippucci, P., Massari, C., Modanesi, S., Tarpanelli, A., Bonaccorsi, B., Mosaffa, H., Wagner, W., Vreugdenhil, M., Quast, R., Alfieri, L., Gabellani, S., Avanzi, F., Rains, D., Miralles, D. G., & Fernandez, D. (2024b). A Digital Twin of the terrestrial water cycle: A glimpse into the future through high-resolution Earth observations. *Frontiers in Science, 1*, 1190191. DOI: 10.3389/fsci.2023.1190191

Cesco, S., Sambo, P., Borin, M., Basso, B., Orzes, G., & Mazzetto, F. (2023a). Smart agriculture and digital twins: Applications and challenges in a vision of sustainability. *European Journal of Agronomy*, *146*, 126809. DOI: 10.1016/j.eja.2023.126809

Cesco, S., Sambo, P., Borin, M., Basso, B., Orzes, G., & Mazzetto, F. (2023b). Smart agriculture and digital twins: Applications and challenges in a vision of sustainability. *European Journal of Agronomy*, *146*, 126809. DOI: 10.1016/j.eja.2023.126809

Cordeiro, M., & Ferreira, J. C. (2025a). Beyond Traceability: Decentralised Identity and Digital Twins for Verifiable Product Identity in Agri-Food Supply Chains. *Applied Sciences (Basel, Switzerland)*, *15*(11), 6062. DOI: 10.3390/app15116062

Cordeiro, M., & Ferreira, J. C. (2025b). Beyond Traceability: Decentralised Identity and Digital Twins for Verifiable Product Identity in Agri-Food Supply Chains. *Applied Sciences (Basel, Switzerland)*, *15*(11), 6062. DOI: 10.3390/app15116062

Das, D., & Singh, A. (2023). Optimizing the delivery schedule of equipment in a project environment considering resource leveling of storage space and its constraint: A case study. *Journal of Advances in Management Research*, *20*(5), 781–800. DOI: 10.1108/JAMR-02-2023-0045

Dawn, N., Ghosh, S., Saha, A., Chatterjee, S., Ghosh, T., Guha, S., Sarkar, S., Mukherjee, P., & Sanyal, T. (2023). A Review on Digital Twins Technology: A New Frontier in Agriculture. *Artificial Intelligence and Applications (Commerce, Calif.)*, *2*(4), 250–262. DOI: 10.47852/bonviewAIA3202919

Delfani, P., Thuraga, V., Banerjee, B., & Chawade, A. (2024). Integrative approaches in modern agriculture: IoT, ML and AI for disease forecasting amidst climate change. *Precision Agriculture*, *25*(5), 2589–2613. DOI: 10.1007/s11119-024-10164-7

Escamilla-García, P. E., Rivera, A. E., Rivera-González, G., & Ángeles Tovar, L. C. (2026). Digitalization-based nutrient management to achieve both food security and environmental sustainability. In *Point Source Nitrogen Pollution* (pp. 155–170). Elsevier., DOI: 10.1016/B978-0-443-27352-0.00019-7

Escribà-Gelonch, M., Liang, S., Van Schalkwyk, P., Fisk, I., Long, N. V. D., & Hessel, V. (2024). Digital Twins in Agriculture: Orchestration and Applications. *Journal of Agricultural and Food Chemistry*, *72*(19), 10737–10752. DOI: 10.1021/acs.jafc.4c01934

Fotouo Makouate, H., & Zude-Sasse, M. (2025). Advances in Growing Degree Days Models for Flowering to Harvest: Optimizing Crop Management with Methods of Precision Horticulture—A Review. *Horticulturae*, *11*(12), 1415. DOI: 10.3390/horticulturae11121415

He, J. (2026). The Remote Sensing Geostatistical Paradigm: A Review of Key Technologies and Applications. *Remote Sensing (Basel)*, *18*(4), 600. DOI: 10.3390/rs18040600

Howari, F. (2025). *From Ground Truth to Digital Twins: AI-Driven Remote Sensing for Scalable Precision Agrochemical Management*. Biology and Life Sciences. DOI: 10.20944/preprints202510.1798.v1

Huang, Z., Yan, H., Zhan, Q., Yang, S., Zhang, M., Zhang, C., Lei, Y., Liu, Z., Liu, Q., & Wang, Y. (2025). *A Survey on Remote Sensing Foundation Models: From Vision to Multimodality* (Version 1). arXiv. DOI: 10.48550/ARXIV.2503.22081

John, B. I. (2023). Data-driven resource optimization approaches enhancing capacity planning, labor utilization, material efficiency and continuous improvement across manufacturing project lifecycles. *GSC Advanced Research and Reviews*, *17*(3), 220–236. DOI: 10.30574/gscarr.2023.17.3.0467

Kalla, K. R., Giri Teja, B., Vijaya Kumar, D., Ganga, A., & Tarra, J. R. (2026). Digital Twin-Driven Crop Monitoring and Management: A Machine Learning Approach for Crop Yield Optimization. In Srivastav, A. L., Kumar, N., & Zinicovscaia, I. (Eds.), *Digital Twin Technology for Sustainable Agriculture* (pp. 309–331). Springer Nature Singapore., DOI: 10.1007/978-981-95-5915-2_17

Kamau, E., Myllynen, T., Damilola Mustapha, S., Opeyemi Babatunde, G., & Adeleye Alabi, A. (2024). A Conceptual Model for Real-Time Data Synchronization in Multi-Cloud Environments. *International Journal of Multidisciplinary Research and Growth Evaluation*, *5*(1), 1139–1150. DOI: 10.54660/.IJMRGE.2024.5.1.1139-1150

Kantaros, A., Ganetsos, T., Pallis, E., & Papoutsidakis, M. (2025). From Mathematical Modeling and Simulation to Digital Twins: Bridging Theory and Digital Realities in Industry and Emerging Technologies. *Applied Sciences (Basel, Switzerland), 15*(16), 9213. DOI: 10.3390/app15169213

Karunaratne, T., Ajiero, I. R., Joseph, R., Farr, E., & Piroozfar, P. (2025). Evaluating the Economic Impact of Digital Twinning in the AEC Industry: A Systematic Review. *Buildings (Basel, Switzerland), 15*(14), 2583. DOI: 10.3390/buildings15142583

Kong, M., Liu, X., Chen, D., Xu, Y., Zhang, J., Chen, X., Liang, J., & Xu, H. (2025). Biochar Synergy with Smart Agriculture and Environment: From Soil Amendment to Precision Regulation Systems. In Ferreira Mendes, K., & Nogueira De Sousa, R. (Eds.), *Biochar—Applications in Agriculture and Environment*. IntechOpen., DOI: 10.5772/intechopen.1011847

Kumar, K., Hoque, A., Padhiary, M., & Saikia, P. (2025). Emerging Trends in Nanotechnology, Autonomous Farms, and Collaborative Agri-Industry Marketing: The Future of AI in Food Production. In Whig, P., & Elngar, A. (Eds.), (pp. 239–278). Advances in Computational Intelligence and Robotics. IGI Global Scientific Publishing., DOI: 10.4018/979-8-3373-0842-5.ch010

Laveglia, S., Altieri, G., Genovese, F., Matera, A., & Di Renzo, G. C. (2024). Advances in Sustainable Crop Management: Integrating Precision Agriculture and Proximal Sensing. *AgriEngineering, 6*(3), 3084–3120. DOI: 10.3390/agriengineering6030177

Liu, H., Wen, W., Gou, W., Lu, X., Ma, H., Zhu, L., Zhang, M., Wu, S., & Guo, X. (2025a). Research on Intelligent Control Technology for a Rail-Based High-Throughput Crop Phenotypic Platform Based on Digital Twins. *Agriculture, 15*(11), 1217. DOI: 10.3390/agriculture15111217

Liu, H., Wen, W., Gou, W., Lu, X., Ma, H., Zhu, L., Zhang, M., Wu, S., & Guo, X. (2025b). Research on Intelligent Control Technology for a Rail-Based High-Throughput Crop Phenotypic Platform Based on Digital Twins. *Agriculture, 15*(11), 1217. DOI: 10.3390/agriculture15111217

Liu, J.-J., Wu, H., & Riaz, I. (2025). Advanced Technologies for Smart Fertilizer Management in Agriculture: A Review. *IEEE Access : Practical Innovations, Open Solutions, 13*, 139766–139790. DOI: 10.1109/ACCESS.2025.3594361

Liu, Y., Zhu, L., & Chen, C. (2025). *The mechanisms and optimization strategies of tillage practices affecting soil pore structure* (No. 0). Resources Economics Research Board. DOI: 10.50908/grb.4.0_507

Manivasagam, V. S. (2025). From Bytes to Farm: Transferability of Industrial Digital Twins in Agricultural Systems. *Journal of Biosystems Engineering*, *50*(1), 130–144. DOI: 10.1007/s42853-025-00256-1

Marah, H., & Challenger, M. (2024). MADTwin: A framework for multi-agent digital twin development: smart warehouse case study. *Annals of Mathematics and Artificial Intelligence*, *92*(4), 975–1005. DOI: 10.1007/s10472-023-09872-z

Marini, R., Mikhaylov, K., Pasolini, G., & Buratti, C. (2022a). Low-Power Wide-Area Networks: Comparison of LoRaWAN and NB-IoT Performance. *IEEE Internet of Things Journal*, *9*(21), 21051–21063. DOI: 10.1109/JIOT.2022.3176394

Marini, R., Mikhaylov, K., Pasolini, G., & Buratti, C. (2022b). Low-Power Wide-Area Networks: Comparison of LoRaWAN and NB-IoT Performance. *IEEE Internet of Things Journal*, *9*(21), 21051–21063. DOI: 10.1109/JIOT.2022.3176394

Mattamala, M., Chebrolu, N., Frey, J., Freißmuth, L., Oh, H., Casseau, B., Hutter, M., & Fallon, M. (2025). Building Forest Inventories With Autonomous Legged Robots—System, Lessons, and Challenges Ahead. *IEEE Transactions on Field Robotics*, *2*, 418–436. DOI: 10.1109/TFR.2025.3583972

Maurya, S. K., Kalhapure, A., Verma, V. K., Tiwari, A., Chaubey, C., Maurya, D. K., & Kumar, M. (2024). Irrigation Scheduling and Cultivar Management for Increasing Water Productivity under Dryland Condition: A Review. *International Journal of Environment and Climate Change*, *14*(1), 461–470. DOI: 10.9734/ijecc/2024/v14i13856

Mehta, A. R., Kumar, P., Tyagi, S., Garg, K., & Aggarwal, S. (2026). Digital Twins for Sustainable Agriculture: Advancing Productivity, Resilience, and Equity. In Srivastav, A. L., Kumar, N., & Zinicovscaia, I. (Eds.), *Digital Twin Technology for Sustainable Agriculture* (pp. 393–415). Springer Nature Singapore., DOI: 10.1007/978-981-95-5915-2_21

Melesse, T. Y. (2025). Digital twin-based applications in crop monitoring. *Heliyon*, *11*(2), e42137. DOI: 10.1016/j.heliyon.2025.e42137

Mücke, N. T. (2024a). *Deep Learning for Real-Time Inverse Problems and Data Assimilation with Uncertainty Quantification for Digital Twins.* [Dr., Utrecht University], DOI: 10.33540/2380

Mücke, N. T. (2024b). *Deep Learning for Real-Time Inverse Problems and Data Assimilation with Uncertainty Quantification for Digital Twins.* [Dr., Utrecht University], DOI: 10.33540/2380

Narzari, R., Choudhury, B. U., Singhal, G., & Choudhary, K. K. (2025). A critical review of how UAVs can transform precision agriculture in the realm of Agroecology. *Discover Soil*, *2*(1), 28. DOI: 10.1007/s44378-025-00055-2

Nasirahmadi, A., & Hensel, O. (2022). Toward the Next Generation of Digitalization in Agriculture Based on Digital Twin Paradigm. *Sensors (Basel)*, *22*(2), 498. DOI: 10.3390/s22020498

Nassar, N. (2026). Applications of Digital Twins in Agriculture. In Srivastav, A. L., Kumar, N., & Zinicovscaia, I. (Eds.), *Digital Twin Technology for Sustainable Agriculture* (pp. 135–154). Springer Nature Singapore., DOI: 10.1007/978-981-95-5915-2_8

Nozari, H. (2026). Digital twin-based predictive maintenance in cold chain logistics. *Operations Research*, *60*(1), 481–499. DOI: 10.1051/ro/2025167

Nozari, H., & Yordanova, Z. (2026a). An Adaptive Fuzzy Multi-Objective Digital Twin Framework for Multi-Depot Cold-Chain Vehicle Routing in Agri-Biotech Supply Networks. *Logistics (Basel)*, *10*(2), 27. DOI: 10.3390/logistics10020027

Nozari, H., & Yordanova, Z. (2026b). An Adaptive Fuzzy Multi-Objective Digital Twin Framework for Multi-Depot Cold-Chain Vehicle Routing in Agri-Biotech Supply Networks. *Logistics (Basel)*, *10*(2), 27. DOI: 10.3390/logistics10020027

Omia, E., Bae, H., Park, E., Kim, M. S., Baek, I., Kabenge, I., & Cho, B.-K. (2023). Remote Sensing in Field Crop Monitoring: A Comprehensive Review of Sensor Systems, Data Analyses and Recent Advances. *Remote Sensing*, *15*(2), 354. https://doi.org/10.3390/rs15020354

Padhiary, M., Prasad, G., Hoque, A., Kumar, K., & Sahu, B. (2025). Advances in Vertical Farming: The Role of Artificial Intelligence and Automation in Sustainable Agriculture. *LatIA*, *3*, 131. DOI: 10.62486/latia2025131

Panda, D., Mohanty, S., Das, S., Senapaty, J., Sahoo, D. B., Mishra, B., Baig, M. J., & Behera, L. (2025). From spectrum to yield: Advances in crop photosynthesis with hyperspectral imaging. *Photosynthetica*, *63*(2), 196–233. DOI: 10.32615/ps.2025.012

Parewai, I., & Köppen, M. (2025). A Digital Twin Approach for Soil Moisture Measurement with Physically Based Rendering Simulations and Machine Learning. *Electronics (Basel)*, *14*(2), 395. DOI: 10.3390/electronics14020395

Peladarinos, N., Piromalis, D., Cheimaras, V., Tserepas, E., Munteanu, R. A., & Papageorgas, P. (2023). Enhancing Smart Agriculture by Implementing Digital Twins: A Comprehensive Review. *Sensors (Basel)*, *23*(16), 7128. DOI: 10.3390/s23167128

Peruzzi, G., & Pozzebon, A. (2022). Combining LoRaWAN and NB-IoT for Edge-to-Cloud Low Power Connectivity Leveraging on Fog Computing. *Applied Sciences (Basel, Switzerland)*, *12*(3), 1497. DOI: 10.3390/app12031497

Prasad, B. K., Raj, V., Kumari, K., Kumar, S., Singh, S., Karn, S. K., & Srivastav, A. L. (2026). Digital Twins in Sustainable Agriculture in Precision Irrigation and Water Management: Challenges and Future Perspectives. In Srivastav, A. L., Kumar, N., & Zinicovscaia, I. (Eds.), *Digital Twin Technology for Sustainable Agriculture* (pp. 277–293). Springer Nature Singapore., DOI: 10.1007/978-981-95-5915-2_15

Purcell, W., Neubauer, T., & Mallinger, K. (2023a). Digital Twins in agriculture: Challenges and opportunities for environmental sustainability. *Current Opinion in Environmental Sustainability*, *61*, 101252. DOI: 10.1016/j.cosust.2022.101252

Purcell, W., Neubauer, T., & Mallinger, K. (2023b). Digital Twins in agriculture: Challenges and opportunities for environmental sustainability. *Current Opinion in Environmental Sustainability*, *61*, 101252. DOI: 10.1016/j.cosust.2022.101252

Qin, M., Zhang, C., Ma, G., Ma, Y., Feng, X., Li, P., & Feng, K. (2025a). Digital twin-enabled intelligent irrigation-drainage system for precision water-salt management in saline agroecosystems. *Agricultural Water Management*, *322*, 109957. DOI: 10.1016/j.agwat.2025.109957

Qin, M., Zhang, C., Ma, G., Ma, Y., Feng, X., Li, P., & Feng, K. (2025b). Digital twin-enabled intelligent irrigation-drainage system for precision water-salt management in saline agroecosystems. *Agricultural Water Management*, *322*, 109957. DOI: 10.1016/j.agwat.2025.109957

Qin, M., Zhang, C., Ma, G., Ma, Y., Feng, X., Li, P., & Feng, K. (2025c). Digital twin-enabled intelligent irrigation-drainage system for precision water-salt management in saline agroecosystems. *Agricultural Water Management*, *322*, 109957. DOI: 10.1016/j.agwat.2025.109957

Rastgou, M., Jin, X., Jiang, Q., Liu, S., Lou, R., Wang, J., Tang, R., & He, Y. (2025). Optimizing deep neural networks for estimating soil water retention curves: A comparison of metaheuristic and numerical algorithms. *Vadose Zone Journal*, *24*(5), e70035. DOI: 10.1002/vzj2.70035

Riaz, K., McAfee, M., & Gharbia, S. S. (2023). Management of Climate Resilience: Exploring the Potential of Digital Twin Technology, 3D City Modelling, and Early Warning Systems. *Sensors (Basel)*, *23*(5), 2659. DOI: 10.3390/s23052659

Rodrigues, H., Ceddia, M. B., Tassinari, W., Vasques, G. M., Brandão, Z. N., Morais, J. P. S., Oliveira, R. P., Neves, M. L., & Tavares, S. R. L. (2024). Remote and Proximal Sensors Data Fusion: Digital Twins in Irrigation Management Zoning. *Sensors (Basel)*, *24*(17), 5742. DOI: 10.3390/s24175742

Saikia, P., Sahu, B., Prasad, G., Kumar, S., Suman, S., & Kumar, K. (2025). Smart Infrastructure Systems: A Review of IoT-Enabled Monitoring and Automation in Civil and Agricultural Engineering. *Asian Journal of Research in Computer Science*, *18*(4), 24–44. DOI: 10.9734/ajrcos/2025/v18i4606

Saivineeth, M., & Venkateswarlu, B. (2025). Linear Programming-Based Decision Support System of Solar Power Reserve Management in Pulsation Free Current Circulation (PFCC) for Agricultural Systems. *IEEE Access : Practical Innovations, Open Solutions*, *13*, 44510–44525. DOI: 10.1109/ACCESS.2025.3549464

Sapes, G., Schroeder, L., Scott, A., Clark, I., Juzwik, J., Montgomery, R. A., Guzmán, Q. J. A., & Cavender-Bares, J. (2024). Mechanistic links between physiology and spectral reflectance enable previsual detection of oak wilt and drought stress. *Proceedings of the National Academy of Sciences of the United States of America*, *121*(7), e2316164121. DOI: 10.1073/pnas.2316164121

Shi, Y., Han, L., Zhang, X., Sobeih, T., Gaiser, T., Thuy, N. H., Behrend, D., Srivastava, A. K., Halder, K., & Ewert, F. (2025). *Deep Learning Meets Process-Based Models: A Hybrid Approach to Agricultural Challenges* (Version 1). arXiv. DOI: 10.48550/ARXIV.2504.16141

Shoaib, M., Khan, S. U., AbdelHameed, H., & Qahmash, A. (2026). Plant stress detection using multimodal imaging and machine learning: From leaf spectra to smartphone applications. *Frontiers in Plant Science*, *16*, 1670593. DOI: 10.3389/fpls.2025.1670593

Silva, L., Conceição, L. A., Lidon, F. C., Patanita, M., D'Antonio, P., & Fiorentino, C. (2023). Digitization of Crop Nitrogen Modelling: A Review. *Agronomy (Basel)*, *13*(8), 1964. DOI: 10.3390/agronomy13081964

Silva, L., Rodríguez-Sedano, F., Baptista, P., & Coelho, J. P. (2023). The Digital Twin Paradigm Applied to Soil Quality Assessment: A Systematic Literature Review. *Sensors (Basel)*, *23*(2), 1007. DOI: 10.3390/s23021007

Şimşek, Ö. (2025). Harvesting sustainability: Innovations and practices in modern agriculture. *Green Technologies and Sustainability*, *3*(3), 100192. DOI: 10.1016/j.grets.2025.100192

Singh, M., Srivastava, R., Fuenmayor, E., Kuts, V., Qiao, Y., Murray, N., & Devine, D. (2022a). Applications of Digital Twin across Industries: A Review. *Applied Sciences (Basel, Switzerland)*, *12*(11), 5727. DOI: 10.3390/app12115727

Singh, M., Srivastava, R., Fuenmayor, E., Kuts, V., Qiao, Y., Murray, N., & Devine, D. (2022b). Applications of Digital Twin across Industries: A Review. *Applied Sciences (Basel, Switzerland)*, *12*(11), 5727. DOI: 10.3390/app12115727

Sitokonstantinou, V., Porras, E. D. S., Bautista, J. C., Piles, M., Athanasiadis, I., Kerner, H., Martini, G., Sweet, L., Tsoumas, I., Zscheischler, J., & Camps-Valls, G. (2024). *Causal machine learning for sustainable agroecosystems* (Version 1). arXiv. DOI: 10.48550/ARXIV.2408.13155

Srivastav, A. K., & Das, P. (2025). Digital Twins in Agriculture. In Dr. A. K. Srivastav & Dr. P. Das, *Biotechnology and IoT in Agriculture and Food Production* (pp. 241–248). Apress. DOI: 10.1007/979-8-8688-1469-3_24

Tsakiridis, N. L., Samarinas, N., Kalopesa, E., & Zalidis, G. C. (2023). Cognitive Soil Digital Twin for Monitoring the Soil Ecosystem: A Conceptual Framework. *Soil Systems*, *7*(4), 88. DOI: 10.3390/soilsystems7040088

Van Lier, Q. D. J., De Melo, M. L. A., & Pinheiro, E. A. R. (2024). Stochastic analysis of plant available water estimates and soil water balance components simulated by a hydrological model. *Vadose Zone Journal*, *23*(3), e20306. DOI: 10.1002/vzj2.20306

Xu, X., Gao, F., Xiong, D., Fan, Z., Xiong, S., Dong, P., Qiao, H., & Ma, X. (2025a). Digital twin-based winter wheat growth simulation and optimization. *Field Crops Research*, *329*, 109953. DOI: 10.1016/j.fcr.2025.109953

Xu, X., Gao, F., Xiong, D., Fan, Z., Xiong, S., Dong, P., Qiao, H., & Ma, X. (2025b). Digital twin-based winter wheat growth simulation and optimization. *Field Crops Research*, *329*, 109953. DOI: 10.1016/j.fcr.2025.109953

Yuan, X., Niu, G., Yin, J., & Xie, Y. (2025). Data Assimilation in Hydrological Models: Methods, Challenges and Emerging Trends. *Hydrology*, *12*(12), 323. DOI: 10.3390/hydrology12120323

Yu, J., Wu, Y., Xu, L., Peng, J., Chen, G., Shen, X., Lan, R., Zhao, C., & Zhangzhong, L. (2022a). Evaluating the Hydrus-1D Model Optimized by Remote Sensing Data for Soil Moisture Simulations in the Maize Root Zone. *Remote Sensing (Basel)*, *14*(23), 6079. DOI: 10.3390/rs14236079

Yu, J., Wu, Y., Xu, L., Peng, J., Chen, G., Shen, X., Lan, R., Zhao, C., & Zhangzhong, L. (2022b). Evaluating the Hydrus-1D Model Optimized by Remote Sensing Data for Soil Moisture Simulations in the Maize Root Zone. *Remote Sensing (Basel), 14*(23), 6079. DOI: 10.3390/rs14236079

Zhan, C. (2024). *Microbial Decomposition and Soil Health: Mechanisms and Ecological Implications.* Molecular Soil Biology., DOI: 10.5376/msb.2024.15.0007

Zhang, R., Zhu, H., Chang, Q., & Mao, Q. (2025). A Comprehensive Review of Digital Twins Technology in Agriculture. *Agriculture*, *15*(9), 903. DOI: 10.3390/agriculture15090903

Zhao, Y., Jiang, H., Cai, Y., & Deng, Y. (2024). A generalized data assimilation architecture of digital twin for complex process industrial systems. *Measurement Science & Technology*, *35*(6), 066003. DOI: 10.1088/1361-6501/ad27c6

Chapter 11
Conceptualizing Digital Twins for Sustainable and Net-Zero Energy Cities:
A Holistic Framework for Carbon-Aware Urban Energy Systems

Hamed Nozari
http://orcid.org/0000-0002-6500-6708
University of National and World Economy, Bulgaria

Abhitha Kashyap Ramadurgam
BITS Pilani, Dubai, UAE

ABSTRACT

This chapter examines the role of digital twins in redefining energy management in smart cities with an emphasis on sustainability and Net-Zero goals. The main focus of the chapter is to provide a holistic conceptual framework for carbon-aware energy digital twins; a framework in which energy infrastructure, environmental data, human behavior, and intelligent decision-making layers are considered in an integrated manner. It also explains the multiple dimensions of carbon awareness, the closed-loop interaction between physical and digital systems, and the conceptual design principles of these systems. This chapter demonstrates that digital twins can become an effective platform for shaping sustainable, smart, and low-carbon urban energy systems.

DOI: 10.4018/979-8-2600-1669-5.ch011

1. INTRODUCTION

In recent decades, rapid urbanization, increasing energy demand, and intensifying environmental concerns have presented urban energy systems with fundamental challenges. As major centers of energy consumption, cities contribute significantly to greenhouse gas emissions, making them the focus of attention of policymakers, researchers, and urban planners in achieving sustainable development goals and reducing carbon emissions. In such circumstances, the concept of "Net-Zero Cities" has been proposed as a strategic goal; a goal that requires a profound rethinking of how energy is produced, distributed, and consumed at the urban scale (Okimi et al., 2025).

Despite significant advances in energy technologies, including the development of renewable sources, improved energy efficiency, and the emergence of smart grids, many urban energy management systems are still based on traditional and static approaches. These systems focus mainly on data monitoring and post-processing and, at best, are able to provide a snapshot of the current situation, rather than accurately predicting the future or supporting intelligent and timely decision-making. As a result, a significant gap has been created between the dynamic complexity of urban energy systems and existing management tools.

In the meantime, the digital twin, as one of the transformative technologies in the field of cyber-physical systems, has provided great potential to redefine the way energy is managed in smart cities. By creating a virtual and dynamic representation of physical systems, the digital twin enables continuous data synchronization, simulation of different scenarios, and analysis of system behavior under various conditions. These features make the digital twin a powerful tool for a deeper understanding of urban energy systems and improving decision-making processes (Nozari & Yordanova, 2026).

However, most of the current applications of digital twins in the energy sector remain at the level of data monitoring and display, and have not moved towards integrating the concepts of sustainability, carbon awareness, and context-based intelligence. In other words, although digital twins have been able to provide better visibility into the performance of energy systems, they have not yet fully served the macro-environmental goals such as reducing carbon emissions and achieving Net-Zero. This indicates the need to develop new conceptual frameworks that can elevate digital twins from a purely analytical tool to an intelligent and integrated platform for sustainable energy management.

In this regard, one of the key concepts that has received attention in recent years is "Carbon Awareness" in energy systems. This concept emphasizes that decisions related to energy production and consumption should be made not only based on economic and technical criteria, but also taking into account carbon and environmental consequences. Integrating this perspective into the digital twin structure can lead to the formation of a new generation of urban energy systems that are able to dynamically monitor carbon impacts and act to reduce them (Nozari, 2026).

On the other hand, the inherent complexity of urban energy systems (which includes the interaction between physical infrastructure, user behavior, environmental conditions and macro policies) requires a comprehensive and multi-layered approach to modeling and analysis. In such a context, the use of a holistic conceptual framework that can consider these different dimensions together becomes particularly important. The digital twin, if properly designed, can act as the central core of this framework and establish the connection between data, knowledge and decision at the urban scale.

Accordingly, the aim of this chapter is to provide a comprehensive conceptual framework for digital twins in urban energy systems with a focus on sustainability and carbon awareness. This framework attempts to redefine the role of digital twins in energy management and to explain the connections between different layers of the system (from physical infrastructure to intelligent and interactive layers) in an integrated manner. It also seeks to show how, by using a holistic approach, digital twins can be an effective tool in the realization of sustainable and low-carbon cities (Nozari & Yordanova, 2026).

Ultimately, this rethinking of the role and function of digital twins can be an important step towards developing a new generation of urban energy systems that are not only more efficient and intelligent, but also aligned with environmental goals and sustainable development requirements.

2. FROM CONVENTIONAL ENERGY MANAGEMENT TO CARBON-AWARE URBAN INTELLIGENCE

Energy management in cities has been largely based on traditional approaches based on predefined controls over the past decades. In these approaches, the main focus has been on sustainable energy supply, cost reduction and, in some cases, improving efficiency. Although these goals remain important, the complex and dynamic nature of today's urban energy systems has shown that these approaches no longer meet new needs, especially in the field of environmental sustainability and carbon emission reduction. In other words, energy management is no longer simply a technical or economic issue, but has become a multidimensional challenge that also includes environmental, social and even behavioral dimensions (Nozari et al., 2025).

In traditional systems, data is often collected and analyzed periodically and decisions are made based on historical patterns or fixed rules. Meanwhile, energy consumption in cities is influenced by multiple factors such as climate change, user behavior, demand fluctuations and infrastructure developments. Such dynamics require systems that can understand the situation in real time and provide appropriate and intelligent responses. This is where the concept of "city intelligence" in the energy sector comes into play; a concept that goes beyond monitoring and control to deep understanding, prediction and active interaction with the system (Nweye et al., 2025).

However, even many so-called smart systems still focus on traditional metrics such as energy efficiency or cost reduction and pay less attention to the environmental impacts of decisions. This is where the concept of "carbon awareness" comes in as a new paradigm. Carbon awareness means that every energy-related decision) whether production, distribution or consumption) must be evaluated in terms of carbon emissions and their impact on the environment. This shift in perspective is a kind of redefining the goals of urban energy systems; Success is no longer measured solely by economic metrics, but also by reducing the carbon footprint as a key indicator.

The move towards carbon-aware intelligence requires a transformation in the way data is collected, analyzed, and used. In this context, energy data must be combined with environmental information, such as the carbon intensity of different energy sources, weather conditions, and even user behavior patterns. This data integration allows for a more comprehensive view of the state of the system and paves the way for more accurate and responsible decision-making. For example, in such a system, the right time to consume energy or use a specific resource is determined not only by price or availability, but also by its carbon emissions.

In the meantime, the digital twin can play a key role in realizing this transition. By providing a virtual and dynamic model of the energy system, the digital twin enables the simultaneous analysis of various indicators (including consumption, cost, and carbon emissions). This capability allows the system to assess the carbon consequences of various decisions before implementation and identify more optimal options. In other words, the digital twin can serve as a platform for the practical implementation of the concept of carbon awareness at the city scale.

Another important point in this transition is to pay attention to the role of humans as one of the main components of the energy system. Consumer behavior, energy use patterns, and even their level of awareness of environmental issues have a direct impact on the overall performance of the system. In a carbon awareness framework, these factors also need to be included in modeling and analysis. This requires the development of systems that can effectively interact with users and guide them towards adopting more sustainable behaviors (Bayasgalan et al., 2024).

Ultimately, the transition from traditional energy management to carbon-aware smart cities is not only a technological change, but also a conceptual transformation. This transformation requires redefining the goals, tools, and even the roles of different actors in the urban energy system. The digital twin, if developed as a comprehensive and holistic conceptual framework, can serve as one of the main pillars of this transformation and pave the way for achieving sustainable and low-carbon cities.

3. FOUNDATIONS OF DIGITAL TWINS IN URBAN ENERGY CONTEXTS

Digital twin has emerged as a key concept in the evolution of cyber-physical systems in recent years and has gradually established itself in various fields, including industry, transportation, and energy. In its simplest definition, a digital twin is a virtual, dynamic, and synchronized representation of a physical system that, using real-time data, is capable of reflecting the state, behavior, and changes of that system over time. However, in the context of urban energy systems, this concept takes on more complex dimensions and requires a deeper understanding of its components, interactions, and different layers.

In a smart city, the energy system does not simply include electricity generation and distribution networks, but also a set of interconnected infrastructures that include renewable energy sources, transmission networks, buildings, storage systems, and even user consumption patterns. In such an environment, the digital twin must be able to model this diversity and complexity in an integrated manner. This requires the definition of a multi-layer framework in which each layer plays a specific role in the representation and analysis of the system (Nozari & Yordanova, 2025).

In the physical layer, the real components of the urban energy system are located; from power plants and solar panels to distribution networks and consumer equipment. This layer is the main source of data and its behavior is the basis for the formation of the digital twin. In the digital layer, these components are reconstructed as virtual models. These models can include mathematical models, dynamic simulations or even data-driven representations, each of which in some way tries to reflect the behavior of the real system.

Between these two layers, there is another crucial layer that can be called the data synchronization layer. This layer is responsible for collecting, transferring and integrating data from different sources. In urban energy systems, data is obtained through sensors, smart metering systems, and communication infrastructure. The quality, accuracy, and timeliness of this data have a direct impact on the performance of the digital twin. Any delay or error in this process can lead to deviation of the digital model from reality, thereby reducing the validity of analyses and decisions.

One of the important features of the digital twin in this context is its dynamism. Unlike static models that are only valid for a certain period of time, the digital twin is continuously updated and can reflect changes in the system in or near real time.

This feature allows for continuous monitoring, trend analysis, and even simulation of different scenarios. As a result, urban managers and decision makers can have a more accurate view of the current and future state of the energy system.

However, one of the fundamental challenges in implementing digital twins in urban energy systems is the integration of data and models from heterogeneous sources. Different energy systems are often managed by different organizations and institutions and do not have the same data standards. This heterogeneity can be a serious obstacle to creating a comprehensive and integrated digital twin. Therefore, developing conceptual frameworks that can manage this diversity and facilitate the interaction between different components is of great importance.

On the other hand, many current digital twin applications in urban energy still remain at the level of monitoring and data visualization. In these applications, the main focus is on displaying the system status and providing analytical dashboards, without establishing a deep connection between data, analysis and decisions. This is while the true potential of digital twins goes beyond this level and can act as an interactive platform for understanding, analyzing and even guiding the behavior of the system (Yordanova & Nozari, 2026).

In this context, one of the important developments is the move towards intelligent, context-aware digital twins that are able not only to process data, but also to interpret it in a meaningful context. In urban energy systems, this means taking into account environmental factors, consumption patterns, and even sustainability goals in the analysis. Such an approach transforms the digital twin from a purely technical tool to a key component in the urban decision-making ecosystem.

Ultimately, it can be said that the digital twin in the context of urban energy systems is not only a technology, but also a conceptual framework for representing and understanding complex systems. This framework, if properly designed, can seamlessly connect the physical and digital worlds and pave the way for the development of smart, sustainable, and environmentally aware energy systems.

4. SUSTAINABILITY AND NET-ZERO PARADIGMS IN SMART CITIES

In recent years, the concept of sustainability has become one of the main pillars of urban planning. The growth of urban populations, increasing energy demand and pressures from climate change have made it more and more obvious that traditional urban development methods need to be reconsidered. In this context, smart cities, as a platform for integrating technology, data and urban infrastructure, play an important role in achieving sustainability goals. However, the real realization of sustainability in these cities requires moving beyond purely technological approaches and moving towards frameworks that can consider environmental, economic and social dimensions in an integrated manner.

One of the key concepts proposed in this direction is "Net-Zero Cities". In this paradigm, the goal is to make the total amount of greenhouse gas emissions in a city equal to the amount of absorption or compensation, so that ultimately the net carbon footprint approaches zero. This goal cannot be achieved simply by improving energy efficiency or using renewable resources, but requires a fundamental transformation in the way urban energy systems are designed, operated and managed (Al-Jamal et al., 2026).

In the Net-Zero framework, the urban energy system is no longer considered as a linear and one-way structure, but rather as a dynamic and interactive system in which production, consumption, storage and even user behavior act simultaneously and in interaction with each other. This perspective requires that energy decisions are made not only based on technical and economic criteria, but also by considering environmental consequences, especially carbon emissions. For this reason, the concept of "carbon awareness" is raised as one of the key elements in this paradigm.

In traditional systems, the main focus has been on sustainable energy supply and cost reduction. Even in more advanced versions of these systems, energy efficiency is considered an important indicator. But in the Net-Zero paradigm, these approaches are not enough. What is important is a deep understanding of the carbon impacts of each decision and efforts to reduce these impacts at all levels of the system. This shift in perspective is a kind of redefining the goals of urban energy systems and requires new tools and frameworks for analysis and decision-making.

In this regard, a comparison between traditional systems and carbon-aware systems can help to better understand this transformation. Table 1 shows some of the conceptual differences between these two approaches:

Table 1. Comparative Conceptual Characteristics of Traditional vs. Carbon-Aware Energy Systems

Dimension	Traditional Systems	Carbon-Aware Systems
Objective	Energy efficiency	Carbon minimization and sustainability
Data Usage	Periodic and static	Real-time, dynamic, and integrated
Decision Logic	Rule-based and predefined	Context-aware and adaptive
System Scope	Local and isolated	Multi-scale and interconnected
Environmental Focus	Limited or indirect	Central and explicitly modeled

As the table shows, the difference between the two approaches is not simply at the level of tools or technologies, but at the level of attitude and conceptual framework. In traditional systems, data is often used in a limited and delayed manner, while in carbon-aware systems, data is collected and analyzed in real time from various sources. This allows for a more accurate understanding of the state of the system and timely decisions to be made.

On the other hand, the logic of decision-making has also evolved. In traditional approaches, decisions are usually made based on predefined rules, while in new systems, decision-making is done in a context-based and adaptive manner. In other words, the system is able to choose the best option from among various options, given the current situation; the option that is not only technically and economically suitable, but also has the lowest carbon footprint (Nozari & Yordanova, 2026).

Another important point is the expansion of the system scope from the local level to the multi-scale level. In the Net-Zero paradigm, one cannot only look at a single building or a limited network, but also needs to consider the interaction between different components at different levels, from individual equipment to the entire city. This holistic view paves the way for the development of frameworks that can model and manage these complex interactions.

In the meantime, the digital twin can serve as a key tool for realizing this paradigm. By providing an integrated representation of the urban energy system and integrating different data, the digital twin allows for the simultaneous analysis of energy and carbon indicators. This capability allows the system to make more informed and effective decisions on the way to achieving Net-Zero goals.

Overall, the Net-Zero paradigm and the concept of carbon awareness represent a fundamental shift in the way we look at urban energy systems. This change requires redefining existing concepts, tools, and frameworks, and the digital twin, if designed properly, can play a pivotal role in this transformation.

5. A HOLISTIC CONCEPTUAL FRAMEWORK FOR CARBON-AWARE ENERGY DIGITAL TWINS

Given the increasing complexity of urban energy systems and the need to align them with sustainability and Net-Zero goals, the need for frameworks that can model these systems in an integrated and multidimensional way is increasingly felt. In the meantime, the digital twin, as a conceptual and technological platform, has the potential to become the central core of such a framework. However, to effectively exploit this potential, it is necessary to move away from insular and one-dimensional views and move towards a holistic approach in which all components of the system (from data and infrastructure to people and the environment) are considered in interaction with each other.

In the proposed framework, the energy digital twin is considered not only as a virtual model of the physical system, but also as an intelligent core that is able to receive data from various sources, interpret them in the appropriate context, and provide meaningful insights for decision-making. The framework consists of several main layers, each of which plays a specific role in the overall performance of the system.

At the lowest level, there is the "environmental layer", which includes natural and environmental factors such as weather conditions, temperature, humidity, and solar radiation intensity. This layer plays an important role in determining energy production and consumption patterns, and is particularly important in analyzing carbon impacts. For example, the carbon emissions of an energy source can depend on environmental conditions and the time of its use (Nozari et al., 2025).

Next to it is the "energy infrastructure layer", which includes the physical components of the system such as power plants, renewable resources, transmission networks, and consumer equipment. This layer is the main source of operational data and its behavior forms the basis for the formation of the digital twin. The interaction between this layer and the environmental layer is one of the key factors in understanding the dynamics of the urban energy system.

At the center of this framework is the "digital twin core", which is responsible for integrating data, modeling the system and creating a dynamic representation of the real situation. Using input data, this core reconstructs the current state of the system and allows for the analysis of its behavior under different conditions. The important point in this section is the ability to continuously synchronize with the physical system and maintain the accuracy and validity of the model (Gassar & Jafar, 2025).

On top of this core is the "intelligence layer", which is responsible for advanced data analysis and extracting actionable insights. In a carbon-aware framework, this layer should be able to analyze the carbon impacts of various decisions in addition to energy indicators. In other words, intelligence here does not simply mean optimizing performance, but also includes a deep understanding of environmental consequences (Nozari, 2025).

Finally, there is the "interaction and decision layer," which connects the digital system with human users or management systems. This layer can include analytical dashboards, decision-support tools, and user interfaces that help decision-makers take appropriate actions based on the information provided. In a holistic framework, this interaction should be two-way; in such a way that not only does the system convey information to the user, but the user's feedback is also taken into account in updating the model and future decisions.

Figure 1. Holistic Conceptual Architecture of Carbon-Aware Energy Digital Twin in Smart Cities

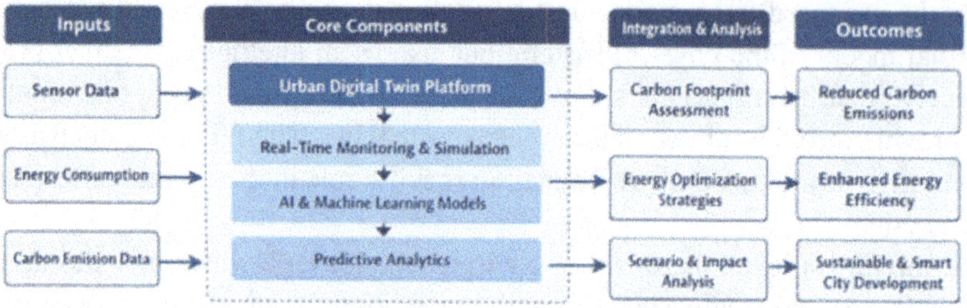

As can be seen in the Figure 1, this framework is designed as a layered structure in which data and information flow from the bottom layers to the top and vice versa. Data from the environmental and infrastructure layers are transferred to the digital twin core and, after processing in the intelligence layer, are provided to the decision layer in the form of usable insights. In turn, decisions made can also be applied as feedback to the physical system, thus forming a closed loop between the real and digital worlds.

One of the important features of this framework is the simultaneous attention to different dimensions of the system. Unlike traditional approaches that mainly focus on a specific aspect (such as energy efficiency), an attempt has been made here to consider energy, carbon, environment and human behavior in an integrated manner. This perspective allows for more comprehensive analysis and more informed decision-making (Nozari et al., 2025).

Overall, the conceptual framework presented can be used as a basis for developing a new generation of digital twins in the field of urban energy; twins that not only reflect the state of the system, but also actively act towards achieving sustainability goals and reducing carbon emissions.

6. MULTI-DIMENSIONAL CARBON AWARENESS IN DIGITAL TWIN SYSTEMS

With the spread of the Net-Zero paradigm and the increasing importance of environmental considerations in the design and operation of urban energy systems, the concept of "carbon awareness" has emerged as a key element in the evolution of these systems. However, in many existing studies and applications, carbon awareness is considered unidimensionally, mainly as a general indicator of greenhouse gas emissions. In reality, carbon emissions are a complex and multifaceted phenomenon that is influenced by temporal, spatial, behavioral, and systemic factors. Therefore, for a deeper understanding and more effective management, it is necessary to adopt a multidimensional approach that can consider these different dimensions in an integrated manner.

In the context of the digital twin, it is possible for carbon awareness to go beyond the level of a simple indicator and become a multilayered analytical mechanism. In other words, a digital twin can provide a platform in which carbon emissions data can be collected, analyzed, and used in decision-making processes at different times. This approach not only increases the accuracy of the analysis, but also allows for the identification of hidden patterns and complex relationships between different factors.

One important dimension in this context is the temporal dimension. Carbon emissions in urban energy systems are highly time-dependent and can vary significantly at different times of the day, days of the week, or seasons. For example, the use of certain energy sources during peak hours may lead to a significant increase in carbon emissions, while the same consumption at other times may have a lower environmental impact. In a digital twin equipped with temporal carbon awareness, these changes are dynamically monitored and can be used to optimally schedule energy consumption or production (Michailidis et al., 2025).

The spatial dimension also plays an important role in carbon analysis. Within a city, different areas may differ significantly in terms of energy sources, consumption patterns, and carbon emissions. For example, an industrial area may contribute more to carbon emissions than a residential area. By taking these spatial differences into account, a digital twin can help city planners design more targeted policies to reduce emissions in different areas.

The behavioral dimension refers to the impact of user behavior and consumption patterns on carbon emissions. Consumers' everyday decisions (such as when to use electrical appliances, adjust the temperature, or choose energy sources) can have a significant impact on emissions. In a digital twin framework, these behaviors can be included as inputs into models and analyzed. This allows for the design of behavioral interventions and incentive policies that lead to a reduction in high-carbon consumption (Nozari & Nassar, 2025).

Finally, the systemic dimension refers to the interactions between different components of the energy system and their cumulative effects on carbon emissions. In this dimension, the focus is on understanding the relationships between energy production, distribution, storage, and consumption on a macro scale. By modeling these interactions, a digital twin can identify the cascading effects of different decisions and prevent unintended consequences.

To summarize these dimensions, Table 2 provides an overview of the different dimensions of carbon awareness in digital twin-driven energy systems:

Table 2. Dimensions of Carbon Awareness in Digital Twin–Driven Energy Systems

Dimension	Description	Role in Digital Twin Systems
Temporal	Time-dependent variation in carbon emissions	Enables dynamic monitoring and forecasting
Spatial	Location-based differences in emissions	Supports urban zoning and planning
Behavioral	Impact of user consumption patterns	Facilitates demand-side management
Systemic	Interactions across the entire energy system	Enables holistic system optimization

As the table shows, each of these dimensions plays a different role in understanding and managing carbon emissions. In an advanced digital twin system, these dimensions are considered simultaneously and in interaction with each other. For example, simultaneous analysis of the temporal and behavioral dimensions can show which consumption patterns have the greatest carbon impact at which times and how they can be modified.

Another important point is the ability to integrate these dimensions into a single analytical framework. By providing a common environment for collecting and processing data, the digital twin allows for the combination of these dimensions and helps decision makers have a more comprehensive picture of the state of the system. This is especially important in situations where decisions must be made quickly and in response to environmental changes.

Overall, a multidimensional approach to carbon awareness is an important step towards elevating the role of the digital twin from an analytical tool to an intelligent platform for sustainable urban energy management. This approach not only increases the accuracy and depth of analyses, but also enables the design of more effective policies and interventions towards reducing carbon emissions.

7. CLOSED-LOOP INTERACTION BETWEEN PHYSICAL AND DIGITAL ENERGY SYSTEMS

One of the fundamental concepts in understanding the true role of the digital twin in urban energy systems is the concept of "Closed-Loop Interaction" between the physical world and the digital space. In traditional approaches, the communication between these two domains has been mainly one-way and based on collecting data from the physical system and transferring it to analytical systems. In such a structure, data is rarely directly returned to the system after analysis, and if there is feedback, it is often applied with a delay and in an unsystematic manner. However, for the effective management of complex and dynamic urban energy systems, there is a need for a continuous, two-way and dynamic interaction between these two spaces.

In the digital twin framework, this interaction is defined as a continuous cycle in which data is collected from the physical system, analyzed in the digital environment, and the results are returned to the physical system in the form of control actions or decision-making recommendations. This cycle is continuously repeated, allowing the system to respond adaptively and intelligently to environmental and operational changes. In other words, the digital twin not only reflects the state of the system, but also actively plays a role in shaping its behavior.

In this closed-loop interaction, several key stages can be identified. First, there is the "sensing" stage, in which data related to the state of the system—including energy consumption, production, environmental conditions, and even user behavior—is collected through sensors and communication infrastructure. This data provides a view of the current state of the system and is the basis for subsequent analysis.

In the next stage, this data is transferred to the "digital twin core," where it is processed and analyzed. This stage involves modeling, simulation, and analysis of various scenarios. In a carbon-aware system, these analyses focus not only on energy performance but also on the carbon impacts of decisions. For example, a digital twin can examine how a change in consumption patterns or the use of a particular energy source will affect carbon emissions.

After analysis, the results are produced in the form of "Insights" or "Recommended Actions." These results can include suggestions for reducing consumption, changing the timing of resource use, or even exercising direct control over some system components. At this stage, the interaction and decision layer plays an important role, as it must convey this information in a way that is understandable and actionable to users or control systems.

Finally, these decisions or actions are fed back to the physical system and implemented. This feedback can be done automatically (in automated systems) or semi-automated (with human intervention). After these changes are made, the system status is monitored again through sensors and the cycle begins again. This iterative process allows for gradual learning and continuous improvement of system performance.

Figure 2. Closed-Loop Interaction Model for Carbon-Aware Digital Twin Systems

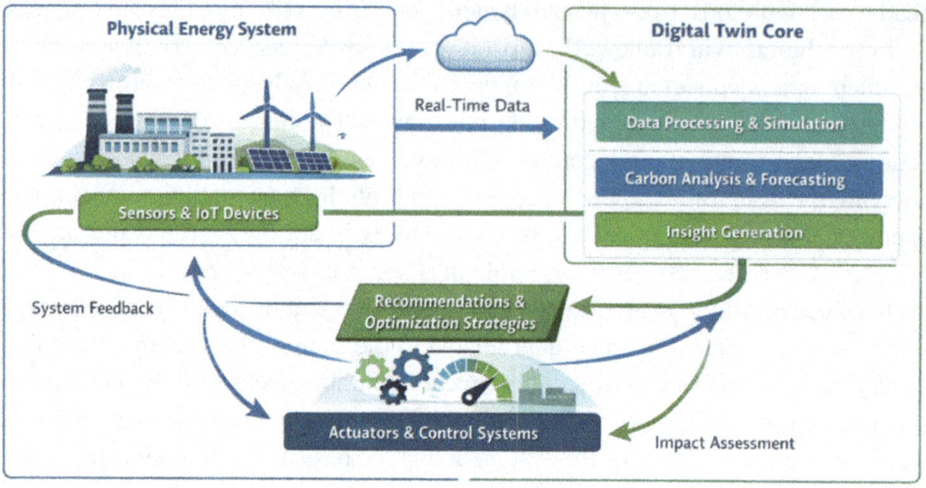

As shown in Figure 2, the interaction between the physical system and the digital twin is designed as a closed and continuous cycle. In this cycle, each step is connected to the next step, and there is no disconnect between data collection, analysis, and action. This continuity is one of the key features of digital twin-based systems that differentiates them from traditional systems.

One of the main advantages of this structure is the increase in speed and accuracy in decision-making. Since data is collected and analyzed in real time, the system can react quickly to changes and prevent undesirable conditions from occurring. In addition, by considering carbon impacts in the analysis process, the decisions made will be more optimal not only from a technical and economic point of view, but also from an environmental point of view.

Another important point is the adaptability of this cycle. In a dynamic system like urban energy, conditions are constantly changing and no one-size-fits-all solution can be right for all situations. Closed-loop interaction allows the system to continuously learn from new data and adapt its behavior to new conditions.

Overall, closed-loop interaction between physical and digital systems is a key element in achieving an effective and efficient digital twin in urban energy. This interaction not only connects data, analysis, and action, but also paves the way for the development of systems that operate intelligently, adaptively, and carbon-consciously.

8. INTEGRATING HUMAN, ENVIRONMENTAL, AND INFRASTRUCTURAL INTELLIGENCE

One of the fundamental challenges in the design and operation of urban energy systems is understanding and managing the interactions between three key components: humans, the environment, and infrastructure. In many common approaches, these three components are examined separately; in such a way that technical analyses focus on infrastructure, environmental studies are limited to natural conditions, and human behavior is often considered in the form of simple assumptions or general models. This separation, although justified in some cases, loses its effectiveness when faced with the real complexities of urban energy systems. In such a context, the digital twin can provide a basis for the synergy of these three types of intelligence as an integrated framework.

Infrastructure intelligence refers to the ability to understand, monitor, and analyze the performance of the physical components of the energy system. This section includes generation, transmission, and distribution networks, renewable energy sources, storage systems, and consumer equipment. In a digital twin, these components are represented as virtual models and their behavior under different conditions can be analyzed. But what is important is to go beyond simply displaying the state of these infrastructures and move towards understanding their internal relationships; relationships that determine how changes in one part can affect others (Nozari & Yordanova, 2026).

Alongside this layer is environmental intelligence, whose role has been increasingly recognized in recent years. Weather conditions, air quality, solar radiation intensity and other environmental factors have a direct impact on energy production and consumption. For example, the amount of solar or wind energy production is highly dependent on weather conditions, and as a result, any analysis or decision-making in the energy sector without taking these factors into account will be incomplete. By integrating environmental data into its models, digital twins enable analyses to be carried out in a more realistic context and closer to real-world conditions.

But perhaps the most complex and yet most decisive component is human intelligence. User behavior, consumption patterns, individual preferences, and even their level of environmental awareness all play a role in shaping the overall energy consumption pattern. Unlike infrastructure and the environment, which are largely predictable, human behavior is inherently dynamic, nonlinear, and sometimes unpredictable. This characteristic makes it even more important to consider it in analytical frameworks.

Integrating these three types of intelligence into a digital twin requires creating meaningful connections between them. In other words, it must be possible to show how environmental changes affect the infrastructure, how user behavior responds to these changes, and ultimately, how these interactions affect the overall performance of the energy system. For example, a rise in temperature on a summer day can lead to increased consumption of cooling systems, which in turn puts more strain on the distribution network and, if fossil fuels are used, increases carbon emissions. In such situations, a digital twin can identify this chain of relationships and allow for the analysis of different scenarios.

One of the key benefits of this integration is the creation of a platform for more informed decision-making. When information about the infrastructure, the environment, and human behavior is available simultaneously, decision-makers can have a more comprehensive picture of the state of the system and adjust their actions based on this complete picture. This is especially important in the context of carbon-conscious systems, since the environmental impacts of decisions are often the result of complex interactions between these three components.

On the other hand, this integration can lead to the development of more effective interaction between humans and the system. In an advanced digital twin, users are considered not only as consumers, but also as part of the system, who can influence its overall performance through their feedback, preferences, and decisions. This view paves the way for designing systems that can guide users' behavior towards sustainable goals without making them feel constrained or constrained.

Finally, the integration of human, environmental, and infrastructure intelligence can be considered as a fundamental step towards the development of next-generation digital twins; twins that not only reflect reality but are also able to shape it towards sustainable and low-carbon goals. This approach draws a new horizon in urban energy management, in which technology, environment, and humans are in a dynamic and purposeful interaction.

9. CONCEPTUAL DESIGN PRINCIPLES FOR CARBON-AWARE DIGITAL TWIN SYSTEMS

Designing digital twins in the context of urban energy systems, especially when the goal is to achieve sustainability and reduce carbon emissions, requires a set of conceptual principles that can structure and manage the complexities of these systems. In contrast to purely technical approaches that focus on tools and algorithms, the emphasis here is on frameworks that guide the way digital twins are thought about, designed, and implemented. These principles provide a roadmap for developing systems that are not only functionally efficient, but also aligned with environmental goals and sustainability requirements.

One of the first principles in this context is "Holistic Design." As mentioned in the previous sections, urban energy systems are a set of interconnected components that include physical infrastructure, environmental conditions, and human behavior. The design of a digital twin should consider these different dimensions in an integrated manner and avoid one-dimensional analyses. This principle is especially important in the context of carbon awareness, because carbon emissions are the result of complex interactions between these components and cannot be attributed to a single factor.

The second principle is "scalability". Digital twins in smart cities must be able to operate at different scales (from a single device or building to the entire city). This requires designing architectures that can adapt to increasing data volumes and system complexity without losing performance. In the context of carbon awareness, this becomes even more important, because analyzing carbon impacts often requires examining data at different scales simultaneously (Mun et al.,2025).

"Interoperability" is also a key principle in this area. Urban energy systems are usually composed of diverse components that are managed by different organizations and platforms. The digital twin must be able to communicate with these systems and receive and integrate data from different sources. This requires the use of common standards and the design of interfaces that allow for effective information exchange.

Another principle is "adaptivity". Environmental conditions, consumption patterns, and even energy policies in cities are constantly changing. Therefore, the digital twin must be able to adapt to these changes and update its models. In a carbon-aware system, this means the ability to understand changes in the intensity of carbon emissions from different sources and respond appropriately to them.

"Transparency" is also an important principle in the design of digital twins. Decisions made in these systems (especially when they affect energy consumption or user behavior) must be explainable and understandable. This not only helps to increase user trust, but also paves the way for their active participation in decision-making processes. In the context of carbon awareness, transparency can help users better understand the environmental impact of their behavior and take action to reduce it.

The principle of "Sustainability Alignment" is also of particular importance. The digital twin should not be considered solely as a tool to improve the technical performance of the system, but should be designed in such a way that the macro-sustainability goals (including reducing carbon emissions, optimizing resource use, and preserving the environment) are taken into account. This principle, in a way, determines the value framework of the system and determines its overall direction.

In addition, "Data-Context Integration" is also an important principle. Data alone has limited value unless it is interpreted in a meaningful context. The digital twin should be able to frame raw data in a way that identifies its relationship to the system's goals (especially carbon goals). This allows data to be transformed into knowledge and ultimately into decisions.

Finally, we can mention the principle of "Human-Centric Design". In many cases, the success or failure of energy systems depends on the behavior of users. Therefore, the design of the digital twin should be such that it puts users at the center of attention and allows them to interact effectively with the system. This interaction can include providing feedback, behavioral suggestions, or even participating in decision-making.

Overall, the conceptual design principles presented provide a framework that can guide the development of carbon-aware digital twins. These principles not only help improve the technical performance of the system, but also pave the way for its alignment with sustainability goals and environmental impact reduction. Such an approach can elevate the role of the digital twin from an analytical tool to a key component in smart and sustainable urban energy management.

10. CONCLUSION

In recent years, rapid developments in the energy sector and increasing environmental concerns have seriously raised the need to rethink urban energy management practices. As the main centers of energy consumption and carbon emission generation, cities play a decisive role in achieving sustainability and Net-Zero goals. In the meantime, digital twin technology, as a new and transformative platform, has the capacity to fundamentally transform the way urban energy systems are understood, analyzed, and managed. However, effective exploitation of this capacity requires moving beyond limited applications and moving towards comprehensive and holistic conceptual frameworks.

This chapter attempts to redefine, from a conceptual and integrated perspective, the role of digital twin in realizing sustainable and carbon-conscious urban energy systems. Initially, it was shown that traditional energy management approaches (which mainly focus on limited monitoring and control) are unable to respond to the complexities and dynamics of today's urban systems. Instead, there is a need for systems that can operate in real-time, context-aware, and adaptive ways, and that make decisions that take into account environmental impacts in addition to technical and economic efficiency.

The concept of carbon awareness was then introduced as a key element in this transformation. By emphasizing the consideration of carbon implications at all stages of decision-making, this concept provides a framework in which sustainability goals are operationally integrated into urban energy systems. Furthermore, a multidimensional approach to carbon awareness (which includes temporal, spatial, behavioral, and systemic dimensions) allows for a deeper understanding of emission patterns and the design of more effective interventions.

One of the most important achievements of this chapter was the presentation of a holistic conceptual framework for energy digital twins, in which different layers of the system (from environment and infrastructure to intelligence and interaction) are considered in an integrated manner. This framework shows how a comprehensive picture of the urban energy system can be created by integrating diverse data and multi-layered analytics and used to support informed decision-making. In the meantime, the closed-loop interaction between physical and digital systems, as one of the key elements, enables a continuous cycle of perception, analysis and action that leads to continuous improvement of system performance.

This chapter also emphasized the importance of integrating human, environmental and infrastructure intelligence. Each of these three components plays an important role in shaping the behavior of the energy system, but only if they are considered simultaneously and in interaction with each other can a true understanding of the dynamics of the system be achieved. By providing a platform for this integration, the digital twin enables the analysis of complex relationships and the design of coordinated solutions.

In addition to these topics, conceptual design principles for the development of carbon-aware digital twins were also examined. Principles such as holistic approach, scalability, interoperability, adaptability, transparency, and alignment with sustainability goals provide a framework that can guide the development of these systems towards greater efficiency and higher environmental impact. These principles are a kind of bridge between theoretical concepts and practical implementation and can be used as a guide for designers and decision-makers in this field.

Finally, it can be said that the digital twin, if developed in the form of a comprehensive and carbon-aware conceptual framework, can become one of the key tools in smart and sustainable urban energy management. This technology not only enables more accurate representation of complex systems, but also, by connecting data, knowledge and decisions, paves the way for systems that actively work towards reducing carbon emissions and achieving Net-Zero goals.

This evolution is not just a technological change, but also represents a paradigm shift in how we view urban energy systems; one in which sustainability, intelligence and environmental responsibility are considered integral components of the design and operation of these systems.

REFERENCES

Al-Jamal, M. Q., Alsarhan, A., Aljamal, Q., AlJamal, M., Khassawneh, B. S., Al Nuaim, A., & Al Nuaim, A. (2026). AI-Enabled System-of-Systems Decision Support: BIM-Integrated AI-LCA for Resilient and Sustainable Fiber-Reinforced Façade Design. *Information (Basel), 17*(2), 126. DOI: 10.3390/info17020126

Bayasgalan, A., Park, Y. S., Koh, S. B., & Son, S.-Y. (2024). Comprehensive Review of Building Energy Management Models: Grid-Interactive Efficient Building Perspective. *Energies, 17*(19), 4794. DOI: 10.3390/en17194794

Gassar, A. A. A., & Jafar, R. (2025). Artificial Intelligence-Enabled Heating, Ventilation, and Air Conditioning Systems Toward Zero-Emission Buildings: A Systematic Review of Applications, Challenges, and Future Directions. *Applied Sciences (Basel, Switzerland), 15*(19), 10497. DOI: 10.3390/app151910497

Michailidis, P., Michailidis, I., & Kosmatopoulos, E. (2025). Reinforcement Learning for Electric Vehicle Charging Management: Theory and Applications. *Energies, 18*(19), 5225. DOI: 10.3390/en18195225

Mun, H.-S., Lagua, E. B., Hong, S.-K., Ryu, S.-B., Sharifuzzaman, M., Hasan, M. K., Kim, Y.-H., & Yang, C.-J. (2025). Energy-Efficient Technologies and Strategies for Feasible and Sustainable Plant Factory Systems. *Sustainability (Basel), 17*(7), 3259. DOI: 10.3390/su17073259

Nozari, H. Integrating Neuromorphic Intelligence with Fuzzy Multi-Objective Models in Resilient Green Supply Chain Design. Process Integr Optim Sustain (2025). https://doi.org/DOI: 10.1007/s41660-025-00598-5

Nozari, H. (2026). Digital twin-based predictive maintenance in cold chain logistics. *Operations Research, 60*(1), 481–499. DOI: 10.1051/ro/2025167

Nozari, H., Abdi, H., Szmelter-Jarosz, A., & Rahmaty, M. (2025). A digital twin-based neutrosophic multi-objective optimization model for smart supply chains in the food industry. *Journal of Intelligent Manufacturing*. Advance online publication. DOI: 10.1007/s10845-025-02689-z

Nozari, H., & Nassar, S. (2025). *Real-time data-driven multi-objective optimization in B2B marketing using digital twins*. OPSEARCH., DOI: 10.1007/s12597-025-01044-3

Nozari, H., Nassar, S., & Szmelter-Jarosz, A. (2025). Fuzzy Multi-Objective Optimization Model for Resilient Supply Chain Financing Based on Blockchain and IoT. *Digital, 5*(3), 32. DOI: 10.3390/digital5030032

Nozari, H., Szmelter-Jarosz, A., & Weiland, D. (2025). A Fuzzy Multi-Objective Sustainable and Agile Supply Chain Model Based on Digital Twin and Internet of Things with Adaptive Learning Under Environmental Uncertainty. *Applied Sciences (Basel, Switzerland), 15*(19), 10399. DOI: 10.3390/app151910399

Nozari, H., & Yordanova, Z. (2025). An intelligent digital twin approach for optimizing multi-channel supply chains in uncertain environments. Green Technologies and Sustainability, 100309.

Nozari, H., & Yordanova, Z. (2026). A fuzzy–digital twin optimization framework for simultaneous management of waste and energy consumption in sustainable manufacturing. *An International Journal of Optimization and Control: Theories & Applications, 16*(1), 160–175. DOI: 10.36922/IJOCTA025360152

Nozari, H., & Yordanova, Z. (2026). A Self-Deciding Adaptive Digital Twin Framework Using Agentic AI for Fuzzy Multi-Objective Optimization of Food Logistics. *Algorithms, 19*(3), 218. DOI: 10.3390/a19030218

Nozari, H., & Yordanova, Z. (2026). An Adaptive Fuzzy Multi-Objective Digital Twin Framework for Multi-Depot Cold-Chain Vehicle Routing in Agri-Biotech Supply Networks. *Logistics (Basel), 10*(2), 27. DOI: 10.3390/logistics10020027

Nozari, H., & Yordanova, Z. (2026). Fuzzy multi-objective digital twin framework for dynamic pricing and assortment optimization in metaverse-based retail environments. *Journal of Revenue and Pricing Management.* Advance online publication. DOI: 10.1057/s41272-026-00586-5

Nweye, K., Kaspar, K., Buscemi, G., Fonseca, T., Pinto, G., Ghose, D., Duddukuru, S., Pratapa, P., Li, H., Mohammadi, J., Lino Ferreira, L., Hong, T., Ouf, M., Capozzoli, A., & Nagy, Z. (2025). CityLearn v2: Energy-flexible, resilient, occupant-centric, and carbon-aware management of grid-interactive communities. *Journal of Building Performance Simulation, 18*(1), 17–38. DOI: 10.1080/19401493.2024.2418813

Okimi, T. O. (2025). Human centric digital twins for sustainable supply chain in construction. *Smart Constr. Sustain. Cities, 3*(1), 31. DOI: 10.1007/s44268-025-00078-2

Yordanova, Z., & Nozari, H. (2026). An Agentic Digital Twin Framework for Fuzzy Multi-Objective Optimization in Dynamic Humanitarian Logistics. *Algorithms, 19*(3), 198. DOI: 10.3390/a19030198

Chapter 12
Finding Leadership Value With Human Digital Twins and Extended Reality in Industry 5.0 Transformation

Kaloyan Dimitrov
http://orcid.org/0009-0006-3752-3413
University of National and World Economy, Bulgaria

Liliya Elenkova
University of National and World Economy, Bulgaria

Dimityr Panchukov
University of National and World Economy, Bulgaria

ABSTRACT

Human Digital Twins (HDTs) and Extended Reality (XR) are key emerging technologies for human-machine interaction within digital transformation of recent years. This chapter evaluates how the integration of these two technologies fosters leadership development and human-centric transformation in the context of Industry 5.0. Within this framework, the conceptual architectures of HDTs and XR are identified, and then the impact of these technologies on redefining leadership capabilities is analyzed. By synthesizing emerging discourse and technological frameworks, the findings emphasize that modern leadership values have transitioned from pure machine optimization to complex socio-technical systems. Finally, the role of the

DOI: 10.4018/979-8-2600-1669-5.ch012

leader as a facilitator of collaborative intelligence is also redefined, establishing the foundational "flight controls" necessary for navigating contemporary industrial volatility.

1. INTRODUCTION

The contemporary convergence of artificial intelligence, cyber-physical systems, and immersive technologies is currently precipitating a transformative shift in industrial production on a global scale. While the initial momentum of Industry 4.0 was predominantly sustained by automation and digital connectivity, a palpable shift in academic and policy discourse has transitioned toward the more nuanced paradigm of Industry 5.0. This new era is not merely a technical upgrade but a fundamental re-centering of industrial logic around human well-being, sustainability, and societal resilience. In this context, the manufacturing industry finds itself at a crossroads, moving away from techno-centric, automation-driven models toward value-driven, human-centric frameworks. However, this transition is fraught with complexity; business leaders are now required to orchestrate intricate socio-technical systems and oversee hybrid human-machine workforces in environments defined by extreme uncertainty.

The optimal trajectory for Industry 5.0 is predicated on a "worker back in the loop" strategy, wherein technology is designed to augment - rather than supplant - human creativity and expertise. In this ideal state of perfection, tools such as Human Digital Twins (HDTs) and Extended Reality (XR) have the potential to empower workers, reduce cognitive loads, and cultivate a culture of resilient collaboration. Unfortunately, the current reality falls short. A significant Digital Transformation (DTx) gap has emerged, particularly regarding how leadership manages the integration of these complex technologies without compromising the human element. Traditional command-and-control models are proving increasingly inadequate for these challenges. Furthermore, many organizations suffer from "Twinwash" - a phenomenon where the term "digital twin" is used as a buzzword, leading to stakeholder mistrust and conflicting system designs. When leaders fail to bridge this gap, the consequences are tangible: high levels of worker fatigue, cognitive overload, and a persistent "cog in the machine" mentality that has plagued industrial revolutions since the 18th century.

This study argues that a holistic approach is necessary, one that utilizes XR as the intuitive interface to bridge physical environments with the deep data insights provided by HDTs. By adopting the lens of strong structuration theory, this research explores the complex interplay between human agency and digital structures. The purpose of this chapter is to evaluate how synergistic integration of HDTs and XR facilitate leadership value and human-centric transformation in Industry 5.0 environments. Specifically, this study aims to achieve the following three key objectives: (1) to identify the conceptual architectures required to fuse HDTs with XR in the context of Industry 5.0; (2) to analyze the impact of these technologies on worker safety, psychological well-being, and performance; and (3) to evaluate how these immersive, data-driven environments redefine leadership capabilities in volatile contexts.

Understanding HDTs and XR synergy matters because it provides the "flight controls" for modern industrial leaders. From a policy perspective, it aligns with European Commission pillars of human-centricity and sustainability. Practically, it offers a way to preserve intergenerational knowledge as experienced workers retire, documenting their expertise within a digital twin environment for the next generation.

2. INDUSTRY 5.0 AND HUMAN-CENTRIC APPROACH

The concept of Industry 5.0 places significant emphasis on the collaboration between humans and intelligent machines with a view to achieving sustainable and resilient production systems. The focus of technological development is no longer on replacing workers; rather, technologies are now being designed to augment human creativity, expertise and decision-making capacity. This paradigm necessitates that organizations achieve a balance between technological innovation and social responsibility, ethical governance, and workforce empowerment (Breque et al., 2021). The historical trajectory of industrialization is characterized by four major shifts prior to the current era.

2.1. Brief Overview of Industrial Revolutions

The First Industrial Revolution, which originated in Britain circa 1760, introduced the concepts of steam power and mechanization, thereby transitioning labor from manual production to machine-based factory systems. The era, distinguished by James Watt's enhancements to the steam engine, witnessed significant efficiency gains. However, it also established a paradigm in which human beings were frequently regarded as mere extensions of the machinery (Pham & Li, 2026; Fogaça et al., 2025).

The Second Industrial Revolution (1870-1914) witnessed the transition from steam to electricity, paving the way for the adoption of standardized mass production methodologies through the utilization of assembly lines. This period is particularly notable for Henry Ford's contributions, which significantly advanced the automotive industry. Although this period gave rise to novel opportunities for the middle class, it also served to further solidify the "cog in the machine" mentality among those engaged in frontline roles (Pham & Li, 2026; Islam et al., 2025).

The Third Industrial Revolution, which commenced in the 1970s, introduced electronics and computer integration, leading to the automation of repetitive tasks through Programmable Logic Controllers (PLCs) and robotics. This transition resulted in a shift in the workforce towards roles that were IT-focused and strategic. Concurrently, it initiated a process of disconnection between the human element and the physical production process (Pham & Li, 2026; Fogaça et al., 2025; Islam et al., 2025).

The advent of the Fourth Industrial Revolution, circa 2010, precipitated the utilization of the Internet of Things (IoT), voluminous data sets, and cyber-physical systems to engender "smart factories" wherein machines communicate autonomously. Nevertheless, the overemphasis on digitalization and efficiency in Industry 4.0 has been the subject of mounting criticism on account of its failure to take into account social fairness, environmental sustainability, and the well-being of the workforce (Yordanova & Nozari, 2025; Fogaça et al., 2025; Islam et al., 2025).

The Fifth Industrial Revolution has been described as a corrective evolution, building upon the digital foundations of Industry 4.0 while re-centering the human element as the primary driver of innovation and resilience. In this paradigm, the objectives of industry extend beyond the pursuit of profit and productivity to encompass societal contribution and the preservation of the planet. In the contemporary era, effective leadership demands a dual emphasis on technological precision and human creativity. This necessitates the establishment of a collaborative environment in which machines and humans' function in partnership rather than in opposition (Pham & Li, 2026; Fogaça et al., 2025).

This conceptual evolution has been accompanied by a series of distinct technological breakthroughs that have fundamentally redefined production methods, societal structures, and the role of the human worker within the economic machine. As illustrated in Table 1, the following section outlines some of the most significant differences between these industrial eras.

Table 1. Evolution of Industrial Revolutions and Supporting Technologies

IR Phase	Primary Focus	Key Enabling Technologies	Human Role
Industry 1.0	Mechanization through water and steam power	Steam engine, water power, mechanical manufacturing facilities (textile, steel, cement)	Labor - intensive manual work; replacement of manual production; workers faced hardship without social protection
Industry 2.0	Mass production, electrification, and energy transition	Electricity, assembly lines, machine tools, railway, petrochemicals, generators, signal processing	Division of labor; assembly line workers; automation of factory environments utilizing machinery and tools
Industry 3.0	Automation, computerization, and information systems	Electronics, Information Technology (IT), computers, Programmable Logic Controllers (PLCs), robotics, basic sensors	Machine operators and supervisors; contraction of blue-collar job market; widespread automation
Industry 4.0	Cyber-Physical Systems (CPS), intelligent manufacturing, and digitization	IoT/IIoT, Big Data analytics, Cloud Computing, Additive Manufacturing, smart sensors, fuzzy logic	Task-based automation; supervisors or operators; human intervention moved "out of the loop"; fear of job security
Industry 5.0	Human-centricity, resilience, and sustainability	Collaborative Robots (Cobots), Generative AI (LLMs), Extended Reality (XR), Blockchain, BCI, Human Digital Twins (HDT)	Human-machine collaboration; worker "back in the loop" as a collaborator, creator, and supervisor of AI; focus on well-being and ethics

2.2. The Conceptual Shift to Industry 5.0

The transition into Industry 5.0 is not merely an incremental technological update, but rather a profound ontological shift that addresses the systemic shortcomings of previous eras. In order to comprehend the leadership value intrinsic to the present transformation, it is imperative to contextualize the historical evolution of industry from mechanization to the digital-centric model of Industry 4.0 and finally to the human-centric value system of Industry 5.0 (Sacavém et al., 2025).

Industry 5.0 is not merely a refinement of Industry 4.0, but rather a paradigm shift in the very essence of industry. The conceptual shift is primarily defined by a move from a technology-driven approach toward a value-driven philosophy (Tooth et al, 2024). While I4.0 optimized efficiency and productivity, I5.0 prioritizes the synergy between human creativity and machine precision. This shift was accelerated by global disruptions, such as the COVID-19 pandemic, which highlighted the limitations of purely automated systems in terms of resilience and adaptability. The emerging paradigm emphasizes "worker back in the loop" strategies, where technology is designed to augment human capabilities rather than replace them (Islam et al., 2025). Moreover, this transformation redefines the role of industry from being an instrument of purely economic growth to a contributor to societal well-being and environmental health.

The modern industry is undergoing a period of significant transformation, with a shift in focus from the techno-centric and automation-driven models of Industry 4.0 to the value-driven, human-centric frameworks of Industry 5.0. In the course of this transition, a considerable digital transformation gap has been identified, particularly with regard to the manner in which organizational leadership manages the integration of complex emerging technologies without compromising the role of the human workforce (Nozari & Yordanova, 2025).

Figure 1 shows the industrial evolution between I4.0 and I5.0. As demonstrated in the accompanying figure, the transition from Industry 4.0 to 5.0 is characterized by the technological foundation of internet of things, big data analytics, cyber-physical system, digital twins, additive manufacturing and robotic systems.

Figure 1. Industrial Evolution: Transition from Industry 4.0 to 5.0

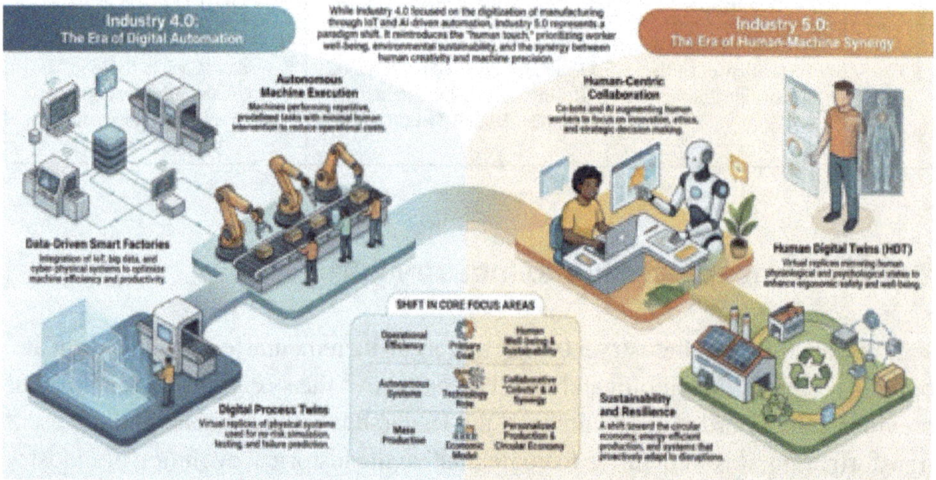

2.3. Core Pillars of Industry 5.0 Transformation as Value Drivers

The Industry 5.0 Transformation has the potential to fundamentally alter the nature of leadership in organizations. In the modern industrial environment, business leaders are required to orchestrate intricate socio-technical systems, oversee hybrid human-machine workforces, and make pivotal decisions in contexts characterized by uncertainty and accelerated transformation. In the context of contemporary environmental challenges, traditional command-and-control models are increasingly inadequate.

The strategic framework of Industry 5.0 is anchored by three interconnected values that serve as the primary metrics for leadership success in the modern age. Nair et al. (2024) provide a comprehensive characterization of the core of Industry 5.0, which is based on three main pillars that redefine the strategic imperatives for modern organizations. These pillars are not merely operational goals, but rather a new system of values that necessitates a corresponding evolution in leadership. The European Commission has defined the three core pillars - *human-centricity, sustainability, and resilience* - represent a shift from exclusively economic indicators to a balanced triple-bottom-line performance (Barata & Kayser, 2024).

Human-Centricity and the Re-Integration of the Worker

This pillar is predicated on the notion that human needs and capabilities should be prioritized in the domain of technological development. The process entails the conceptualization of systems that are designed to safeguard the autonomy, dignity and psychological well-being of workers (Bucci et al., 2024). Human-centricity in Industry 5.0 is defined by the re-evaluation of the worker as a primary asset. The creativity, problem-solving abilities and emotional intelligence of the worker are considered to be indispensable to the industrial process. In contradistinction to preceding paradigms that sought to displace manual labor through automation, Industry 5.0 aspires to augment human capabilities through intelligent tools (Davila-Gonzalez & Martin, 2024). This approach acknowledges that, whilst artificial intelligence and machines are capable of repetitive and precise tasks, human intelligence continues to prevail in critical thinking and innovation.

Sustainability and Environmental Stewardship

In the context of Industry 5.0, sustainability should be expanded to encompass not only environmental concerns, but also social dimensions; has evolved from being a discretionary corporate social responsibility initiative to a fundamental operational pillar (Awouda et al., 2024). The promotion of a circular economy, characterized by the reuse and recycling of resources, is central to the company's mission, alongside the provision of inclusive and fulfilling work environments. It is incumbent upon business leaders to implement circular economy principles, utilizing advanced technologies such as AI and additive manufacturing to minimize waste, optimize energy consumption, and reduce the environmental footprint of production. The integration of emerging digital technology facilitates the precise modelling of energy flows and resource utilization, thereby enabling real-time adjustments that align industrial activities with the planet's production limits (Chichi & Mamad, 2025).

Resilience Amidst Global Volatility

This value refers to the industry's capacity to maintain operations and adapt swiftly during crises. Resilience is the capacity of an industrial system to recover from disturbances, whether they be geopolitical shifts, natural disasters, or supply chain shocks (Barata & Kayser, 2024). The concept of Industry 5.0 underscores the imperative for the development of agile, decentralized, and flexible infrastructure, capable of expeditious adaptation to uncertainty. Leaders can develop robust contingency plans and respond with speed and precision by utilizing emerging digital technology to simulate various disruption scenarios, such as shipping delays or equipment failures (Chichi & Mamad, 2025).

Recent scholarship has explored these pillars through various technological and strategic lenses. Gaffley and Pelser (2025) investigate how manufacturing leaders navigate the I4.0 to I5.0 transition using a mixed-methods approach, including a survey of 2500 South African leaders. Their study identifies a "digital transformation gap" and highlights the importance of resilience and human-machine collaboration as critical success factors. While robust in its empirical reach, the study's focus on the South African context may limit the immediate generalizability of its strategic model to different global regulatory environments.

Islam et al. (2025) provide a bibliometric analysis of the I4.0 to I5.0 transition, identifying three primary clusters: technology (IoT/AI), sustainability, and human-centric approaches. It has been established that, while I5.0 builds on the automation of its predecessor, it introduces ethical considerations as a central tenet. A potential limitation of this study is its reliance on historical patterns, which may not fully capture the rapid, non-linear trajectories of generative AI advancements in the last year.

Overall, the quality of the literature is high in conceptualizing the ideals of Industry 5.0, effectively establishing the significance of human-centricity, sustainability, and resilience as core value drivers. The extant literature aligns firmly with the research aims of defining the shift from I4.0.

3. DIGITAL TWINS AND EMERGENCE OF HUMAN DIGITAL TWINS (HDTS)

The Fourth Industrial Revolution placed particular emphasis on cyber-physical systems, the Internet of Things, and automation driven by artificial intelligence (AI). In contrast, IR5 adopts a human-centric approach, emphasizing environmental sustainability and system resilience. This approach treats humans not merely as operators, but as central components whose well-being and creativity guide the manufacturing process (Barata & Kayser, 2024).

The transition from the techno-centricity of Industry 4.0 to the human-centric paradigm of Industry 5.0 has necessitated a fundamental re-evaluation of digital simulation technologies. Central to this evolution is the digital twin, a technology that has progressed from a mere mirroring of physical assets to a complex, bidirectional socio-technical system. There is a broad consensus that I5.0 is essentially a sociotechnical system, requiring integration of human factors and advanced data analytics. A recurrent theme pertains to the function of DT as a "key enabling technology" for the convergence of physical and digital domains.

3.1. Traditional Digital Twin (DTs)

Traditional DTs originated as advanced cyber-physical systems designed to create dynamic, virtual replicas of physical objects, such as machines, production lines, and supply chains. Therefore, DTs predominantly focus on machine productivity and treat humans as either external supervisors or mere users, failing to integrate the complexity of human needs, cognitive limits, and behaviors.

The concept of the digital twin has historically been anchored in the digitization of physical assets to optimize operational efficiency and predictive maintenance. Lv (2023) highlights that traditional DTs project the physical world through data accumulation and AI, feeding simulation results back to optimize physical machine performance. Similarly, Awouda et al. (2024) demonstrate that DTs effectively decouple physical flows from planning, enabling virtual scenario testing and remote management before real-world implementation.

Tooth et al. (2024) reveal through interviews with 51 experts that the term "digital twin" is often used as a buzzword ("Twinwash"), leading to stakeholder mistrust and conflicting system designs. Moreover, while some studies view the shift as a necessary social evolution, others like Rojek et al. (2025) emphasize that the shift to Industry 6.0 is already on the horizon, driven by self-organizing autonomous ecosystems. This creates a tension between the immediate need to solve I5.0's social sustainability goals and the rapid push toward further autonomy.

Figure 2 shows the evolution of digital twins from Industry 4.0 to 5.0, characterized by a shift in value, human-machine integration and human-centric metadata.

Figure 2. Digital Twins: The core of Industry 5.0

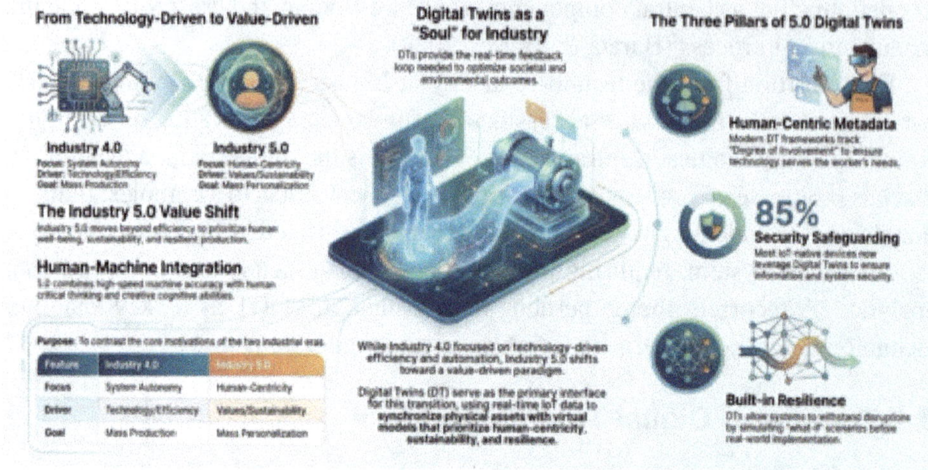

However, recent critiques suggest that traditional DT models are insufficient for the Industry 5.0 era. Barata and Kayser (2024) argue through the lens of strong structuration theory that DTs have been overly focused on technical agency, often neglecting the human agent's role within the system. Their bibliometric analysis reveals that while DTs have successfully shaped physical-digital convergence, they often operate as closed-loop systems that prioritize machine autonomy over human-system integration. Similarly, Islam et al. (2025) note that while Industry 4.0 achieved high levels of automation, it created a "C" where human factors were treated as external variables rather than core components. The limitation of the traditional DT, therefore, lies in its inability to capture the nuance of human cognition, health, and behavior in real-time, necessitating a shift toward more holistic modeling.

3.2. Human Digital Twins (HDTs)

The emergence of HDTs addresses the glaring limitations of traditional DTs by digitally replicating the biological, psychological, and emotional characteristics of human workers. Therefore, it represents an ontological shift from Digital Human Modeling (DHM) to dynamic, real-time human replicas. He et al. (2024) provide a comprehensive review of this evolution, noting that while traditional DHM supported ergonomic analysis, it lacked the real-time data integration and bidirectional interaction required for modern manufacturing. The HDT addresses this by integrating multi-source data and AI to enable continuous monitoring and personalized services.

The significance of HDTs lies in their potential to transform worker safety and well-being. Davila-Gonzalez and Martin (2024) propose a holistic approach that incorporates emotional analytics and advanced AI. Their research highlights that HDTs can go beyond physical monitoring to include psychological and emotional states, effectively "humanizing" the digital replica. This is echoed by Bucci et al. (2024), who argue that HDTs serve as a "human-centric cornerstone", allowing for a more nuanced approach to smart manufacturing where the system adapts to the human, rather than the human adapting to the system.

Bucci et al. (2024) conducted a systematic review revealing that HDTs facilitate the real-time monitoring of fatigue, stress, and posture, allowing for proactive interventions that drastically reduce workplace injuries and mitigate cognitive overload. Furthermore, Barata and Kayser (2024) emphasize that HDTs extend beyond the individual worker to encompass broader social sustainability, utilizing strong structuration theory to understand the complex interplay between human agency and digital structures.

Overall, the Human Digital Twin is the technological cornerstone of the Industry 5.0 transformation, representing a sophisticated digital replica of a human being that mirrors biological, psychological, and emotional characteristics. Whilst traditional Digital Twins have been utilized in sectors such as aerospace and automotive manufacturing to monitor mechanical systems, HDTs extend this capability to the most complex entity in the production line: the human worker. Therefore, traditional DTs mirror physical machinery, HDTs replicate human characteristics, incorporating physiological and psychological data to optimize ergonomics, predict fatigue, and tailor workstations to individual needs.

3.3. Conceptual and Technical Architecture of the Human Digital Twin (HDTs)

The architecture of the HDT is significantly more complex than that of its machine-based predecessors. He et al. (2024) propose a unified framework comprising three layers: the physical twin (the human worker), the virtual twin (the digital model), and the linkage (the data exchange mechanism). This framework is unique because the virtual twin incorporates not only geometric and physical data but also "*model-data-hybrid-enabled simulation*", allowing for predictive insights into human fatigue and cognitive load (Yordanova & Nozari, 2026; Nozari & Yordanova, 2026).

Building upon this, Awouda et al. (2024) introduce an IoT-based framework for DTs in the Industry 5.0 era, utilizing the IoT Architectural Reference Model (IoT-ARM). Their technical architecture emphasizes sustainability and resilience, arguing that a standardized collection of architectural instruments is necessary for interoperability. This is a critical observation, as the lack of a standardized framework has been cited as a primary barrier to widespread HDT adoption.

The Three-Layer Framework of HDT

The architecture of a human digital twin is comprised of three distinct but interconnected layers: the physical twin, the virtual twin, and the synchronization thread (Scataglini & Truijen, 2023; He et al, 2024; Davila-Gonzalez & Martin, 2024):

- ***The Physical Twin*** is defined as the individual worker operating in the industrial environment. The data is collected from the physical twin using an array of sensors, including wearable devices (*e.g. smartwatches and biosensors*), video logs for facial expression analysis, and environmental sensors.
- ***The Virtual Twin*** is defined as the digital representation hosted in a virtual environment. The system integrates human modelling, AI engines and machine learning models to simulate the worker's physical and cognitive states. The virtual twin is "experimentable", i.e. business leaders are able to analyses "what-if" scenarios regarding worker fatigue or task allocation in a risk-free digital space.
- ***The Synchronized Thread***: This model thus represents the bi-directional communication channel between the physical and virtual worlds. The transmission of data occurs in real-time, with the virtual model subsequently processing the information and providing feedback to the operative or the operations manager, frequently through XR interfaces.

The Hierarchical Framework of HDT Profiles

Other modern research into human digital twin suggests using an integrated system to capture data across three main domains: *physical, mental, and emotional* (Table 2). These data points are often stored in a unified "*Human Digital Twin Profile*", typically in a structured JSON format, to facilitate real-time interoperability between systems (Asad et al., 2023; Davila-Gonzalez & Martin, 2024; Jeong et al., 2025):

- ***The Physical Domain:*** This captures kinematics and physiological metrics. Technologies such as wearable inertial measurement units (IMUs) and RGB-D sensors are used for posture estimation and ergonomic performance analysis. Real-time metrics include heart rate variability (HRV), sleep quality, and GPS tracking to ensure worker safety and optimize physical tasks.
- ***The Mental Domain:*** This level focuses on psychological states and cognitive load. Through personality tests and behavioral analytics, the HDT identifies potential stressors and assesses a worker's professional characteristics, such as information preferences and skill levels.

- ***The Emotional Domain:*** Utilizing "Emotion Artificial Intelligence" (EAI), HDTs analyze facial expressions, speech patterns, and text-based sentiment from written conversations. Empathic chatbots, trained using Natural Language Processing (NLP), evaluate these emotional cues to identify early signs of anxiety or depression before they escalate into mental health crises

Table 2. HDT Data Domains and Enabling Technologies

Domain	Key Metrics	Enabling Technologies
Physical	Kinematics, Heart Rate, HRV, Sleep, Fatigue	IMUs, RGB-D Sensors, Microsoft Kinect, Wearables
Mental	Cognitive load, stress levels, personality traits	Behavioral analytics, cognitive tests, BCIs
Emotional	Anxiety, depression, engagement, satisfaction	EAI, NLP, Speech-to-Emotion algorithms, Video logs

The role of artificial intelligence (specifically Generative AI) is increasingly central to this architecture. Rojek et al. (2025) explore how GenAI can enhance DTs by providing autonomous decision-making and synthetic data generation for training human-centric models. They suggest that the evolution toward Industry 6.0 will see HDTs becoming even more proactive, utilizing AI to predict and mitigate risks before they manifest in the physical twin.

Moreover, the integration of advanced AI paradigms, including Natural Language Processing (NLP) for the analysis of emotional intents in written conversations and video logs for speech analysis, enables the HDT to achieve a profound comprehension of mental mindsets and potential stressors. It is imperative to acknowledge the pivotal role of this capability in attaining the objective of "0-accident" goals. This is achieved by means of the identification of emotional factors, such as anxiety and depression, which have the potential to compromise safety.

The functionality of human digital twins is categorized in a range of ways, from simple data storage to complex multi-objective optimization. The following hierarchy (Table 3) enables leaders to ascertain the requisite level of technological maturity for their specific organizational requirements.

Table 3. Functional Taxonomy and Categorization of Human Digital Twins

Functional Level	Capability	Industrial Application
Level 1: Store	Data Integration	Career history and basic medical records
Level 2: Analyze	Pattern Recognition	Identifying ergonomic risks or performance trends
Level 3: Personalize	User Customization	Tailoring machine interfaces to worker skill levels
Level 4: Predict	Forecasting	Predicting burnout, attrition, or future skill needs
Level 5: Control	Real-time Feedback	Triggering safety alerts or adjusting light/temp based on stress
Level 6: Optimize	Dynamic Reconfiguration	Real-time job rescheduling based on workforce disturbances

4. EXTENDED REALITY (XR) AND HUMAN DIGITAL TWINS (HDTS) INTEGRATION FOR LEADERSHIP VALUE

The digital transformation of organizations is an extensive process which often necessitates a fundamental shift that extends beyond the mere implementation of technology. In most cases, this necessitates profound changes in organizational leadership approaches, management structures, and the integration of people and systems. In the context of Industry 5.0, the focus is shifting from automation-driven metrics to human-centric, sustainable and resilient operations. Two vital catalysts in this shift are Extended Reality and Digital Twins, and more specifically the emerging concept of Human Digital Twins. The combination of these technologies has the potential to transform organizational leadership by facilitating immersive, data-driven environments for complex decision-making, team management and fostering empathy. Moreover, the future of manufacturing is becoming increasingly characterized by increased immersion, intelligence and human-centred design.

4.1. Definition of XR and Human-Machine Interaction

In general, immersive technologies are closely aligned with Industry 4.0 technologies, which, however, are evolving toward Industry 5.0 and prioritize sustainable, human-centred industrial applications. This technology has the capacity to leverage the principles of Industry 5.0 by emphasising dynamic content and rapid interactions between humans and machines. Consequently, this involves a digital ecosystem where the physical and virtual worlds converge, driven by artificial intelligence, the Internet of Things, and digital twins. As demonstrated in the study conducted by Wang et al. (2024), this interconnected space facilitates the real-time exchange of data, predictive maintenance procedures, and optimized workflows. Consequently, it is instrumental in promoting innovation, enhancing efficiency, and ensuring sustainability.

Extended Reality, a term used to denote the integration of Virtual Reality (VR), Augmented Reality (AR), and Mixed Reality (MR), has emerged as the predominant interface through which humans interact with the data-rich environments of Industry 5.0. The integration of the physical and virtual domains, facilitated by XR, fosters a more natural and intuitive human-machine collaboration. This paradigm shift enables leaders to visualize intricate systems and workers to receive real-time, context-aware guidance, thereby enhancing operational efficiency and safety (Sthapit & Olbina, 2026). Therefore, XR acts as an intuitive interface bridging physical environments with the digital twin data. AR and MR enhance real-time situational awareness by superimposing digital insights onto physical workspaces, transforming passive observation into active human-machine collaboration.

In the context of Industry 5.0, XR fundamentally transforms Human-Machine Interaction (HMI) by translating complex, high-volume operational data into intuitive, spatial visualizations. XR leverages human perceptual and motor capabilities, reducing cognitive load and ergonomic strain while significantly enhancing situational awareness (Manoli et al., 2026). Consequently, XR facilitates a more natural, collaborative, and safer interaction paradigm between human operators and cyber-physical systems, empowering individuals to intuitively grasp complex system behaviors.

The conceptual boundary of XR has undergone significant refinement in recent years. Girginova (2026) argues that XR should no longer be viewed merely as a visual overlay but as a fundamental "redrawing of the conceptual map" where the digital and physical are inextricably linked. This perspective aligns with the requirements of Industry 5.0, where Human-Machine Interaction (HMI) evolves from simple command-response cycles to intuitive, bidirectional collaborations.

A critical component of this interaction is "human-centric" design. According to Leirmo (2024), the "holy grail" of Industry 5.0 lies at the intersection of sustainability and human-centricity. Here, technology is designed to compensate for the limitations of the human mind and facilitate superior creative thinking. This concept is realised through sophisticated HMI interfaces. For example, Manoli et al. (2026) present a modular metaverse framework that uses ROS2 and Unity to enable gesture-based control in manufacturing. Their study shows that XR interfaces that immerse users in the virtual environment can improve human-robot collaboration (HRC) by making interactions more natural and less cognitively demanding.

4.2. XR and HDTs Integration

The integration of XR and HDTs represents the technical convergence of real-time data streaming and immersive visualization. Whilst traditional DTs serve as high-fidelity, bidirectional virtual replicas of physical assets, the concept of the Human Digital Twin (HDT) extends this digital architecture to the human operator (Manoli et al., 2026). HDTs are defined as dynamic digital representations of human physical, physiological and cognitive characteristics. The integration of extended reality with high-definition telepresence technology establishes the foundational framework for the conceptualization of the “industrial metaverse”. This emerging paradigm encompasses an interactive spatial computing environment, wherein the realms of the digital and physical domains are synchronized, enabling users to seamlessly monitor and manipulate processes in real-time (Solmaz et al., 2026).

Stacchio et al. (2022) advance this integration through the Human Collaborative Intelligence empowered Digital Twin (HCLINT-DT) framework. This study is significant because it moves beyond mere visualization, integrating human annotations (textual and vocal) into the DT to preserve organizational knowledge. Their methodology, involving user testing and Cronbach’s alpha validation, suggests that XR-DT integration fosters a “collaborative intelligence” where the DT becomes an active resource for training and remote assistance.

Contemporary cyber-physical systems (CPS) are undergoing an evolution that encompasses the integration of Human Digital Twins (Nozari & Yordanova, 2025) - virtual replicas that monitor physiological, cognitive, and psychological states - and Extended Reality interfaces that facilitate the convergence of physical and digital domains. It is imperative that leadership recognizes the necessity to master the application of these tools, not solely for the purpose of operational efficiency, but also to empower workers, reduce cognitive loads, and cultivate a culture of resilient, sustainable human-machine collaboration.

In addition, HDTs differ from standard DTs by incorporating human-specific data, such as ergonomic markers, health metrics, and cognitive loads. Asad et al. (2023) provide a comprehensive review of enabling technologies, noting that the “human-in-the-loop” approach requires advanced sensor fusion and AI-driven analytics to create a meaningful digital counterpart. Furthermore, using XR interfaces to interact with DTs naturally, enabling immersive training, real-time remote assistance, and collaborative intelligence (Nozari & Yordanova, 2025).

Immersive Training and Knowledge Transfer

XR and HDTs are a technology that is transforming the field of workforce development by offering safe and repeatable simulations for tasks that are considered to be high-risk or complex. This is of particular importance in sectors such as healthcare, aviation, energy, and manufacturing, where errors can be costly and safety is of the utmost importance. This training programme enables employees to rehearse emergency procedures and engage with virtual apparatus without causing interruption to actual production processes or jeopardizing their own physical well-being (Jeong et al., 2025).

Real-Time Operational Support and Troubleshooting

In addition to training, XR and HDTs provide essential frontline support through the utilization of AR-guided maintenance and remote expert assistance. Technicians have the capability of utilizing AR headsets to overlay step-by-step instructions or real-time diagnostic data onto physical machinery, thereby reducing the time required for troubleshooting and repair. In complex assembly environments, technology applications offer ergonomic cyber-representations of production processes (Feddoul et al., 2023). These applications have been shown to improve quality and flexibility, while simultaneously enhancing safety.

Figure 3. Human Digital Twins: The Human-Centric of Industry 5.0

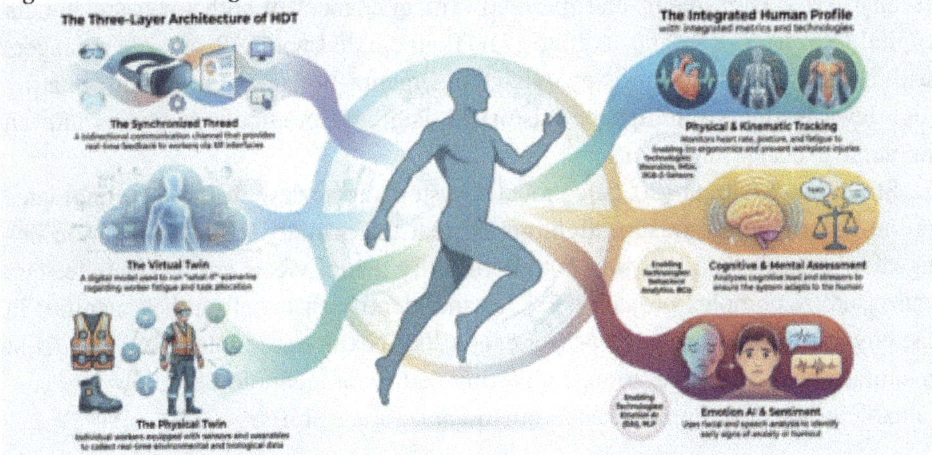

Overall, despite these advancements, the extant literature reveals a persistent challenge, namely the digitization of the environment itself. Solmaz et al. (2026) address this gap by proposing a methodology to reconstruct 3D models from shop floor scans when native CAD models are unavailable. The researchers' findings emphasize that the value of an XR-HDT system is contingent upon the fidelity and modularity of the data pipelines. In the event of a disparity between the digital environment and the physical shop floor, the "twin" is rendered incapable of providing reliable insights for leadership.

4.3. XR and HDTs as the Immersive Interface for Leadership Value Proposition

While Digital Twins were traditionally relegated to the monitoring of physical assets, the emergence of HDTs allows for the digital replication of human physiological, cognitive, and behavioral states. When coupled with XR - encompassing VR, AR, and MR - these technologies provide an immersive environment for the purpose of developing next-generation leadership capabilities. Therefore, the intersection of XR and HDTs offers a profound value proposition for new organizational leadership. The human factor, which encompasses human intelligence, emotional capabilities, ethical thinking, and adaptive learning, represents both the most valuable asset and the most vulnerable element in digital transformation initiatives (Attah et al., 2023).

The term "Leadership 5.0" was coined by Warner-Søderholm and Kuoppamäki (2025) to denote a specific concept in the field of leadership. This leadership style prioritizes resilience and human potential over raw output. Within this framework, XR and HDTs function as the immersive interface that enables leaders to engage in "scenario risk assessment" and "predictive maintenance" of both assets and human capital (Nozari & Yordanova, 2026). HDTs integrated with XR provide managers with a comprehensive "birds-eye view" of organizational systems, consolidating high-level business intelligence, historical data, and predictive analytics into an immersive, accessible format.

Sthapit and Olbina (2025) conduct a systematic review of these technologies in the construction industry, highlighting that the primary value proposition lies in safety training and real-time monitoring. By using XR as an interface, leaders can visualize complex project lifecycles and identify risks before they manifest in the physical world. Similarly, Mayer et al. (2025) explore the role of XR and AI in manufacturing reconfiguration, suggesting that these technologies allow for "sustainable urban development" and agile organizational shifts.

Figure 4. Leadership 5.0: Unlocking Value thought XR and Human Digital Twins

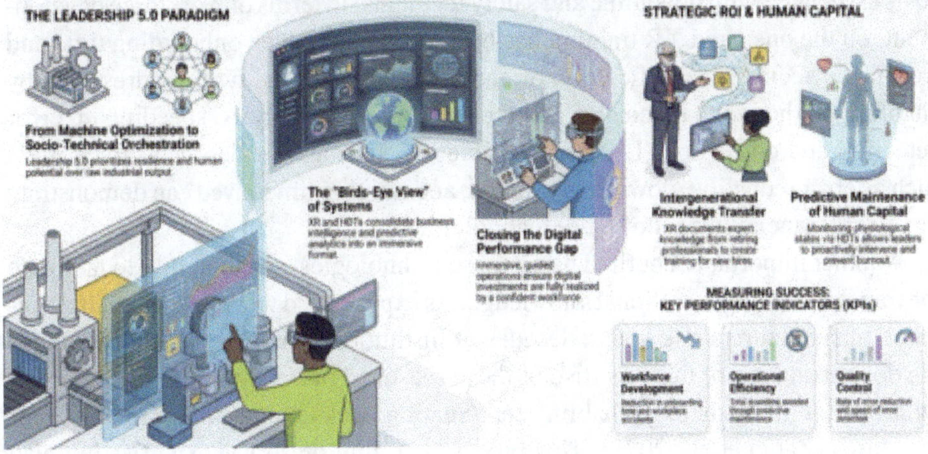

A significant challenge for today business leaders is the gap between smart system investments and frontline performance. The value of leadership in Industry 5.0 is found in closing this gap through hands-on enablement that "actually sticks". By using digital twins and extended reality to provide immersive, guided operations, leaders can ensure that digital investments are fully realized by a confident and skilled workforce (Belfadel et al., 2026). Moreover, digital leadership in the modern era requires navigating extreme ambiguity, managing decentralized virtual teams, and utilizing data-driven insights for strategic agility. The prevailing paradigm of leadership value has undergone a significant shift, transitioning from the realm of machine optimization to the orchestration of socio-technical systems. These systems are characterized by a strategic prioritization of human creativity, well-being, and environmental resilience.

In this regard, a critical gap exists in how "Leadership Value" is measured. Most studies, such as those by Manoli et al. (2026) and Stacchio et al. (2022), focus on technical performance metrics (e.g., latency, accuracy, user satisfaction scores) rather than long-term strategic value or organizational ROI. While the literature establishes that XR and HDTs improve collaboration, it often fails to articulate how this translates into leadership effectiveness or cultural transformation within the organization. According to Warner-Søderholm & Kuoppamäki (2025), leadership value is realized when workers are freed from rote tasks to focus on strategic and creative roles, leading to more fulfilling and meaningful employment. This transition requires ergonomic workplace designs that prioritize mental health, safety, and worker satisfaction. By using XR and HDTs to monitor physiological and emotional states, business leaders can proactively intervene to prevent burnout and ensure a safe, supportive environment.

In addition, digital twins and extended reality deliver tangible ROI by reducing costs associated with downtime and safety incidents. In terms of workforce development, on the one hand XR training has been shown to reduce onboarding time and workplace accidents significantly; on the other hand, digital twins address this by enabling predictive maintenance and real-time process analysis, speeding up error detection and correction (Lv, 2023). Therefore, business leaders who track KPIs such as error reduction, downtime avoided, and training time saved can demonstrate the clear financial impact of these technologies.

Another important benefit of using these technologies is their ability to facilitate the transfer of intergenerational knowledge. As experienced workers approach retirement, manufacturers risk losing decades of institutional knowledge. XR facilitates the documentation of the expertise of these seasoned professionals within a digital twin environment, thereby enabling the creation of detailed training modules for new hires (Zahid et al., 2025). This process not only preserves expertise but also motivates younger, digitally native workers who prefer technology-driven learning solutions. Conversely, the shift towards sustainability also aligns with the values of digitally native generations, such as Millennials and Gen Z, who priorities environmental preservation. A leadership approach that integrates sustainability into its core strategy fosters long-term resilience and strengthens the brand's social license to operate. Social resilience is equally critical and encompasses the behavioral response of communities and workforces to change. Resilient leadership focuses on building trust and empowering employees to take ownership of organizational goals, ensuring the workforce can adapt and remain confident in the face of technological upheaval.

In summary, extended reality and human digital twins' integration is tailored to the individual context of the worker. It is imperative for business leaders to acknowledge this paradigm shift if they are to unlock latent value by leveraging human digital twins and extended reality. By doing so, they can facilitate the creation of more inclusive, adaptable and ethically responsible industrial ecosystems (Wang et al., 2024). Therefore, the immersive interface of XR and HDTs equips leaders with both the macro-level systemic understanding and the micro-level empathetic insights necessary to drive sustainable digital transformation.

5. CONCLUSION

The study characterizes the emergence of Industry 5.0 as not just the technological successor to Industry 4.0, but also as a profound philosophical shift towards placing greater emphasis on people, sustainability and resilience within a new industrial paradigm. Driven by the need to overcome the limitations of a model focused entirely on efficiency, this evolution has fundamentally changed the requirements for leadership value.

This chapter attempts to evaluate how the synergistic integration of Human Digital Twins and Extended Reality facilitates leadership value and human-centric transformation within Industry 5.0 environments. Specifically, the research aimed to identify the conceptual architectures required for this fusion, analyze their impact on worker well-being and performance, and evaluate how these immersive, data-driven environments redefine leadership capabilities in volatile industrial contexts. By synthesizing emerging discourse and technological frameworks, this paper underscores the transition from the techno-centric efficiency of Industry 4.0 to a value-driven paradigm that prioritizes human-machine synergy.

The findings indicate that the integration of HDTs and XR serves as a critical technological cornerstone for Industry 5.0. HDTs extend traditional digital twin capabilities by replicating biological, psychological, and emotional characteristics, thereby enabling real-time monitoring of fatigue, stress, and cognitive load. When interfaced through XR, these twins provide leaders with an intuitive, spatial environment for "scenario risk assessment" and "predictive maintenance" of human capital. One of the most important findings of this chapter is that leadership value in the modern era is no longer derived solely from machine optimization. Instead, it is derived from the leadership of socio-technical systems that safeguard worker dignity and foster resilient, inclusive industrial ecosystems.

From a theoretical perspective, this research contributes to the burgeoning field of "Leadership 5.0" by reframing the leader's role as a facilitator of collaborative intelligence. It challenges the historical "cog in the machine" mentality by positioning the worker as a primary asset whose creativity is augmented, rather than replaced, by intelligent tools. Furthermore, the chapter integrates strong structuration theory to explain the complex interplay between human agency and digital structures, suggesting that the "industrial metaverse" is not merely a visual tool but a fundamental redrawing of the conceptual map between physical and digital domains.

Future research should focus on the long-term strategic value and organizational ROI of these technologies, moving beyond the current emphasis on technical performance metrics like latency or user satisfaction. There is a pressing need for empirical studies that track the impact of HDTs on cultural transformation and intergenerational knowledge transfer within diverse global regulatory environments.

Despite its insights, this chapter is limited by the rapid, non-linear trajectory of generative AI advancements, which may quickly surpass current historical patterns. Additionally, the fidelity of the "twin" remains contingent upon the modularity of data pipelines; disparities between digital environments and physical shop floors can undermine the reliability of leadership insights. Future research could address these gaps by developing standardized interoperability frameworks and exploring the proactive risk-mitigation capabilities of Industry 6.0 autonomous ecosystems.

In conclusion, the study advances the understanding of leadership in Industry 5.0 by demonstrating that the integration of HDTs and XR provides the necessary "flight controls" for navigating modern industrial volatility. While these technologies offer unparalleled productivity, the ultimate leadership value resides in maintaining the "human touch" - striking a delicate balance between high-fidelity monitoring and the preservation of psychological safety and trust.

ACKNOWLEDGEMENT

The author acknowledges with gratitude funding provided by the Bulgarian National Research Fund under administrative contract KP-06-H65/5 of 12.12.2022.

REFERENCES

Asad, U., Khan, M., Khalid, A., & Lughmani, W. A. (2023). Human-centric digital twins in industry: A comprehensive review of enabling technologies and implementation strategies. *Sensors (Basel), 23*(8), 3938. DOI: 10.3390/s23083938 PMID: 37112279

Attah, R. U., Ogunsola, O. Y., & Garba, B. M. P. (2023). Leadership in the digital age: Emerging trends in business strategy, innovation, and technology integration. *Iconic Research and Engineering Journals, 6*(9), 389–411.

Awouda, A., Traini, E., Bruno, G., & Chiabert, P. (2024). IoT-Based Framework for Digital Twins in the industry 5.0 era. *Sensors (Basel), 24*(2), 594. DOI: 10.3390/s24020594 PMID: 38257686

Barata, J., & Kayser, I. (2024). How will the digital twin shape the future of industry 5.0? *Technovation, 134*, 103025. DOI: 10.1016/j.technovation.2024.103025

Belfadel, A., Lacheroy, T., Creff, S., Hamida, A. B., & Didier, J. Y. (2026). XRTwin-4Industry: Empowering Industrial Digital Twins with XR and Instruction System-A Framework for Standardization and Interoperability. *Future Generation Computer Systems, 182*, 108476. DOI: 10.1016/j.future.2026.108476

Breque, M., De Nul, L., & Petridis, A. (2021). Industry 5.0: Towards a sustainable, human-centric and resilient European industry. European Commission. DOI: 10.2777/308407

Bucci, I., Fani, V., & Bandinelli, R. (2024). Towards human-centric manufacturing: Exploring the role of human digital twins in industry 5.0. *Sustainability (Basel), 17*(1), 129. DOI: 10.3390/su17010129

Chichi, O., & Mamad, M. (2025). Exploring synergies between Industry 4.0 and Industry 5.0 for digital, resilient and sustainable supply chains: a systems thinking perspective. *Continuity & Resilience Review*, 1-35. https://doi.org/10.1108/CRR-05-2025-0014

Davila-Gonzalez, S., & Martin, S. (2024). Human digital twin in industry 5.0: A holistic approach to worker safety and well-being through advanced AI and emotional analytics. *Sensors (Basel), 24*(2), 655. DOI: 10.3390/s24020655 PMID: 38276347

Feddoul, Y., Ragot, N., Duval, F., Havard, V., Baudry, D., & Assila, A. (2023). Exploring human-machine collaboration in industry: A systematic literature review of digital twin and robotics interfaced with extended reality technologies. *International Journal of Advanced Manufacturing Technology, 129*(5), 1917–1932. DOI: 10.1007/s00170-023-12291-3

Fogaça, D. R., Grijalvo, M., & Sacomano Neto, M. (2025). What are Industry 4.0 and Industry 5.0 all about? An integrative institutional model for the new industrial paradigms. *Administrative Sciences, 15*(4), 118. DOI: 10.3390/admsci15040118

Gaffley, G. R., & Pelser, T. G. (2025). A digital transformation strategy model for leadership in manufacturing: Considering the technological innovations to advance industry 5.0 in smart manufacturing. *South African Journal of Business Management, 56*(1), 1–16. DOI: 10.4102/SAJBM.v56i1.5449

Girginova, K. (2026, February 13). When XR Meets Digital Twins: Redrawing the Conceptual Map of Extended Reality. *Frontiers in Virtual Reality, 7*, 1762601. DOI: 10.3389/frvir.2026.1762601

He, Q., Li, L., Li, D., Peng, T., Zhang, X., Cai, Y., Zhang, X., & Tang, R. (2024). From digital human modeling to human digital twin: Framework and perspectives in human factors. *Chinese Journal of Mechanical Engineering, 37*(1), 9. DOI: 10.1186/s10033-024-00998-7

Islam, M. T., Sepanloo, K., Woo, S., Woo, S. H., & Son, Y. J. (2025). A review of the industry 4.0 to 5.0 transition: Exploring the intersection, challenges, and opportunities of technology and human–machine collaboration. *Machines (Basel), 13*(4), 267. DOI: 10.3390/machines13040267

Jeong, H., Chen, K. B., Jeon, M., & Stanley, L. (2025). Human-Centered eXtended Reality for Occupational Applications in the Era of Industry 5.0: Introduction to the Special Issue. *IISE Transactions on Occupational Ergonomics and Human Factors, 13*(2), 75–78. DOI: 10.1080/24725838.2025.2512681 PMID: 40464284

Leirmo, T. L. (2024). Digital twins for Industry 5.0: Unlocking the human potential. *Procedia CIRP, 130*, 761–766. DOI: 10.1016/j.procir.2024.10.161

Lv, Z. (2023). Digital twins in industry 5.0. *Research, 6*, 0071. DOI: 10.34133/research.007

Manoli, E., Mattera, G., Caggiano, A., Marzano, A., & Nele, L. (2026). From Digital Twins to Immersive Manufacturing: XR and Gesture-Based Control for Enhancing Human–Robot Collaboration. *IET Collaborative Intelligent Manufacturing*, *8*(1), e70055. DOI: 10.1049/cim2.70055

Mayer, A., Greif, L., Häußermann, T. M., Otto, S., Kastner, K., El Bobbou, S., Chardonnet, J.-R., Reichwald, J., Fleischer, J., & Ovtcharova, J. (2025). Digital twins, extended reality, and artificial intelligence in manufacturing reconfiguration: A systematic literature review. *Sustainability (Basel)*, *17*(5), 2318. DOI: 10.3390/su17052318

Nair, A., Pillai, S. V., & Senthil Kumar, S. A. (2024). Towards emerging Industry 5.0–a review-based framework. *Journal of Strategy and Management*, •••, 1–19. https://doi.org/10.1108/JSMA-04-2023-0067

Nozari, H., & Yordanova, Z. (2025). A fuzzy–digital twin optimization framework for simultaneous management of waste and energy consumption in sustainable manufacturing. *An International Journal of Optimization and Control: Theories & Applications*, 25360152. https://doi.org/10.36922/IJOCTA025360152

Nozari, H., & Yordanova, Z. (2025). An intelligent digital twin approach for optimizing multi-channel supply chains in uncertain environments. *Green Technologies and Sustainability*, 100309. https://doi.org/10.1016/j.grets.2025.100309

Nozari, H., & Yordanova, Z. (2025). Hybrid digital twin and quantum AI with fuzzy multiobjective modeling in supply chain management. *Edelweiss Applied Science and Technology*, *9*(10), 609–628. DOI: 10.55214/2576-8484.v9i10.10482

Nozari, H., & Yordanova, Z. (2026). An Adaptive Fuzzy Multi-Objective Digital Twin Framework for Multi-Depot Cold-Chain Vehicle Routing in Agri-Biotech Supply Networks. *Logistics (Basel)*, *10*(2), 27. DOI: 10.3390/logistics10020027

Nozari, H., & Yordanova, Z. (2026). Digital twin–driven decision support for human capital management under uncertainty. *Green Technologies and Sustainability*, 100353. https://doi.org/10.1016/j.grets.2026.100353

Pham, H. S. T., & Li, S. B. (2026). Towards Industry 5.0: A conceptual model for leading organisational change in digital age. *Journal of Organizational Change Management*, *39*(1), 125–161. DOI: 10.1108/JOCM-03-2025-0264

Rojek, I., Mikołajewski, D., Piszcz, A., Małolepsza, O., & Kozielski, M. (2025). Role of Generative AI in AI-Based Digital Twins in Industry 5.0 and Evolution to Industry 6.0. *Applied Sciences (Basel, Switzerland)*, *15*(18), 10102. DOI: 10.3390/app151810102

Sacavém, A., de Bem Machado, A., dos Santos, J. R., Palma-Moreira, A., Belchior-Rocha, H., & Au-Yong-Oliveira, M. (2025). Leading in the digital age: The role of leadership in organizational digital transformation. *Administrative Sciences, 15*(2), 43. DOI: 10.3390/admsci15020043

Scataglini, S., & Truijen, S. (2023, July). Critical appraisal of using digital human model, virtual human, human digital twin and digital twin. In *International conference on digital human modeling* (pp. 154-158). Cham: Springer Nature Switzerland. DOI: 10.1007/978-3-031-37848-5_17

Solmaz, S., Mahy, W., & Zogopoulos, V. (2026). Shop floor environment digital twin scan for extended reality applications. *Procedia CIRP, 138*, 1127–1132. DOI: 10.1016/j.procir.2026.01.194

Stacchio, L., Angeli, A., & Marfia, G. (2022). Empowering digital twins with extended reality collaborations. *Virtual Reality & Intelligent Hardware, 4*(6), 487–505. DOI: 10.1016/j.vrih.2022.06.004

Sthapit, I., & Olbina, S. (2026). Digital Twin (DT) and Extended Reality (XR) in the Construction Industry: A Systematic Literature Review. *Buildings (Basel, Switzerland), 16*(3), 517. DOI: 10.3390/buildings16030517

Tooth, J., Watson, J. D. M., & Tuptuk, N. (2024). Transdisciplinary Perspectives on Human-Centric Digital Twins in Industry 4.0 & 5.0. *Transdisciplinary Perspectives on Human-Centric Digital Twins in Industry, 4*. http://dx.doi.org/10.2139/ssrn.5011257

Wang, B., Zhou, H., Li, X., Yang, G., Zheng, P., Song, C., Yuan, Y., Wuest, T., Yang, H., & Wang, L. (2024). Human Digital Twin in the context of Industry 5.0. *Robotics and Computer-integrated Manufacturing, 85*, 102626. DOI: 10.1016/j.rcim.2023.102626

Warner-Søderholm, G., & Kuoppamäki, M. (2025). Beyond Industry 5.0: Leadership 5.0—Driving Future-Ready Organizations. *Businesses, 5*(4), 56. DOI: 10.3390/businesses5040056

Yordanova, Z., & Nozari, H. (2025). Harnessing cyber-physical systems and digital twins for industrial innovation and economic resilience. In *Dynamic and Safe Economy in the Age of Smart Technologies* (pp. Yordanova, Z., & Nozari, H. (2026). An Agentic Digital Twin Framework for Fuzzy Multi-Objective Optimization in Dynamic Humanitarian Logistics. *Algorithms, 19*(3), 198. DOI: 10.3390/a19030198

Zahid, A., Ferraro, A., Petrillo, A., & De Felice, F. (2025). Exploring the role of digital twin and industrial metaverse technologies in enhancing occupational health and safety in manufacturing. *Applied Sciences (Basel, Switzerland)*, *15*(15), 8268. DOI: 10.3390/app15158268

Chapter 13
Hybrid Digital Twin and AI Frameworks for Climate-Resilient Agriculture

Khatere Rafiei
http://orcid.org/0009-0005-4471-6929
Independent Researcher, UAE

ABSTRACT

Increasing climate instability and limited natural resources have created serious challenges for the sustainability of agricultural production. In such circumstances, the use of advanced digital technologies can play an important role in improving the management of agricultural systems. This research examines the role of hybrid frameworks based on digital twins and artificial intelligence in the development of climate-resilient agriculture. In this framework, smart data ecosystems, digital models, and predictive analytics tools are considered to support adaptive decision-making in farm management. The results show that the integration of digital twins with data analytics can significantly improve resource efficiency, decision-making accuracy, and sustainability of agricultural systems.

1. INTRODUCTION

Agriculture has faced extensive developments in technology, data, and production systems management in recent decades. At the same time, one of the most important challenges that threatens the sustainability of agricultural production is the increase in climate instability and environmental changes. Rising global temperatures,

DOI: 10.4018/979-8-2600-1669-5.ch013

changing rainfall patterns, the occurrence of long-term droughts, and the increase in the intensity of extreme climatic phenomena have seriously limited the traditional structure of farm management. These conditions have made management approaches based on experience and fixed decisions no longer respond to the complexities of modern agricultural systems. In such circumstances, the development of intelligent frameworks that can analyze environmental data and adjust management decisions dynamically and adaptively has become a fundamental necessity (Barzigar et al., 2025).

In response to these challenges, the concept of smart agriculture has been proposed as one of the most important new approaches in the management of food production systems. Smart agriculture is based on the use of digital technologies, sensor networks, the Internet of Things, and environmental data analysis, which enable accurate monitoring of farm conditions and rapid response to environmental changes. However, many existing smart agriculture systems still focus mainly on data monitoring and have limited ability to dynamically model and predict the behavior of biological systems. Therefore, the development of more advanced models that can analyze the relationship between environmental data, biological processes, and management decisions in an integrated manner has become one of the main research areas in the field of digital agriculture (Hurmat et al., 2026).

In the meantime, digital twin technology has been considered as one of the most innovative tools for modeling and analyzing complex systems. A digital twin is actually a virtual representation of a physical system that is continuously updated using real data, simulation models, and analytical algorithms. In agriculture, this technology can enable accurate modeling of farm processes, monitoring of environmental conditions, and prediction of biological system behavior. In other words, a digital twin can act as an analytical platform that combines environmental data, sensor data, and biological models into a unified framework, providing a comprehensive picture of the state of the agricultural system. This capability allows decision makers to assess the potential effects of management decisions in the digital environment before making changes to the real environment (Yordanova & Nozari, 2026).

Despite the significant potential of digital twins in agricultural systems analysis, the effectiveness of this technology largely depends on its ability to analyze data and extract decision-making patterns. Here, the role of artificial intelligence as an advanced analytical engine becomes important. Artificial intelligence algorithms can identify complex patterns that are not visible in traditional analytical methods by analyzing huge volumes of environmental and biological data. Combining digital twins with artificial intelligence models can enable the development of systems that not only monitor the current state of the farm, but also predict future trends and provide management recommendations. Such frameworks can play an important role in improving resource efficiency, reducing production risks, and increasing the sustainability of agricultural systems (Suman et al., 2025).

Given the increasing complexity of agricultural systems and the pressures caused by climate change, the development of hybrid frameworks that can combine biological systems modeling, environmental data analysis, and intelligent decision-making in an integrated platform has become particularly important. Hybrid frameworks based on digital twins and artificial intelligence can enable the simultaneous analysis of various factors affecting agricultural production and help create adaptive management systems. In such an approach, data collected from the farm, predictive models, and decision-making algorithms are put into a feedback loop, continuously improving the system performance.

Despite recent advances in the field of digital agriculture, many existing frameworks still lack an integrated structure to combine the capabilities of digital twins and artificial intelligence in climate risk management. For this reason, the development of new conceptual models that can systematically analyze the interaction between environmental data, predictive models, and decision-making processes can play an important role in advancing future research in this area. This chapter attempts to examine the role of these technologies in increasing the resilience of agricultural systems to climate change by focusing on the development of hybrid frameworks based on digital twins and artificial intelligence. The presented approach, with its emphasis on data integration, analytical models, and intelligent decision-making mechanisms, provides a basis for the development of a new generation of adaptive agricultural systems that are capable of maintaining productivity and production sustainability in changing climatic conditions.

2. CLIMATE CHALLENGES AND THE NEED FOR INTELLIGENT AGRICULTURAL SYSTEMS

One of the most important challenges that agricultural production systems have faced in recent decades is the increase in climate instability and the complexity of environmental conditions. Climate change has not only affected rainfall and temperature patterns, but also caused phenomena such as prolonged droughts, flash floods, severe temperature fluctuations, and increased biological stresses in farms. These changes have made many traditional patterns of cultivation and farm management no longer as efficient as they used to be. In fact, agricultural systems that were designed in the past based on relatively stable climatic conditions are now faced with an environment in which the level of uncertainty has increased significantly. These conditions have made it more evident than ever before that there is a need to review agricultural management practices and move towards intelligent and data-based systems (Tariqul Islam et al., 2026).

One of the important consequences of climate change in the agricultural sector is the increase in production risk and the decrease in predictability in agricultural processes. In many parts of the world, the timing of growing seasons, rainfall patterns, and even soil quality are changing under the influence of climate change. This makes decisions about planting time, irrigation rates, crop types, and nutrient management more difficult. In such circumstances, farmers and agricultural system managers need tools that can accurately analyze environmental data and predict the potential consequences of management decisions. Without such tools, decisions are often based on limited experience or information that may not be effective in changing climate conditions (Nozari, 2026).

In addition to the challenges associated with climate change, the increase in global population and the growing demand for food products have also put additional pressure on agricultural systems. The demand for food production is expected to increase significantly in the coming decades, while the natural resources used in agriculture, especially water and soil, are facing serious constraints. This situation has made the need for optimal use of resources and increased productivity in agricultural systems a strategic priority. In such circumstances, efficient resource management will be difficult without the use of advanced technologies and intelligent decision-making systems. Future agricultural systems must be able to balance economic productivity, environmental sustainability, and resilience to climate change (Al-Asady, n.d.).

In response to these challenges, the concept of smart agriculture has been proposed as a new approach to farm and agricultural production management. Smart agriculture emphasizes the use of digital technologies, sensor networks, environmental monitoring systems, and big data analysis. The main goal of this approach is to create systems that can monitor the state of the farm in real time and adjust management decisions based on real-time data. For example, data related to soil moisture, air temperature, solar radiation, and plant growth status can be analyzed simultaneously to optimize decisions related to irrigation or plant nutrition. Such an approach allows for more precise resource management and reduced energy and water waste.

Despite these advances, many smart farming systems are still in their early stages of development and focus mainly on data collection. The main challenge in this area is to convert scattered and voluminous data into knowledge that can be used for management decision-making. Environmental data alone cannot be an effective guide for decision-making unless it is used in the form of analytical and predictive models. Therefore, developing systems that can convert data into information and then into operational decisions is one of the most important requirements of smart farming.

In the meantime, smart farming systems must be able to adapt to environmental uncertainties and sudden changes in climate conditions. This feature, often known as climate resilience, refers to the ability of the system to maintain its performance in the face of environmental stresses and shocks. Achieving such a capability requires the use of advanced analytical models, predictive algorithms, and adaptive decision-making mechanisms that can dynamically analyze environmental conditions and propose appropriate management strategies. For this reason, much recent research has focused on developing frameworks that can combine environmental data analysis, biological systems modeling, and management decision-making in an integrated framework (Hossain et al., 2026).

From a systems perspective, modern agriculture can be considered as a complex, multi-layered system in which various factors, including climatic conditions, soil characteristics, crop type, resource management, and economic factors, interact with each other. Effective management of such a system requires tools that can model these complex interactions and predict the consequences of different decisions. Without such tools, long-term planning in the field of agriculture will be accompanied by significant uncertainties. For this reason, the move towards smart agricultural systems that can use data, models, and digital technologies to improve decision-making has become one of the main axes of transformation in the agricultural sector.

Overall, the new climate conditions and increasing pressure on natural resources indicate that the future of agriculture will be highly dependent on the development of intelligent, technology-based management systems. These systems must be able to continuously analyze environmental data, predict changes in climate conditions, and suggest appropriate management solutions. Such an approach can pave the way for the creation of agricultural production systems that, in addition to increasing productivity, also have a higher level of resilience to climate fluctuations.

3. EVOLUTION OF DIGITAL TWIN APPLICATIONS IN AGRICULTURE

The evolution of digital technologies in recent decades has led to the emergence of new approaches to the management of agricultural systems. One of the most important of these approaches is the use of the concept of digital twins in the modeling and analysis of complex agricultural production processes. Digital twins essentially refer to the creation of a virtual representation of a physical system that is continuously updated using real data and analytical models. This technology was initially developed in advanced industries such as aerospace and industrial manufacturing, but its applications have gradually expanded to other areas, including agriculture. In the agricultural sector, digital twins can reconstruct the real state of the field, environmental conditions, plant behavior, and management processes in a digital environment, allowing the analysis of different management scenarios (Nozari & Yordanova, 2026b).

In the early stages of the development of digital agriculture, the main focus was on collecting environmental data and monitoring the condition of the fields. The use of soil sensors, local weather stations, and remote monitoring systems provided access to more accurate data on environmental conditions. However, these data were often used independently of each other and there was no unified framework for their analysis. For this reason, many management decisions were still made based on limited analysis or individual experiences. With the advancement of computing technologies and the increase in data processing power, it became possible to develop more advanced models for analyzing this information, paving the way for the emergence of the concept of digital twins in agriculture (Mehta et al., 2026).

The first generation of digital twin applications in agriculture focused more on simulating farm processes. At this stage, digital models were used to represent some aspects of the agricultural system, such as plant growth, water consumption, or temperature changes. These models allowed for the prediction of some system behaviors, but were often designed statically and had limited ability to interact with real-time data. Over time, with the expansion of sensor networks and the development of the Internet of Things in agriculture, it became possible to directly connect digital models to real-time data. This evolution has transformed digital twins from simple simulation tools into dynamic analytical systems capable of continuously representing the real-world situation on the farm.

In the next stage of the evolution, digital twins were used as part of smart farm management systems. At this stage, environmental data, soil information, climate conditions, and crop growth status were combined into a single digital framework. This framework enabled the simultaneous analysis of multiple factors affecting agricultural production and allowed farm managers to examine the effects of different management decisions before they were implemented. For example, different irrigation or nutrient management scenarios could be simulated in a digital environment to select the best option for specific farm conditions. Such an approach has become particularly important in a situation where climate change has increased uncertainty in agricultural production (Alexander et al., 2025).

In recent years, the application of digital twins in agriculture has entered a new phase in which this technology is combined with big data analytics, predictive models, and decision support systems. At this stage, digital twins are no longer just simulation tools, but also serve as analytical platforms for managing complex agricultural systems. These systems can analyze various data, including meteorological information, remote sensing data, soil information, and crop yield data, in an integrated environment and provide a more comprehensive view of the state of the agricultural system. Such a capability allows for better prediction of environmental changes and optimization of management decisions (Nozari & Yordanova, 2025a). In order to better understand the evolution of digital twin applications in agriculture, the evolution of this technology is presented in the form of a conceptual framework in Figure 1.

Figure 1. Evolution of digital twin applications in agricultural systems

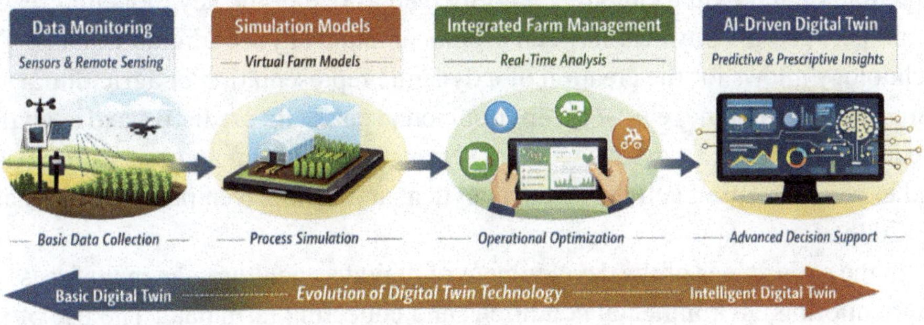

This figure shows that digital twin applications in agriculture started from the stage of data monitoring and basic modeling and gradually moved towards advanced analytical and decision-support systems. In the first stage, the focus was on collecting environmental data and monitoring the condition of the farm. In the next stage, simulation models were developed to analyze some biological and environmental processes. In the next stages, these models were integrated with real-time data and farm management systems, allowing for the analysis of management scenarios. Finally, by combining digital twins with advanced data analysis technologies, it has become possible to create intelligent and adaptive agricultural systems.

The evolution of digital twin applications in agriculture shows that this technology has gradually transformed from a modeling tool to a strategic infrastructure for managing food production systems. As agriculture faces multiple challenges, including climate change, resource constraints, and increasing food demand, the use of such tools can play an important role in improving management decisions and increasing the productivity of agricultural systems. From this perspective, digital twins not only enable more accurate analysis of the state of farms, but also pave the way for the development of a new generation of agricultural systems that are able to continuously adapt to changing environmental conditions and optimize their performance.

4. FROM FARM MONITORING TO INTELLIGENT FARMING: THE EVOLUTION OF DIGITAL TWIN AGRICULTURE

Technological developments in recent decades have led to fundamental changes in the management and production methods in the agricultural sector. One of the most important of these developments is the gradual transition from simple farm monitoring-based agriculture to smart, data-driven agricultural systems. In this direction, digital twin technology has emerged as one of the key tools for creating a link between the physical farm environment and analytical digital models. This technology allows for the creation of a dynamic representation of agricultural systems that can combine real-world environmental data with predictive and analytical models. As a result, farm management is transforming from an observation-based and experience-based activity to an analytical and data-driven process (Mălina et al., 2025).

In the early stages of the development of digital agriculture, the main focus was on monitoring environmental conditions and collecting farm data. The use of soil sensors, local weather stations, moisture monitoring systems, and remote sensing technologies allowed for a more accurate recording of environmental conditions. These data provided valuable information about soil conditions, temperature, moisture, and plant growth, but were often used independently of each other. In many cases, the data were used simply to observe the state of the field or record environmental changes, and were rarely used as a basis for advanced analysis or management decisions. Therefore, although the data monitoring stage was an important step towards the digitalization of agriculture, there was still a long way to go before creating intelligent management systems (Suthar et al., 2026).

With the development of digital infrastructure and the increase in data processing power, a new phase in the transformation of agriculture began, in which data collected from the field were systematically analyzed. At this stage, simulation models were developed to analyze biological and environmental processes. These models were able to predict the behavior of some components of the agricultural system, such as plant growth or water consumption, under different conditions. However, many of these models were still dependent on historical data and did not directly relate to real-time farm conditions. As a result, although these models could provide useful insights into production trends, they had limited ability to support day-to-day operational decisions.

In the wake of this trend, the concept of the digital twin emerged as a solution to integrate real-world data and analytical models. By creating a virtual version of a farm or production system, the digital twin enables simultaneous analysis of environmental data and predictive models. These systems are continuously updated with real-world data and can more accurately represent the current state of the agricultural system. In such a framework, data related to soil conditions, water status, climate conditions, and plant growth are combined in a digital environment, allowing for the analysis of different management scenarios. This evolution has transformed digital twins from mere simulation tools to analytical platforms for managing agricultural systems (Nozari & Yordanova, 2025b).

The next step in the evolution of digital agriculture was the move towards smart agriculture and data-driven decision-making. At this stage, digital systems not only monitor field data, but also analyze this data and provide management recommendations. For example, data on soil moisture, ambient temperature, and plant growth stage can be analyzed simultaneously to determine the best time to irrigate or feed the plant. Such an approach allows management decisions to be made based on accurate information and scientific analysis, and reduces the error rate in resource management.

In recent years, the development of data analysis technologies and predictive models has led to digital twins playing a broader role in managing agricultural systems. These systems can analyze a variety of data, including meteorological information, remote sensing data, field sensor data, and historical production information, in an integrated environment. Such a capability allows for the identification of complex patterns in agricultural systems and the prediction of future trends. As a result, farm management is transformed from a reactive process to a predictive process in which decisions are made based on the analysis of future trends (Sharma & Kaur, 2025).

The evolution from simple farm monitoring to smart agriculture shows that digital technologies can play an important role in improving the efficiency of agricultural production systems. In this direction, digital twins act as a bridge between environmental data and management decisions, enabling more accurate analysis of complex agricultural systems. This technology not only helps improve resource management, but can also play an important role in increasing the resilience of agricultural systems to climate change.

Overall, the evolution of digital agriculture from farm monitoring to the development of smart farming systems shows that the future of the sector will be highly dependent on the ability to use data, analytical models, and digital technologies. With the expansion of data infrastructure and the development of advanced analytical tools, digital twins are expected to become a key pillar of agricultural management in the coming decades, paving the way for more sustainable and efficient food production systems.

5. HYBRID DIGITAL TWIN–AI ARCHITECTURE FOR CLIMATE-RESILIENT AGRICULTURE

Increasing climatic uncertainties and the complexity of agricultural production systems have made traditional methods of farm management less efficient than they used to be. In a situation where environmental factors such as temperature, rainfall, soil quality, and water availability are constantly changing, management decisions must be based on accurate data analysis and predictive models. In such a context, the combination of digital twin and artificial intelligence has been proposed as an advanced framework for agricultural systems management. This approach allows for the creation of a dynamic analytical environment in which real-world farm data is combined with predictive models and decision-making algorithms to provide optimal management solutions (Awais et al., 2025).

In a digital twin-based agricultural system, a virtual representation of the farm or production system is first created. This digital representation includes information related to soil conditions, climate, crop type, plant growth patterns, and resource management. Environmental data are collected through field sensors, weather stations, remote sensing systems, and climate databases and are continuously updated in the digital model. As a result, the digital twin can represent the real state of the agricultural system with high accuracy and provide a platform for analyzing various management scenarios (Nozari, 2025a).

However, the ability of the digital twin to provide advanced analytics largely depends on the use of analytical algorithms. Here, the role of artificial intelligence as an engine for data analysis and predicting system behavior becomes important. Machine learning algorithms and predictive models can analyze environmental data, historical production data, and information related to climate conditions and identify hidden patterns in the agricultural system. These analyses can be used to predict crop yield, assess climate risks, and suggest management solutions. In fact, the combination of digital twin and artificial intelligence transforms the agricultural system from a simple monitoring system to an intelligent decision-making system (Chaudhary et al., 2025).

The combined architecture of digital twin and artificial intelligence usually consists of several main layers, each of which has a specific task in managing the agricultural system. The first layer is the physical layer, which includes the actual farm environment, agricultural products, water resources, soil, and production infrastructure. In this layer, sensors and environmental monitoring equipment collect data related to the actual conditions of the farm. The second layer is the data layer, where the collected data is stored, processed, and integrated. This data can include sensor information, climate data, satellite images, and historical production data.

The third layer is the digital twin modeling layer, where a virtual model of the agricultural system is created. This model can simulate the behavior of different system components and analyze possible changes in environmental or management conditions. The fourth layer is the artificial intelligence layer, where machine learning algorithms, predictive models, and data analysis methods are used to extract knowledge from the data. This layer plays an important role in identifying environmental patterns, predicting climate change, and optimizing management decisions. Finally, there is the management decision-making layer, which translates the results of the analyses into practical recommendations for farm management (Manivasagam, 2025). To better understand the structure of this framework, the proposed architecture of the hybrid digital twin and artificial intelligence system is presented in Figure 2.

Figure 2. Hybrid digital twin and artificial intelligence architecture for climate-resilient agriculture

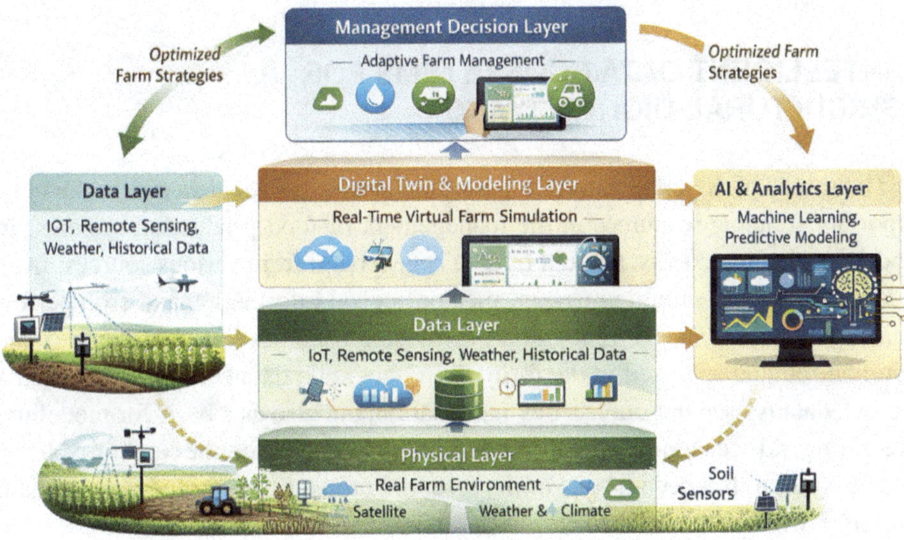

This figure shows how the different layers of the system are interconnected, from the physical environment of the farm to the management decision-making level. Environmental data is first collected in the physical layer and then processed and integrated in the data layer. This data is transferred to the digital twin model to represent the state of the agricultural system in a virtual way. Next, AI algorithms analyze the data and simulation models and transfer the results to the management decision-making layer. At this stage, the system can provide suggestions for resource management, adjusting irrigation schedules, choosing the right planting time, or assessing climate risks.

One of the important features of this architecture is the existence of a feedback loop between the real environment and the digital model. New data collected from the farm is continuously updated in the digital model, improving the accuracy of the analysis. This feedback loop allows the system to respond to environmental changes and dynamically adjust management strategies. Such a capability plays an important role in increasing the resilience of agricultural systems to climate fluctuations.

Overall, the hybrid architecture of digital twin and AI can be used as a strategic infrastructure for managing agricultural systems in conditions of climate change. By integrating environmental data, simulation models, and analytical algorithms, this framework enables more accurate analysis of agricultural systems and provides effective management solutions. As a result, the use of such architectures can pave the way for the development of a new generation of agricultural systems that are able to maintain production efficiency under changing environmental conditions while optimizing the use of natural resources.

6. INTELLIGENT DATA ECOSYSTEMS FOR AGRICULTURAL DIGITAL TWINS

The spread of digital technologies in agriculture has made data one of the most important strategic resources in the management of food production systems. In modern agriculture, a vast amount of data is generated from various sources; from field sensor data to satellite imagery, meteorological information and crop performance data. However, the real value of this data becomes apparent when it can be organized and analyzed in the form of an integrated data ecosystem. In such a context, data is used not only as raw information, but also as a basis for modeling, forecasting and decision-making in agricultural systems. Hence, the creation of smart data ecosystems has become one of the essential prerequisites for the development of digital twins in agriculture (Nozari et al., 2025a).

The agricultural data ecosystem can be considered as a network of data sources, processing infrastructures, and analytical tools that work together to collect, store, process, and analyze data. In this ecosystem, data comes from a variety of sources. One of the most important sources of data is environmental sensors installed in fields that provide information about soil moisture, ambient temperature, solar radiation, and plant growth conditions. In addition, remote sensing systems and satellite images provide valuable information about the state of vegetation, crop health, and environmental changes on a large scale. Climatic data, obtained from meteorological stations or global databases, also play an important role in analyzing environmental trends and predicting future conditions (Abdulameer et al., 2025).

One of the main challenges in creating an agricultural data ecosystem is the diversity and dispersion of data sources. Agricultural data is often stored in different formats and in separate systems, and integrating them for advanced analytics requires appropriate data infrastructure. In this context, the use of cloud-based data platforms and data management systems can provide the ability to store and process large volumes of information. These platforms allow farmers, researchers, and agricultural system managers to manage different data in an integrated environment and use analytical tools to extract knowledge from them.

In the digital twin framework, the data ecosystem plays the role of an infrastructure layer that establishes a connection between the real farm environment and digital models. Data collected from various sensors and sources is transferred to the digital model and used to update the system status. This process enables the digital twin to provide an accurate picture of the real farm situation and reflect environmental changes in real time. As a result, the digital model becomes a dynamic tool for analyzing agricultural system conditions that can predict system behavior under different scenarios (Huang et al., 2025).

Another important feature of smart data ecosystems is their ability to analyze data and transform raw information into usable knowledge. In agricultural systems, data is often generated in a scattered and large volume, and it is difficult to extract useful patterns from them without the use of advanced analytical methods. Here, the use of data analysis methods, predictive models, and big data processing tools becomes important. These tools can identify complex relationships between environmental factors, soil conditions, and crop yield and provide valuable information for farm management.

In a smart data ecosystem, data is processed in several stages. First, raw data is collected from different sources and stored in databases. Next, data cleaning and integration processes are performed to correct inconsistent or incomplete data. The data is then prepared for advanced analytics, and analytical algorithms are used to extract important patterns and trends. The results of these analytics can be used to predict crop yields, assess climate risks, and optimize resource management.

In recent years, the development of communication technologies and data infrastructures has made it possible to create extensive data ecosystems in the agricultural sector. These ecosystems can combine data from different farms, different geographical regions, and long time periods into a unified analytical environment. Such a capability allows researchers and agricultural system managers to analyze long-term trends and design appropriate management strategies for future conditions (Nozari et al., 2025b).

From a strategic perspective, the creation of smart data ecosystems can play an important role in the development of digital twin-based agriculture. These ecosystems not only provide the infrastructure for data collection and analysis, but also provide the basis for the creation of advanced decision-support systems. In such a framework, environmental data, analytical models, and decision-making tools are placed in a feedback loop that continuously improves the performance of the agricultural system.

Ultimately, the success of digital twins in agriculture depends largely on the quality and efficiency of their data ecosystem. The more accurate, extensive, and integrated the data used in these systems, the more accurate the digital models will be able to provide accurate analyses. Therefore, the development of smart data ecosystems is one of the most important steps towards creating sustainable, efficient, and climate-resilient agricultural systems.

7. ADAPTIVE DECISION-MAKING FOR CLIMATE-RESILIENT FARMING SYSTEMS

Managing agricultural systems in a changing climate requires approaches that can adapt to environmental uncertainties and dynamically adjust management decisions. In many regions of the world, climate change has caused traditional patterns of agricultural production to face serious challenges. Increased temperature fluctuations, changes in the timing and amount of precipitation, the emergence of water stresses and the increase in the occurrence of extreme climatic events have created conditions in which fixed decision-making based on predetermined plans is not effective. In such conditions, agricultural systems must be able to continuously modify their management decisions based on new data and environmental changes. This approach, known as adaptive decision-making, is one of the key elements in the development of climate-resilient agricultural systems (Jan & Jan, 2025).

Adaptive decision-making in agriculture means the use of data, analytical models and forecasting tools to continuously adjust management strategies. In this framework, farm management decisions are made not only based on current conditions but also taking into account future trends. For example, information on climate forecasts, soil conditions, moisture levels, and plant growth stages can be analyzed simultaneously to determine the best time to irrigate, fertilize, or harvest. Such an approach allows the agricultural system to respond quickly to environmental changes and prevent potential damage (Gupta et al., 2025).

One of the important features of adaptive decision-making systems is the use of real-time data to assess the state of the system. In modern agriculture, environmental sensors and digital monitoring systems have made it possible to collect accurate data on farm conditions. These data include information on air temperature, soil moisture, nutrient status, solar radiation, and plant health. When this data is continuously processed in analytical systems, it is possible to quickly assess farm conditions and adjust management decisions. In this way, decision-making is transformed from a periodic process into a continuous and data-driven process.

In the context of digital twin-based agriculture, adaptive decision-making plays an important role in the management of production systems. By creating a virtual model of the farm, the digital twin allows for the analysis of different management scenarios. In this digital environment, system managers can examine the possible effects of different decisions before implementing them in the real environment. For example, it is possible to examine how a change in the irrigation schedule or the use of a specific type of fertilizer will affect crop growth and resource consumption. This capability helps farm managers make their decisions based on accurate analyses and scientific simulations (Ma et al., 2025).

Another important aspect of adaptive decision-making in agriculture is the ability to predict climate risks. Sudden changes in weather conditions can have significant impacts on crop yields. Therefore, management systems must be able to assess the likelihood of such changes using historical data and forecasting models. For example, predicting the occurrence of drought or an increase in temperature in a certain period can lead to changes in irrigation schedules or crop selection. Such measures can prevent damage caused by adverse climatic conditions and increase the sustainability of the agricultural system.

In addition to predicting environmental conditions, adaptive decision-making can also play an important role in optimizing resource use. Resources such as water, energy, and soil nutrients are among the vital factors in agricultural production, and their efficient management has a direct impact on the productivity of the system. In an adaptive management system, the use of these resources is adjusted based on the actual needs of the farm and environmental conditions. For example, if soil moisture data shows that the amount of water in the soil is at an appropriate level, the system can temporarily reduce the irrigation schedule. Such an approach reduces resource consumption and increases system efficiency.

Adaptive decision-making can also play an important role in the long-term management of agricultural systems. Analysis of historical data and climate trends allows the identification of long-term patterns in agricultural production systems. This information can be used to plan crops for future years, select crops suitable for specific climatic conditions, and design resource management strategies. In this way, agricultural decision-making is transformed from a short-term process to a strategic process that also considers the future of the production system (Nozari et al., 2025c).

Overall, the development of adaptive decision-making systems can play an important role in increasing the resilience of agricultural systems to climate change. These systems, using environmental data, analytical models, and forecasting tools, allow for rapid responses to changes in environmental conditions and allow farm managers to dynamically adjust their management decisions. As a result, agriculture can transform from an activity dependent on constant environmental conditions into a flexible and intelligent management system that is able to maintain its productivity and sustainability even in changing climatic conditions.

8. TOWARD AUTONOMOUS AND SELF-LEARNING AGRICULTURAL SYSTEMS

Recent developments in digital technologies and data analytics have led to the concept of smart agriculture gradually moving towards the formation of automated and self-learning agricultural systems. In such systems, management decisions are not only made based on data collected from the farm, but the system is also able to continuously improve its performance using past experiences and new data. This approach, which is based on continuous learning and automated decision-making, can fundamentally change the way farms and agricultural production are managed. In a situation where climate change and increasing food demand have put significant pressure on agricultural systems, the development of such systems can play an important role in increasing productivity and sustainability of production (Liang et al., 2025a).

Automated agricultural systems are usually based on a combination of several key technologies. The first element of these systems is the environmental monitoring infrastructure that provides accurate data about the state of the farm. Soil sensors, weather stations, remote sensing systems, and plant health monitoring equipment provide information about environmental conditions and crop growth status. This data is used as the main input for analytical systems. The second element is digital models and simulations that allow for a virtual representation of the agricultural system. These models can analyze the behavior of the system under different environmental conditions and evaluate different management scenarios.

Along with these elements, learning and data analysis algorithms play an important role in transforming raw data into usable knowledge. These algorithms are able to identify complex relationships between environmental factors and crop performance and extract important patterns in the data. For example, analyzing multi-year data can show how a specific combination of temperature, soil moisture, and planting time affects crop yield. Such information can be used to optimize management decisions. When this data analysis process is performed continuously, the system will be able to gradually gain more knowledge about the farm's performance and make more accurate decisions (Cohen-Manrique et al., 2025).

A key feature of self-learning agricultural systems is the existence of a feedback loop between data, analytical models, and management decisions. In this loop, new data collected from the farm are used to update the analytical models. The results of the analyses are then converted into management decisions, and these decisions are implemented in the real farm environment. The results of the implementation of these decisions are fed back to the system and used as new data in the learning process. This feedback loop allows the system to continuously evaluate its performance and improve management strategies.

In such a framework, agricultural systems can gradually move towards automating some management processes. For example, in a smart farming system, the irrigation schedule can be automatically adjusted based on soil moisture data and weather forecasts. Similarly, the system can determine the right time to apply fertilizer or other agricultural inputs. This type of automation not only reduces the need for human intervention, but also increases the accuracy of management decisions.

In addition to resource management, automated systems can also play an important role in predicting and managing agricultural risks. Sudden changes in climatic conditions can cause significant losses to agricultural production. In a self-learning system, climatic and environmental data are continuously analyzed to predict the likelihood of adverse conditions. If such conditions are identified, the system can suggest preventive measures such as changing the irrigation schedule or adjusting the harvest time. Such a capability allows the agricultural system to actively adapt to changing environmental conditions (Yordanova & Nozari, 2025c).

Another important advantage of self-learning agricultural systems is their ability to improve resource efficiency. In many regions of the world, resources such as water and energy are severely limited, and efficient management of these resources is of great importance. In a learning-based system, data on resource consumption and crop performance are continuously analyzed to identify optimal solutions for using these resources. This process can help reduce resource consumption and increase the productivity of the production system.

Despite the significant advantages of automated and self-learning agricultural systems, the development of these systems is also accompanied by challenges. One of the most important challenges is the need for extensive and reliable data infrastructure. In order for the system to learn correctly and make accurate decisions, the data used must be of high quality and accuracy. In addition, integrating different data and designing appropriate analytical models requires the development of advanced data mining and modeling tools (Liang et al., 2025b).

Overall, the move towards automated and self-learning agricultural systems represents a new phase in the evolution of digital agriculture. By combining environmental monitoring technologies, digital models, and learning algorithms, these systems can intelligently adjust management processes and continuously improve their performance. In the future, such systems are expected to play an important role in creating sustainable agricultural production systems and pave the way for the development of a new generation of smart farms that can maintain their productivity and resilience in changing climate conditions.

9. IMPLICATIONS FOR SUSTAINABLE AND CLIMATE-RESILIENT AGRICULTURE

Technological developments in the field of digital agriculture, especially the emergence of digital twins and data-driven analytics, have opened up new perspectives for sustainable resource management and increased resilience of food production systems. In a context where climate change, declining natural resources, and increasing global demand for food have placed significant pressure on the agricultural sector, the use of smart technologies can play an important role in creating sustainable production systems. One of the most important consequences of the development of these technologies is the possibility of more accurate and efficient management of natural resources. In traditional agriculture, many management decisions were made based on general estimates or past experiences, while digital technologies have enabled detailed analysis of environmental conditions and the adjustment of management strategies based on real data (Omrany & Al-Obaidi, 2025).

One of the most important aspects of sustainability in agriculture is the optimal management of water resources. In many regions of the world, water resources are facing serious limitations, and increasing temperatures and changing rainfall patterns have further complicated this issue. Using data-driven farming systems can help reduce water use and increase irrigation efficiency. When data on soil moisture, ambient temperature, and climate conditions are analyzed simultaneously, irrigation schedules can be adjusted to meet the actual needs of plants and prevent excessive water consumption. Such an approach not only conserves water resources, but can also have a positive impact on soil health and crop productivity.

In addition to water resource management, digital technologies can also play an important role in reducing the consumption of agricultural inputs. Excessive use of chemical fertilizers and pesticides is one of the major challenges in modern agriculture, which can have significant negative impacts on the environment and the quality of food products. In smart farming systems, data on soil conditions, plant growth, and environmental conditions are continuously analyzed to determine the exact amount of inputs needed. This process reduces the use of fertilizers and other chemicals to the minimum necessary amount, thereby reducing the environmental impact of agricultural activities (Henriksen et al., 2022).

Another important consequence of the development of digital agriculture is the increased ability of agricultural systems to cope with climate change. Climate resilience means the ability of a system to maintain its performance in the face of environmental stresses and shocks. In agriculture, this concept refers to the ability of a farm to cope with conditions such as drought, increased temperatures, or sudden changes in rainfall patterns. Agricultural systems based on data analysis and predictive models can help farmers make more appropriate management decisions by identifying climate trends and predicting environmental changes. For example, predicting the occurrence of a drought period can lead to changes in cropping plans or water resource management and prevent possible losses.

From an economic perspective, the use of smart technologies in agriculture can also have positive consequences. Accurate analysis of production and resource consumption data helps farmers reduce their operational costs and increase the efficiency of the production system. In many cases, the use of digital technologies can reduce costs related to water, energy, and agricultural inputs. In addition, more accurate crop performance predictions can help improve production planning and supply chain management. Such developments can increase the economic sustainability of agricultural systems and enhance food security at a larger scale (Banciu & Florea, 2026).

In addition to environmental and economic benefits, the development of smart agriculture can also have significant social consequences. Increasing the efficiency of production systems and reducing climate risks can improve farmers' livelihoods and increase the sustainability of rural communities. When farmers have access to advanced analytical tools, they can make more informed decisions about farm management and prevent damage from environmental change. Such a capability can help reduce farmers' economic vulnerability and increase the stability of agricultural production.

Despite these advantages, the full realization of sustainable agriculture based on digital technologies requires the development of appropriate data and technological infrastructure. Access to accurate data, stable communication infrastructure, and advanced analytical tools are among the key factors in the success of these systems. In addition, farmer education and the development of digital skills also play an important role in the adoption and effective use of these technologies. Without the establishment of such infrastructure, the full exploitation of the potential of digital agriculture will face limitations (Cravero et al., 2026).

Overall, the development of agricultural systems based on digital twins and data analytics can provide a new path to achieving sustainable and climate-resilient agriculture. These technologies can help improve the efficiency of food production systems by enabling accurate analysis of environmental conditions, optimizing resource use, and predicting climate change. In the future, the combination of digital technologies, data analytics, and predictive models is expected to play an important role in shaping a new generation of agricultural systems that are able to ensure sustainable and efficient production even under changing environmental conditions.

10. CONCLUSION

Rapid climate change and increasing pressure on natural resources have caused agricultural production systems to face unprecedented challenges. Changes in rainfall patterns, rising temperatures, extreme weather events, and water resource limitations are among the factors that threaten the sustainability of agricultural production in many regions of the world. In such circumstances, the continued use of traditional management approaches, which are mainly based on experience and fixed decision-making patterns, cannot meet the complexities of modern agricultural systems. Therefore, the development of new approaches based on digital technologies and data analysis has become one of the basic necessities in sustainable agricultural management.

This chapter attempted to examine the role of hybrid frameworks based on digital twins and artificial intelligence in creating agricultural systems resilient to climate change. By creating a virtual representation of agricultural systems, digital twins allow for more accurate analysis of production processes and evaluation of different management scenarios. In this digital environment, real-world field data, climate conditions, soil characteristics, and plant growth status are analyzed in an integrated manner, and system managers can examine their potential effects in a simulated environment before implementing management decisions in the real environment. Such a capability can help reduce management risks and increase the accuracy of decisions in agricultural systems.

On the other hand, the use of artificial intelligence algorithms alongside digital twins can significantly increase the capacity to analyze data and predict the behavior of agricultural systems. Analytical algorithms are able to analyze a large volume of environmental, climate, and production data and identify complex patterns in agricultural systems. These analyses can be used to predict crop performance, assess climate risks, and optimize resource management. Combining these capabilities with digital models allows agricultural systems to move from a reactive management approach to a predictive and intelligent approach.

One of the most important aspects examined in this chapter was the role of smart data ecosystems in supporting agricultural digital twins. Data collected from various sources such as field sensors, remote sensing systems, meteorological data, and historical production information form the main foundation of digital models. When this data is stored and analyzed in an integrated data infrastructure, it allows for the creation of more accurate models and advanced analytics. Such data ecosystems can serve as the main infrastructure for the development of smart agricultural systems and pave the way for data-driven decision-making.

Next, the concept of adaptive decision-making was examined as one of the key elements in the development of resilient agricultural systems. In an adaptive management system, farm management decisions are continuously updated based on new data and predictive analytics. This approach allows for rapid response to environmental changes and dynamic adjustment of management strategies. Such a capability is particularly important in situations where agricultural systems are faced with climate uncertainties.

This chapter also discusses the trend of agricultural systems moving towards automated and self-learning systems. In these systems, environmental data and the results of management decisions are continuously analyzed, and the system is able to improve its performance using past experiences. The existence of feedback loops between data, analytical models, and management decisions allows agricultural systems to continuously learn and optimize their management strategies. Such an approach can play an important role in increasing the efficiency of resource use and reducing the vulnerability of agricultural systems to climate change.

Overall, the combination of digital twins, artificial intelligence, and data ecosystems can provide a powerful framework for managing agricultural systems in a changing climate. These technologies enable more accurate analysis of environmental data, predict climate change, and optimize management decisions. The use of such frameworks can increase resource efficiency, reduce water and input consumption, and ultimately contribute to the development of sustainable food production systems.

Given current trends in digital technologies, the role of digital twins and advanced analytical systems in agricultural management is expected to expand even more in the coming years. The development of data infrastructures, the advancement of data analysis algorithms, and the expansion of environmental sensor networks can pave the way for the creation of a new generation of agricultural systems that are capable of intelligently and adaptively adapting to changing environmental conditions. Such developments can be considered an important step towards realizing sustainable agriculture and ensuring food security on a global scale.

REFERENCES

Abdulameer, L., Al-Khafaji, M. S., Al-Awadi, A. T., Al Maimuri, N. M., Al-Shammari, M., Al-Dujaili, A. N., & DhiyaAl-Jumeily, . (2025). Artificial intelligence in climate-resilient water management: A systematic review of applications, challenges, and future directions. *Water Conservation Science and Engineering*, *10*(1), 44. DOI: 10.1007/s41101-025-00371-2

Al-Asady, H. A. J. (n.d.) A Cognitive Digital Twin Framework for Resilient Water Infrastructure under Climate Change. In *Smart Technologies for Sustainable Development Goals* (pp. 321–350). CRC Press. DOI: 10.1201/9781003714774-18

Alexander, R., Janet, A. M. V., Sri, R. V., Sasirekha, S., Kathyani, A., & Priya, V. S. (2025, September). A Multilingual LLM-Driven Digital Twin Framework for Climate Resilient Precision Agriculture. In 2025 6th International Conference on Electronics and Sustainable Communication Systems (ICESC) (pp. 1794-1805). IEEE.

Awais, M., Wang, X., Hussain, S., Aziz, F., & Mahmood, M. Q. (2025). Advancing precision agriculture through digital twins and smart farming technologies: A review. *AgriEngineering*, *7*(5), 137. DOI: 10.3390/agriengineering7050137

Banciu, C., & Florea, A. (2026). AIoT at the Frontline of Climate Change Management: Enabling Resilient, Adaptive, and Sustainable Smart Cities. *Climate (Basel)*, *14*(1), 19. DOI: 10.3390/cli14010019

Barzigar, A., Hosseinalipour, S. M., & Mujumdar, A. S. (2025). AI-driven optimization of hybrid solar-wind dryers for climate-resilient agriculture. *Drying Technology*, *43*(14), 2018–2019. DOI: 10.1080/07373937.2025.2540731

Chaudhary, P., Kandacharam, S., & Krishnan, B. S. (2025). Climate Change and Agriculture: Emerging Challenges and Far-Reaching Impacts. In Smart Agriculture for Sustainable Practices (pp. 67-90). Chapman and Hall/CRC.

Cohen-Manrique, C. S., Villa-Ramírez, J. L., Camacho-León, S., Solano-Correa, Y. T., Alvarez-Month, A. A., & Coronado-Hernández, O. E. (2025). Simulation and optimisation using a digital twin for resilience-based management of confined aquifers. *Water (Basel)*, *17*(13), 1973. DOI: 10.3390/w17131973

Cravero, A., Sepúlveda, S., Gutiérrez, F., & Muñoz, L. (2026). From Precision Agriculture to Intelligent Agricultural Ecosystems: A Systematic Review of Machine Learning and Big Data Applications. *Agronomy (Basel)*, *16*(5), 516. DOI: 10.3390/agronomy16050516

Gupta, A., Singh, B., & Saini, H. K. (2025). Enhancing Crop Resilience to Climate Change with AI. In *AI and Data Analytics in Precision Agriculture for Sustainable Development* (pp. 133–151). Springer Nature Switzerland. DOI: 10.1007/978-3-031-93087-4_8

Henriksen, H. J., Schneider, R., Koch, J., Ondracek, M., Troldborg, L., Seidenfaden, I. K., Kragh, S., Bøgh, E., & Stisen, S. (2022). A new digital twin for climate change adaptation, water management, and disaster risk reduction (HIP digital twin). *Water (Basel), 15*(1), 25. DOI: 10.3390/w15010025

Hossain, M. I., Hossan, M. R., Shaon, Z. H., & Ferdous, M. N. (2026). Linking digital twin paradigm for urban heat monitoring and policy integration to building smart city climate resilience. *Discover Cities, 3*(1), 1. DOI: 10.1007/s44327-025-00179-8

Huang, Y., Ren, F., & Wang, Y. (2025). Evaluation and pathways for achieving agricultural resilience under the framework of climate-smart agriculture. *Humanities & Social Sciences Communications, 13*(1), 105. DOI: 10.1057/s41599-025-06409-z

Hurmat, K. B., Rani, L., & Srivastav, A. L. (2026). Modelling the Future: Digital Twins in Sustainable and Climate-Aware Farming. In Digital Twin Technology for Sustainable Agriculture: Applications, Implementation and Future Trends (pp. 179-193). Singapore: Springer Nature Singapore.

Jan, A., & Jan, A. U. (2025). Harnessing Digital Agriculture to Combat Climate Change and Feed the Future. In Climate Resilient and Sustainable Agriculture: Volume 2: Social and Transformative Strategies (pp. 313-332). Cham: Springer Nature Switzerland. DOI: 10.1007/978-3-032-04141-8_12

Liang, S., Long, N. V. D., Parvin, M. I., Escribà-Gelonch, M., Lantin, S., & Hessel, V. (2025). Digital twin for lettuce growth in six climate zones ('Climate 2050'). *Computers and Electronics in Agriculture, 238*, 110880. DOI: 10.1016/j.compag.2025.110880

Liang, X., Yu, S., Ju, Y., Wang, Y., & Yin, D. (2025). Multi-scale remote-sensing phenomics integrated with multi-omics: Advances in crop drought–heat stress tolerance mechanisms and perspectives for climate-smart agriculture. *Plants, 14*(18), 2829. DOI: 10.3390/plants14182829 PMID: 41011981

Ma, Z. G., Billanes, J. D., & Jørgensen, B. N. (2025). Climate Resilience and Energy Flexibility in Industrial Systems: A Scoping Review of Concepts, Technologies, Applications, and Policy Links. *Energies, 18*(18), 4985. DOI: 10.3390/en18184985

Mălina , C. D., Matei, F., Pop, I. D., Sălăgean, T., & Mălina , A. (2025). Individual and Synergistic Contributions of GIS, Remote Sensing, and AI in Advancing Climate-Resilient Agriculture. *AgriEngineering*, *7*(7), 230. DOI: 10.3390/agriengineering7070230

Manivasagam, V. S. (2025). From bytes to farm: Transferability of industrial digital twins in agricultural systems. *Journal of Biosystems Engineering*, *50*(1), 130–144. DOI: 10.1007/s42853-025-00256-1

Mehta, A. R., Kumar, P., Tyagi, S., Garg, K., & Aggarwal, S. (2026). Digital Twins for Sustainable Agriculture: Advancing Productivity, Resilience, and Equity. In *Digital Twin Technology for Sustainable Agriculture: Applications, Implementation and Future Trends* (pp. 393–415). Springer Nature Singapore. DOI: 10.1007/978-981-95-5915-2_21

Nozari, H. (2026). Digital twin-based predictive maintenance in cold chain logistics. *Operations Research*, *60*(1), 481–499. DOI: 10.1051/ro/2025167

Nozari, H., Abdi, H., Szmelter-Jarosz, A., & Rahmaty, M. (2025). A digital twin-based neutrosophic multi-objective optimization model for smart supply chains in the food industry. *Journal of Intelligent Manufacturing*, 1–15. DOI: 10.1007/s10845-025-02689-z

Nozari, H., Rahmaty, M., Foukolaei, P. Z., Movahed, H., & Bayanati, M. (2025). Optimizing cold chain logistics with artificial intelligence of things (AIoT): A model for reducing operational and transportation costs. *Future Transportation*, *5*(1), 1. DOI: 10.3390/futuretransp5010001

Nozari, H., Szmelter-Jarosz, A., & Weiland, D. (2025). A Fuzzy Multi-Objective Sustainable and Agile Supply Chain Model Based on Digital Twin and Internet of Things with Adaptive Learning Under Environmental Uncertainty. *Applied Sciences (Basel, Switzerland)*, *15*(19), 10399. DOI: 10.3390/app151910399

Nozari, H., & Yordanova, Z. (2025). A fuzzy–digital twin optimization framework for simultaneous management of waste and energy consumption in sustainable manufacturing. An International Journal of Optimization and Control: Theories & Applications, 25360152.

Nozari, H., & Yordanova, Z. (2025). An intelligent digital twin approach for optimizing multi-channel supply chains in uncertain environments. Green Technologies and Sustainability, 100309.

Omrany, H., & Al-Obaidi, K. M. (2025). Application of digital twin technology for Urban Heat Island mitigation: Review and conceptual framework. *Smart and Sustainable Built Environment*, *14*(7), 2145–2175. DOI: 10.1108/SASBE-05-2024-0189

Sharma, H., & Kaur, S. (2025). Deep learning for sustainable development across climate, energy, agriculture and urban systems. *Discover Sustainability*, *6*(1), 1408. DOI: 10.1007/s43621-025-02186-6

Suman, S., Ray, S., & Prusty, A. K. C, U., Rath, G. P., Patnaik, S., ... & M, L. (2025, July). Digital Twins in Agriculture: Revolutionizing Climate Resilience with AI and IoT. In International Conference on ICT for Sustainable Development (pp. 444-454). Cham: Springer Nature Switzerland.

Suthar, O. P., Parihar, V., Kumar, M., Verma, N. K., & Kashyap, A. K. (2026). Digital Twins and AI: Synergizing Intelligence for Sustainable Farm Management. Digital Twin Technology for Sustainable Agriculture: Applications, Implementation and Future Trends, 347-372.

Tariqul Islam, A. F. M., Saha, K. K., & Mahaboob, M. G. (2026). Computational Intelligence in Climate-Adaptive Agriculture: Pathways to Resilient Food Systems. In *IoT and AI in Agriculture: Strategies to Develop Smart Agricultural Space for In-field and In-house Crop Production* (pp. 67–88). Springer Nature Singapore. DOI: 10.1007/978-981-95-5218-4_5

Yordanova, Z., & Nozari, H. (2026). An Agentic Digital Twin Framework for Fuzzy Multi-Objective Optimization in Dynamic Humanitarian Logistics. *Algorithms*, *19*(3), 198. DOI: 10.3390/a19030198

Chapter 14

Sustainability-Driven Cognitive Digital Twins for Enterprise Transformation

Nandini Bhati
http://orcid.org/0009-0008-1535-7519
Christ University, India

Abhinav Kataria
Christ University, India

Pankaj Dhaundiyal
http://orcid.org/0000-0002-4815-1647
Christ University, India

ABSTRACT

The rapid advancement of digital twin technology, combined with developments in artificial intelligence and sustainability-driven innovation, is transforming enterprise asset management. Traditional approaches rely on periodic monitoring and reactive maintenance, limiting predictive capabilities and long-term environmental and economic optimization. While digital twins provide virtual representations of physical assets, many applications remain focused on monitoring rather than intelligent decision support. This chapter proposes a conceptual framework for Cognitive Digital Twins to enable sustainable enterprise transformation. The framework integrates cognitive capabilities—perception, attention, memory, reasoning, learning, and problem solving—within a multi-layer architecture that supports predictive analytics, lifecycle management, and decision-making. By embedding

DOI: 10.4018/979-8-2600-1669-5.ch014

sustainability principles and circular resource management, the framework aligns digital twin systems with Industry 5.0, enabling resilient, intelligent, and environmentally responsible enterprise systems.

1. INTRODUCTION: SUSTAINABILITY AS THE MISSING CORE OF DIGITAL TWIN TRANSFORMATION-

1.1 Background and Context

The operational monitoring tools have been evolved into digital twins that can help in making strategic decisions. Nevertheless, their possibilities to incorporate sustainability into the decision making in organizations are under-researched. Digital twins have been initially designed as condition operators, performance simulators, predictive maintenance, but currently, it is considered a continuously synchronised cyber-physical system with real-time, simulation and predictive analytics (Fuller et al., 2020; Grieves and Vickers, 2017). These features can help organisations to build dynamic online versions of physical assets and optimise performance throughout their lifecycle. The use of digital twins has proliferated quickly in transport, energy systems and infrastructure management (Qiu et al., 2025). The development of the artificial intelligence system has also reinforced their service by facilitating the process of anomaly detection, predictive maintenance, and decision-support based on the scenarios (Rasheed et al., 2020). Measurable benefits are pointed at by empirical evidence, such as enhanced operational efficiency, streamlined resource use, and less emission, especially in the field of smart-city and industry (Li, 2025). However, despite such developments, current applications of digital twin are mainly aimed at efficiency and performance optimization with little consideration given to the environmental and social aspects. Sustainability indicators tend to be discussed as a monitoring output and not a variety of decisions that are typically embedded in the optimisation processes (Botin-Sanabria et al., 2022). This introduces an underlying disparity between digital transformation efforts and overall sustainability goals.

1.2 The Sustainability Imperative

Sustainability is no longer a voluntary corporate action but a strategy and a regulation. The Corporate Sustainability Reporting Directive (CSRD), Task Force on Climate-related Financial Disclosures (TCFD), and standards that are set by the International Sustainability Standards Board (ISSB) are some of the frameworks that compel organisations to make reporting of environmental performance transparent and continuous. Simultaneously, the Industry 5.0 focuses on the creation of usually human, sustainable, and resilient systems to redefine the objectives of digital change (Xu et al., 2021). Digital twins in this changing context are a great potential to aid the sustainability governance system due to their ability to provide real-time review of the energy consumption, emissions, and resource flow (Alcacer and Cruz-Machado, 2019). Nevertheless, the existing implementations are still in scope. Sustainability measures are normally observed but not incorporated in decision-making processes. Lifecycle assessments are not usually applied in connection with existent working systems, and the principles of the circular economy are hardly reflected in the digital decision-making framework (Ali et al., 2025). The environmental impact of digital infrastructure per se is also a source of extra problems, specifically in the large-scale deployment (Bieser and Hilty, 2018). All these constraints emphasize the necessity of more integrated approach where the concept of sustainability is implemented as one of the design principles of digital twin systems.

1.3 Problem Statement

Although the sphere of digital twins has been evolving rapidly, a number of structural issues restrict the role played by this framework in facilitating transformation of corresponding enterprises in terms of sustainability. First, sustainability is still something marginal to system design. Monitoring layers are frequently added to environmental indicators but not included in decision-making, and thus, affect change outcomes of operations to less degrees (Pater and Stadnicka, 2021). Second, digital twins are usually deployed in functional silos, i.e. production, logistics, or infrastructure systems. This fragmentation inhibits the asset-level intelligence of supply chain sustainability goals, such as emissions in the supply chain and circular economy strategies, on an enterprise-wide basis (Kritzinger et al., 2018). Third, multi-objective optimisation is not well developed. The current digital twins approaches seldom incorporate economic performance, environmental effects, and operational resiliency in the same framework of decision making (Zhang et al., 2024).

These issues give rise to the question of the central research: **What are the ways of designing digital twin systems to be sustainability-based cognition systems to facilitate enterprise-wide change in line with the principles of Industry 5.0?**

To answer this research question, this chapter suggests a theoretical framework of Sustainability-Driven Cognitive Digital Twins (SCDTs), and the objectives are the following: **Objective 1:** Cognitive Systems Redefine digital twins as systems where sustainability has become a fundamental aspect of decision-making.

Objective 2: Have a multi-layered architecture that incorporates lifecycle intelligence, interoperability and transparency in governance.

Objective 3: Find transformation pathways, in which SCDTs can facilitate quantifiable sustainability outcomes at operational, tactical, and strategic level.

Objective 4: suggest a maturity model with sustainability-related KPIs through various phases of digital twin preparation.

Objective 5: Examine the problems of implementing and govern its experience to practitioners and policymakers..

2. THEORETICAL FOUNDATIONS AND CONCEPTUAL ANCHORS

2.1 Evolution of Digital Twin Technology

The Digital Twin (DT) concept was initially developed as a part of the product lifecycle management, with the virtual model being used to simulate and analyse the behaviour of physical assets throughout the technology design, manufacturing, and maintenance process (Grieves and Vickers, 2017). The initial uses were mostly engineering-oriented and was used in condition monitoring, predictive maintenance and simulation of performance. Digital twins have grown and changed to be dynamic cyber-physical systems that incorporate physical models, real-time sensor measurements and service-oriented architectures over time (Tao et al., 2019). Contemporary digital twins can have real-time, predictive, and autonomous optimisation (Fuller et al., 2020). This progression indicates a wider tendency of transformation of fixed digital representations to functional, data-driven, systems that assist in ongoing decision-making. Recent research also points to the shift of digital twins that are no longer asset-level to enterprise-level solutions. According to Jones et al. (2020), this development is characterized as a shift towards a descriptive model of intelligence to predictive and prescriptive intelligence. These developments offer technological basis to cognitive digital systems that have the potential of underpinning the organisational decision processes. Rapid adoption is also evidenced by the sectors of transportation, energy systems, infrastructure management, and others (Qiu et al., 2025). Regardless of this development, most digital twin applications are dedicated to optimisation of operations and efficiency. The adoption of sustainability and governance in consideration is still scarce. The above gap demonstrates the importance of expanding digital twins to explore the world of technical functionality and, more crucially, to enterprise-level and sustainability-oriented decision systems.

2.2 Artificial Intelligence in Enterprise Systems

Artificial intelligence plays a critical role in improving the potential of digital twin systems since it allows superior analytics, predictive modelling, and automated decision support. Machine learning, anomaly detection, and reinforcement learning are some of the most common techniques that are used to reveal patterns, forecast system failures, and optimise resource allocation (Rathore et al., 2021). Strategically, the dynamic capabilities theory offers an effective way of perception of AI-enabled systems. Teece (2018) believes that organisations can attain long-run competitive advantage by their capabilities to sense opportunities, seize them and restructure resources. Digital twins with AI capabilities facilitate these functions by constantly recording the operational data, producing evidence-based insights. Digital twins powered by AI can be used in the industrial setting to facilitate better control and coordination of the enterprise systems (Leng et al., 2021). Nevertheless, issues pertaining to data governance, model transparency, and explainability are still important. Such concerns are also vital in environments where sustainability is being addressed, whereby, ESG reporting needs decision making activities that are transparent and verifiable. These capabilities are further extended to cognitive digital twins. Intizar Ali et al. (2021) note that knowledge representation, reasoning mechanism, and human-machine interaction are not only incorporated in cognitive systems but also essential to them. This allows digital twins to proceed beyond operational monitoring to intelligent decision-systems that can respond to the complex organisational problems. Nonetheless, sustainability goals remain not well implemented in most AI-supported digital twins, as the focus is mostly on efficiency and output in performance. This poses a critical discontinuity in the provision of ESG-oriented decision-making into intelligent enterprise systems.

2.3 Sustainability and Circular Economy Theory and ESG Frameworks

The concepts of sustainability and a circular economy form the theoretical part of including the environmental goals into the digital transformation strategies. The concept of a circular economy features closed-loop production, which traditional linear models of production do not have; reuse, remanufacturing, and recovery of raw materials (Kristoffersen et al., 2020).

Digital twins have proven to be one of the enablers of circular economy practices. They facilitate lifecycle optimisation and better-managed use of resources by offering real-time insights into the status of assets, energy as well as material flows (Ali et al., 2025). As empirical research of smart manufacturing shows, digital twins can enhance energy efficiency and resources utilisation by means of predictive maintenance and smart planning of operation (Warke et al., 2021).

Nevertheless, the sustainability of digital twins has not been fully exploited. The majority of the implementations do not aim at converting sustainability goals into the optimization algorithms and decision making process but instead monitoring the environmental indicators. This has mainly led to the fact that sustainability metrics have been applied to reporting activities instead of being used as operational and strategic decision-making tools.

Though digital twins are well in line with the principles of the circular economy, there is still a distance between this opportunity and its realization. The current systems do not usually facilitate the closed-loop decision-making or even optimise sustainability results.

The other factor that is very much urgent is the environmental impact of digital infrastructure itself. Digitalisation may have a positive or a negative impact on the environment, especially when the resulting efficiency gains translate into additional resource demand (Bieser and Hilty, 2018). Massive digital twin implementation can raise the amount of energy used, which can nullify the sustainability gains unless environmental cost is factored in the system design.

Governance-level reports, including GRI, SASB, ISSB, CSRD, and TCFD reports, have worsened pressure on the needless and up-to-date environmental information. Conventional methods of reporting no longer suffice to satisfy them. When organisations incorporate sustainability in the digital transformation structures, they are in a better position to record better ESG performance (Ben Youssef, 2025).

The mentioned challenges emphasize the importance of digital twin systems with sustainability as a fundamental design and decision-making concept rather than as an additional reporting task.

2.4 Industry 4.0 to 5.0 and Enterprise Cognitive Architecture

Creation of digital twins is strongly associated with more global industrial transformation paradigms. Industry 4.0 brought cyberspace-physical systems, Internet of Things connectivity, and cloud computing systems that were implemented to improve automation and efficiency in operation (Alcacer and Cruz-Machado, 2019). Digital twins in this paradigm served mostly in the area of the industrial process monitoring and the optimisation. The most common classification by Kritzinger et al. (2018) includes digital models, digital shadows, and full digital twins. Although this framework is effective in the increase of technological maturity, the model is mainly concerned with technical aptitudes as opposed to issues of governance and sustainability concerns. The new Industry 5.0 paradigm can be discussed as one that is symbolisvious as it focuses on sustainability, resiliency, and human-centric innovation (Xu et al., 2021). Here, one can assume that digital technologies will not only result in a more efficient operation process but also environmental and societal impacts. According to recent studies, digital twin architecture needs to be improved to correspond to such goals. Obesity researchers believe that the design principles of next-generation systems should be transparency, human oversight, and sustainability (Barata and Kayser, 2024). Such Industry 4.0 to Industry 5.0 transformation requires a radical change in the design of digital twins. Systems should transform to adopt efficiency-based tools in sustainability-based cognitive structures. This change is one of the main conceptual backgrounds of the framework in this chapter.

2.5 Enterprise Architecture and Cognitive Systems

Digital twins have been successfully applied in the system and asset level, however, robust enterprise-level integration is required in achieving large-scale sustainability transformation. Enterprise Architecture (EA) provides a structural architecture that is necessary in connecting digital twin systems of functions and decision making processes across the organisation.

Most implementations of digital twins often exist in functional silos and not global aspects in the absence of EA integration. This fragmentation limits their ability to assist in sustainability governance related to the enterprise, as well as, supply-chain coordination, Scope 3 emissions, and creative circular-economy efforts.

According to Ali et al. (2025), enterprise cognitive system has three key requirements, including semantic interoperability, knowledge representation, and governance architecture. Semantic interoperability aids in meaningful data exchange across the systems whereas the knowledge representation supports multi-domain and complex decision-making. The governing structures encourage transparency, accountability, and organisational goals.

As argued by Rasheed et al. (2020), digital twins should be developed not only as data-aggregation tools but as systems that would be able to facilitate strategic environmental governance. This means that, to make effective, transformation that affects the whole organisation, the inculcation of sustainability in the digital infrastructure of an enterprise is imperative.

In this regard, the inclusion of digital twins in enterprise architecture will be a key move in the direction of ensuring sustainability-driven digital transformation and be one of the core components of the Sustainability-Driven Cognitive Digital Twin model mentioned in this chapter..

3. SYSTEMATIC LITERATURE REVIEW

Digital twin (DT) has gone viral in manufacturing, infrastructures, supply chains, and urban systems. In order to assess the level of sustainability integration of DT architectures, a methodical literature analysis was conducted. The results show similarities: although the notion of sustainability is much debated, it is not always integrated into the decision-making process. Environmental information is also normally monitored and reported, but rarely analysed as part of optimisation logic.

This highlights a major disconnect between the system intelligence measurement and action.

3.1 Review Methodology and Inclusion/Exclusion Criteria

The review focused on peer-reviewed academic journal articles that were included in Scopus and Web of Science and had been published within the time frame 20182025. The keywords included the presence of both domain-specific and thematic keywords that include manufacturing, construction, supply chain, infrastructure, sustainability, circular economy, life-cycle assessment, ESG, and Industry 5.0. Conference proceedings, editorials and non peer reviewed sources were avoided in order to achieve methodological rigor. However, the research that mentioned the concept of sustainability but had not implemented it was retained to help in identifying the gaps. The original search results were more than 600 records, 42 articles of which met the final inclusion criteria.

Table 1. Systematic Literature Review — Inclusion and Exclusion Criteria

Criterion	Inclusion	Exclusion
Publication type	Peer reviewed journal articles	Conference papers, editorials, technical notes
Indexing	Scopus or Web of Science	Non-indexed and grey literature
Time window	2018–2025	Pre-2018 publications
Focus	DT systems, architectures, or applied AI/ sustainability integration	Simulation models without real-time data integration
Evidence standard	Architectural sustainability evidence or explicit gap identification	Opinion essays without methodological basis
Access	Full-text available	Duplicate records; abstract-only sources

Explanation:

The restricted amount of eligible studies implies the fact that sustainability integrated DT research is still young with relatively a few pages giving an insight into the architectural level or decision level integration.

3.2 DT Applications by Domain and AI-Enabled Digital Twins

The review proves that adoption of digital twins is the most developed in the field of manufacturing. These systems are high in operation intelligence especially monitoring processes and efficiency of resources. Ideally, however, the sustainability goals are not taken into account in decision-making structures; the environmental indicators are most commonly observed but not used as optimisation variables.

Digital twins are finding more and more application in the construction sector, where life-cycle assessment and carbon assessment are concerned. Nevertheless, sustainability reports tend to be displayed through dashboards but not through decisions; optimisation lays more emphasis on measures of cost and performance ratios.

Supply-chain applications show more favorable correspondence to sustainability aims, in particular, to conceptual models that consider environmental restrictions. However, there are still few big implementation scales.

Digital twins are used in the infrastructure and city systems to do resilience planning and monitor the resources. So, although these applications send environmental awareness, they are never organized towards sustainability governance at the enterprise level. In general, it can be stated that, as it is shown in the literature, digital twins are widely used in various domains, but the application thereof is operational, not strategically oriented towards the sustainability objective..

3.3 Sustainability Integration in DT Research

The concept of sustainability has acquired growing interest in digital twin studies, especially in terms of energy efficiency, life-cycle assessment, resource management of a circle, and sustainable supply chain. These papers highlight the possibilities of digital twins to improve environmental performance. However, the integration of sustainability is usually superficial; the indicators of environmental sustainability are mostly used to monitor and report and not to be integrated in decisions structures. As a result, real time environmental trade-off analysis or optimisation in digital twin system is hard to come by. The models of dynamic life-cycle assessment that would allow perceiving the environment impact continually during the course of operations are under-explored. Likewise, the integration of the principles of the circular economy with digital twin architecture is not extensive, although there is high alignment between the two concepts. This implication implies that sustainability is largely viewed as a measure of a system to be attained and not a parameter that influences system behaviour.

3.4 ESG Reporting and DT Linkages

The connection between new digital twin systems and ESG reporting systems is still disrupted. The regulatory frameworks like CSRD, ISSB and TCFD have increased the strain on the accuracy and real time, verifiable sustainability data. Nevertheless, the majority of organisations still use manual or semi-automated methodology of generating reports. The digital twins have the ability of overcoming this problem through continuous data to be collected and real time environmental monitoring. Digital twin systems have been shown to produce auditable sustainability data in some of the areas, especially the built environment. However, to bring about this potential, there is a lack of integration between ESG framework and digital twin architectures. Digital twins and ESG reporting are usually divided into two independent structures, resulting in the duplication of data processing and decreased efficiency. This lack of connection makes organisations unable to realise real-time and data-driven sustainability governance.

3.5 Identified Research Gaps

The review mentions five structural gaps that are interconnected and outline the modern research field. Although digital-twin technologies have been developing significantly, the concept of sustainability at the architectural and governance levels did not follow this trend.

Table 2. Identified Research Gaps and SCDT Framework Responses

Research Gap	Description	SCDT Response
Sustainability Peripherality	Environmental metrics monitored but not embedded as optimisation criteria in DT decision architectures	Sustainability-first architecture: environmental objectives govern cognition from Layer 1 through Layer 5
Enterprise Fragmentation	DT deployments confined to functional silos; cross-enterprise sustainability coordination absent	Five-layer architecture spanning operational to strategic enterprise levels with semantic interoperability
Cognitive Framework Deficit	DTs rarely theorised as cognitive systems with embedded sustainability governance	Six-capability SCDT model: sustainability embedded within each cognitive capability
ESG–DT Disconnect	ESG frameworks and DT architectures operate as independent infrastructures	ESG data schemas mapped to DT architecture at Layer 2; regulatory alignment automated
Interoperability Deficits	Semantic misalignment prevents enterprise-scale sustainability governance	Semantic modelling layer enables Scope 3 governance and industrial symbiosis coordination

Explanation:

Table 2 points out to the fact that the majority of the existing research concentrates on operational efficiency, and few studies introduce sustainability as the core system design. The reactions to the gaps identified earlier denote the lack of connectivity between environmental goals and digital decision structures, and thus the need to have systems that put sustainability in the logic of systems as opposed to its being presented as an additional reporting facade.

3.6 Synthesis of Findings and Research Positioning

Although prior research has explored digital twins across multiple domains, most studies emphasise performance optimisation rather than sustainability integration. Furthermore, existing approaches lack structured mechanisms for incorporating environmental constraints into real-time decision-making processes. Another key limitation is the absence of enterprise-level integration. Current implementations remain fragmented and fail to support organisation-wide sustainability governance. Additionally, the disconnect between ESG frameworks and digital twin architectures further limits the practical application of sustainability-driven decision systems. This chapter addresses these limitations by proposing a Sustainability-Driven Cognitive Digital Twin (SCDT) framework in which sustainability is embedded as a core design principle. The framework integrates decision-making, governance, and system architecture to enable real-time, enterprise-level sustainability transformation.

4. CONCEPTUALISING SUSTAINABILITY-DRIVEN COGNITIVE DIGITAL TWINS

The growth in the maturity of digital-twin (DT) systems in technical terms has not been balanced by equivalent developments in conceptual sense. The recent applications have placed most digital twins as surveillance and visualisation systems that help improve the operational visibility with real-time data. Though a good idea in terms of performance optimisation, such systems can hardly serve as decision architectures that are able to facilitate the sustainability-based change in the enterprise level. This displays a larger gap found in the literature: sustainability is often regarded as a reporting or monitoring layer but not as a part of decision logic. Environmental information is received but it is not proactively utilised in order to optimise or make strategic decisions. Thus, an integrated system that incorporates cognitive intelligence, lifecycle conscious, and ESG informed decision making in digital-twins is necessary. To counter, this chapter has conceptualised the Sustainability -based Cognitive Digital Twin (SCDT) as an enterprise-wide system in which sustainability is used as the major goal in directing system cognition. The SCDT plays with real-time environmental signals, knowledge of lifecycle and the use of multi-objective reasoning in one combined decision framework.

4.1 From Monitoring to Cognitive Systems

Early digital twins focused on synchronising physical and virtual systems through real-time data integration and predictive maintenance (Fuller et al., 2020; Grieves and Vickers, 2017). Although these systems improved operational efficiency, they remained largely reactive and limited to performance monitoring.

The transition to cognitive systems requires digital twins to move beyond representation toward decision-making. This involves the ability to evaluate alternatives, reason under constraints, and adapt system behaviour over time. Existing implementations, however, continue to use predictive insights primarily for short-term optimisation rather than long-term sustainability outcomes.

Cognitive digital twins address this limitation by incorporating knowledge representation, reasoning, and adaptive learning. These capabilities enable systems to support complex, sustainability-oriented decision processes rather than isolated operational improvements.

4.2 Six Core Cognitive Capabilities

The SCDT framework is structured in terms of a complex of cognition-capabilities that are interdependent and that offer sustainability to the behaviour of the system. These strengths include perception (real-time acquisition of environmental data), attention (prioritisation of risks that concern ESG), memory (traceability of lifecycle), problem-solving, rationalisation (multi-criteria decision making), and learning (continuity of adaptation).

Table 3. Six Core Cognitive Capabilities of the SCDT and Their Sustainability Dimensions

Capability	Technical Function	Sustainability Dimension
Perception	Multimodal sensor acquisition; IoT integration; edge processing	Real-time environmental sensor - emissions - energy - material flows - is a core duty, not an add-on.
Attention	Anomaly detection; signal prioritisation; threshold alerting	ESG breaches and emissions exceedances of equal urgency as the operational risk signal.
Memory	Historical state retention; lifecycle traceability; knowledge accumulation	lifecycle visibility Lifecycle knowledge memory; Lifecycle audit trail Lifecycle data continuity among asset lifecycles
Rationalization	Multi-criteria decision modelling; hybrid AI-physics models	Multi-objective optimisation: economic, environmental, and operational objectives treated as co-equal criteria
Learning	Predictive model recalibration; reinforcement learning; self-updating architectures	Sustainability-oriented training goals; adaptive ESG compliance; feedback loops in a circular economy.
Problem Solving	Orchestrated synthesis of all capabilities; autonomous decision execution	Enterprise sustainability governance: the synthesis of ESG goals at operational, tactical and strategic levels.

Explanation:

The table indicates that the concept of sustainability cuts across all functions of the system. In contrast to traditional digital twins, where appearances of environmental indicators are active inputs, the SCDT incorporates sustainability into perception, reason and decision-making activities.

4.3 Sustainability as a Cognitive Objective

One such critical conceptualisation is the re-definition of sustainability as a mind-set and not an outcome of monitoring. In the traditional systems, the economic performance is optimised, and the environmental impact is considered separately. On the other hand, the SCDT incorporates environmental constraints within decision reasoning. System recommendations will be assessed on the basis of the ESG criteria before implementation and make sure that the sustainability requirements are actively used in defining the outcome of the decision. This paradigm changes sustainability as a passive measure to a rule of system behaviour.

4.4 Multi-Objective Optimisation with Environmental Priority

Conventional digital twin models are based on single or weak multi-objective optimisation, where the environmental elements are considered as secondary constraints. This strategy restricts their ability to attain balanced sustainability results. The SCDT takes a truly multi etc approach where the balance between economic performance, environmental impact and operational resilience are optimised. One of the characteristics of this model is the imposition of environmental thresholds, in accordance to which the decisions that do not correspond to the conditions of sustainability are not accepted irrespective of the hedonic considerations, which will guarantee adherence to sustainability goals.

4.5 Human–AI Collaborative Intelligence

The proposed SustainabilityBased Cognitive Digital Twin (SCDT) model anticipates human-system collaboration and not systemic automation. Sustainability decisions regularly involve moral considerations, legal considerations, and trade-offs that are long-term and require human judgment. As such, the system has been designed to be transparent, interpretable, and accountable hence, keeping the decision processes consistent with the organisational and regulatory requirements and the system delivers data-driven insights to support better decision-making.

Central Conceptual Claim

A Sustainability Based Cognitive Digital Twin is different, as opposed to a more traditional digital twin that has sustainability capabilities not in the character of the monitoring it performs, but motivations that govern how it functions. The SCDT thinking machine uses the sustainability results as first-order goals: all the working recommendations are screened according to the sustainability requirements before publication. The idea of sustainability is what rules cognition; it is this inverse of architecture that forms the conceptual basis of all later design decisions expressed in this chapter.

5. SUSTAINABILITY-EMBEDDED ARCHITECTURE AND DESIGN PRINCIPLES

The structure of Sustainability-Embedded Cognitive Digital Twin (SCT) is not an objective technical artefact, but the value-oriented framework where sustainability is embedded with design principles. In contrast to the classical digital twins, in which sustainability is confined to the external reporting layer, the SCT incorporates the environmental and social concerns into the information streams, intelligence tools, and decision-making procedures.

To make it more coherent and applicable in practice, the suggested architecture can be cognized as happening between three mutually supportive layers:

(i) Data Layer, which frames operational and environmental data
(ii) Intelligence Layer, which processes the information to come up with ESG-conscious insights
(iii) a Decision Layer, which implements the mcasures and governance requirements that reflect sustainability principles.

A cross-cutting governance framework reinforces these layers to facilitate accountability, transparency, and matching the lifecycle. In a practical sense, these sustainability design modules and principles refer to existing literature related to digital twins (Kristoffersen et al., 2020; Tao et al., 2019), AI-assisted building and infrastructure asset management (Bibri and Huang, 2025), and cognitive system interoperability (Ali, Patel, Breslin, Harik, and Seth, 2021) and the circular economy intelligence (Kristoffersen et al., 2020; Zhang et al., 2022 The suggested architectural schema will be seven layers depending on each other. Sustainability obligations follow the hierarchical pattern of downwards flows of sustainabilities, therefore, each layer of SCT has sustainability obligations and provides quality-enhancing outputs to the higher level. More to the point, governance as a traditionally extrinsic control mechanism is placed as a two-way facet that relates with each of the other layers to make sure that system behaviour is kept within the frame of its claimed sustainability throughout the lifecycle.

5.1 Three-Layer Architectural View of SCT

The SCT architecture can be conceptually structured into three functional layers:

- **Data Layer:**

Captures real-time operational and environmental data through IoT sensors, enterprise systems (e.g., ERP, MES), and lifecycle datasets. This layer establishes the foundation for sustainability visibility by enabling traceability of energy use, emissions, material flows, and waste generation.

- **Intelligence Layer:**

Transforms raw data into sustainability-aware insights through semantic modelling, artificial intelligence, and domain-specific analytics. This includes emissions tracking, resource efficiency analysis, and circularity assessment.

- **Decision Layer:**

Translates intelligence outputs into ESG-constrained actions, ensuring that all recommendations or autonomous decisions comply with sustainability targets, regulatory requirements, and organisational policies.

This layered abstraction simplifies the complexity of the architecture while maintaining integration across enterprise systems.

5.2 Data Acquisition and Environmental Sensing

The data- acquisition layer involves the sensory substrate of the sustainable cognitive twin. Its design determines the operational and sustainability features, with which the twin can check its carbon credentials: a twin which fails to measure its energy power, its pollution emissions, its water consumption, its waste levels or its material flows cannot infer or make reasoning as to its dimensions. In line with this, the SCT paradigm assumes the role of infrastructural environment senses as a first-class service, as opposed to a peripherally attached accessory to operationally focused sensor networks. Environmental sensing, which is used in manufacturing and built-environment design, could include sub-metering of electrical demands at machine and zone level; measures of gaseous and thermal energy, water within the product, ambient air quality at the location (PM, volatile organic compounds, and CO 2) and in extreme designs the attachment of sensors to the material to track physical and chemical properties of the product throughout its life cycle (Bibri 2025, and Huang 2025; Jones et al. 2020). SacotoCabrera PerezTorres, TelloOquendo, Cerrada (2025) document the trend of setting up IoT sensor networks in smart-city environments, which are increasingly being arranged to co-locate operational and environmental sensing the same data infrastructure, so as to lower implementation cost and maximize performance and sustainability goals. The other important design principle of this layer is that of the calibration of time and space granularity. The thin resolutions needed by sustainability metrics can track the emitted or consumed energy and resources to particular activities or sets of assets, but the embodied carbon and maintenance cost of vast sensor proliferation should be taken into account. SCT suggests a hierarchic sensing hierarchy: high-freqency and fine-grained asset sensitization of items with high and tough environmental footprints; lower-frequency aggregate sensitization of stable or low-impact procedures; and periodic sample audit for sluggish and dynamic material and wastes. Tiering of this nature is necessary to have the data infrastructure proportionate, justifiable and energy efficient.

5.3 Data Integration and Semantic Modelling

Raw sensor data, regardless of its richness, does not make sustainability intelligence. The data integration and semantic modelling layer takes heterogeneous streams of operational technology (OT) data streaming out of SCADA and PLC systems, information technology (IT) data streaming out of the ERP, MES, and supply-chain platforms and environmental data streaming out of the IoT and monitoring systems and transforms this information into a machine-interpretable body of knowledge capable of informing advanced inference (Ali et al., 2025). Three artefacts are identified in relation to the semantic modelling of the SCT.

The initial one is the domain ontology that describes the terms of the conceptual terms in the domain of assets such as asset-equipment, processes, materials, energy carriers, emission sources, and waste streams and how they relate to one another each other via formal logic that can be readily interpreted by automated reasoning.

Second, a sustainability ontology is connected to the known sustainability frameworks of greenhouse gas accounting schemes (GHG Protocol, ISO 14064) and life-cycle assessment standards (ISO 14040/44), circular economy indicators (Ellen MacArthur Foundation Material Circularity Indicator), and ESG reporting taxonomies (GRI, ISSB IFRS 1/2 S1/S2).

Third, time-based model describes how both physical condition and sustainability performance will change over time, hence, allowing outcomes changes to be attributed to interventions or conditions. Semantic integration is an engineering issue in the deployment of CDT, and has been the hardest to apply in practice (Ali, Patel, Breslin, Harik, and Seth, 2021). Current standards, including Asset Administration Shelling in manufacturing, Building Information Modelling in construction (IFC schema), and Coordination Information Modelling in an energy system, are partial solutions, with more sustainability-specific data they have added to. The SCT framework provides that semantic interoperability is not only aspirational but is compulsory to be deployed to deployment, as it acknowledges that without it, the cross-domain and cross-boundary rationale on which sustainability intelligence at the ecosystem level relies structurally is impossible.

Design Principle: Sustainability Ontology as a Living Artefact

The sustainability ontology needs to be viewed as a dynamic organisational resource, which gets updated on a regular basis as the regulatory frameworks, scientific knowledge, and expectations of the stakeholders change. Preceded by formal change governance, version-controlled ontology management guarantees that the reasoning on sustainability of the twins is up-to-date and can be audited. The ontology drift, which occurs when the conceptual model of the twin is not as reflected by the regulatory or scientific standards it claims to be, is a type of governance risk just as serious as data drift.

5.4 Sustainability Intelligence Modules

The unique architectural innovation that separates the sustainable cognitive twin as opposed to the traditional CDTs that have sustainability dashboards is the Sustainability Intelligence Module (SIM). Instead of considering the sustainability metrics as deliverables of a separate reporting subsystem, SIM integrates sustainability goals, limits, and indicators as directly as part of the inference and optimisation algorithms of the cognitive engine. As a result, sustainability is ceased to be a retrospective one; it runs simultaneously with the operational reasoning of the twin it allows the identification of trade-offs and Pareto-optimal solutions on both operational and sustainability levels (Rasheed et al., 2020; Zhang et al., 2024). There are four functional sub- modules of the SIM. The Emissions Accounting Engine applies a continuous and activity-based model of Scope 1, 2 and 3 greenhouse gas emissions by giving emissions to particular assets, processes and decisions in near real-time. Understanding how to improve material, energy, and water productivity at their most basic level, the Resource Efficiency Tracker identifies sources of inefficiencies and measures the amount of improvement opportunities. Circular Economy Intelligence sub-module denotes material movement across life cycles of products, discovery of recovery and reuse, assessment of material degradation, and ideal end-of life routing choices (Kristoffersen, Blomsma, Mikalef, and Li, 2020). The ESG Alignment Assessor associates operational performance with ESG commitments that have stakeholders, identifies deviations, as well as forecasts compliance pathways in the event of different operating systems. According to Rathore, Shah, Shukla, Bentafat, and Bakiras (2021), artificial intelligence and machine learning capabilities, especially bridge-model architecture which combines physics-based with data-driven algorithms is crucial because it is the envision of which clothing designs are created based on SIM reasoning involved in SIM-based apparel design execution. Physics-informed models also guarantee that the sustainability analysis of the twin is done in accordance with thermal and material science constraints, thus controlling the systematic underestimation of environmental impact that solely statistical models can bear outside of their training data.

5.5 Decision Execution with ESG Constraints

The action layer converts the twin cognitive results into actionable suggestions or autonomous control imperatives, which includes forecasts, optimal recommendations and situation analyses. In the sustainable cognitive twin (SCT), what defines this layer is that it is mandated that all decision logic includes compliance constraints with ESG rules: without a prior check that decision logic is in compliance with the sustainability pledges and regulations stated by an organisation, no autonomous action can be taken, no advice can be given. The execution of decisions subject to ESG as a multi objective optimisation architecture is where sustainability constraints appear as a hard constraint (e.g. regulation of emissions) and soft constraints (e.g. continuing minimisation of energy used per unit of output). Since the sustainability agent that utilizes this limited goal space inherently entails the sustainability performance into the decision-making criteria, sustainability is not viewed as an external control post-factum (Bibri and Huang, 2025; Intizar Ali et al., 2021). The human-in-the-loop structure is necessary in the decision layer; therefore, great carefulness is taken in involving this layer. Major-stake decisions, including changing the capital equipment operations, changing the supply-chain partners, or changing internal processes, need (i) the human approbation of the recommended decision of the CDT before execution and (ii) the verification of motions occurrence, whether by the adopted decision or by the approval of one or more contingency plans. The activation of contingency plans provides the human decision-maker with a decision brief that describes the action recommended, an approximation of operational and sustainable product performance, the available consideration and trade-offs of each option, and the realization of the potential outcomes. Therefore, this framework maintains the complementary character of the human and machine thinking, such that end result of making the most best suited action lies on a human agent, informed by all the facts and information at the moment of decision-making.

Practical Illustration: Smart Manufacturing Application

In a smart manufacturing environment, the SCT can continuously monitor machine-level energy consumption and emissions across production systems. When predefined sustainability thresholds are exceeded, the system can recommend corrective actions such as rescheduling energy-intensive operations, adjusting machine configurations, or redistributing workloads.

By embedding ESG constraints directly into decision logic, the system ensures that productivity improvements are achieved without compromising environmental performance. This demonstrates the practical applicability of the SCT framework in real-time industrial contexts.

5.6 Lifecycle Assessment and Carbon Accounting Integration

Traditional digital twin designs are primarily associated with monitoring the operational performance of physical assets during their utilisation phase. In contrast, the SCT model extends the analytical scope to encompass the full lifecycle of materials and assets, including extraction, component manufacturing, implementation, maintenance, refurbishment, and eventual dismantling. This lifecycle perspective is enabled through the integration of hybrid lifecycle assessment (LCA) frameworks and carbon accounting standards within the knowledge base and logical architecture of the twin.

Lifecycle-capable CDTs maintain a flexible, asset-level environmental bill of materials, representing a structured record of the embodied carbon, water footprint, and material composition of each physical asset. This record is continuously updated as assets are modified, refurbished, or replaced with new components. Such capability allows the twin to estimate aggregate lifecycle environmental impacts rather than focusing solely on isolated operational effects, thereby supporting sustainability-informed decisions related to maintenance strategies, technology upgrades, and capital replacement.

Lifecycle-integrated CDTs also enable the evaluation of trade-offs between operational efficiency and lifecycle environmental performance. For instance, improvements that reduce operational carbon emissions may involve components with higher embodied carbon, and such trade-offs can be made visible and measurable within the lifecycle-integrated CDT (Jones et al., 2020; Kristoffersen et al., 2020).

Carbon accounting integration further allows the emission reporting of the twin to align with organisational greenhouse gas accounting systems, ensuring continuity in tracking Scope 1 and Scope 2 emissions while linking relevant Scope 3 emissions to supply-chain activities. As sustainability disclosure requirements expand under emerging regulatory frameworks such as the **Corporate Sustainability Reporting Directive (CSRD)**, the **SEC Climate Disclosure Rules**, and standards issued by the **International Sustainability Standards Board (ISSB)**, lifecycle-integrated CDTs can support both regulatory compliance and strategic sustainability management. The lifecycle phases, associated CDT data inputs, LCA outputs, and decision-support applications are summarised in **Table 4**.

Table 4. Lifecycle-integrated CDT: data inputs, outputs, and decision applications by phase

Lifecycle Phase	CDT Data Inputs	LCA/Carbon Output	Decision Support Application
Material & Manufacturing	Bill of materials; supplier emissions factors; transport logistics	Embodied carbon; material footprint	Supplier selection; specification optimisation
Commissioning	Energy demand curves; baseline metering; calibration data	Operational carbon baseline	Efficiency gap analysis; benchmark setting
Operation & Maintenance	Real-time sensor data; maintenance logs; energy metering	Dynamic Scope 1/2 emissions; maintenance impact	Continuous optimisation; predictive maintenance
Refurbishment	Component condition data; upgrade options; energy modelling	Lifecycle extension vs. replacement trade-off	Investment prioritisation; retrofit specification
End of Life	Material composition; recovery pathway data; logistics	Circular recovery rate; avoided Scope 3 emissions	Circular routing; take-back programme optimisation

5.7 Circular Economy Mechanisms and Resource Loop Optimisation

The most structurally ambitious sustainability goal integrated into the SCT framework is the one that aims at the eradication of waste at the design level keeping materials, components, and products at their highest value and in use as long as possible. The concept of digital twin is perhaps the only technology with the capacity to operationalise principles of a circular economy at scale, due to the fact that circular material flows produce the type of complex and dynamic, multi-party logistics and quality management issues that CDT cognition is precisely designed to handle (Kristoffersen et al., 2020; Zhang et al., 2025; Ali et al., 2022).

There are four functional capabilities within the circular economy mechanisms layer of SCT architecture. Material Passport Management keeps a digitally validated list of the material composition, quality history, treatment history, and recovery potential of the physical assets and components, which allows making informed decisions regarding the reuse, remanufacturing, recycling, or energy recovery upon end of life. Closed-Loop Quality Tracking will observe the state and results of recovered materials and parts when returning into processes and make sure that the circular input is up to spec and that any quality degradation is identified and controlled before it spreads down the supply chain. Reverse Logistics Optimisation uses ML-based route optimisation to optimise the collection, sorting, and redistribution of end-of-life materials to reduce the transport carbon and cost overhead of the circular flows. The Facilitation of Industrial Symbiosis finds a chance to have waste stream of one organisation to act as an input resource of another organisation, which necessitates cross-boundary data sharing made possible through the interoperability architecture of the two twins.

The review of CDT applications in the construction industry with a specific focus on the mechanisms of a circular economy, Zhang et al. (2025) identify that one of the most valuable CDT applications related to sustainability is the use of digital material passports and lifecycle tracking, but it is one of the least used ones, mostly because of data governance barriers. The SCT framework fills this gap by defining material passport management as a primitive and not a sophisticated ability- one that needs to be installed at initial stages of deployment so as to ensure continuity of data needed by subsequent stages of the circular economy optimisation.

Circular Economy Design Principle

Circular economy mechanisms cannot be retrofitted to a twin designed solely for linear production optimisation. Material flow visibility, lifecycle data continuity, and multi-party data sharing must be designed into the twin's architecture from the outset. Organisations that defer circular economy architecture to a later deployment phase will face substantial re-engineering costs and data discontinuities that compromise the quality of routing decisions.

5.8 Governance: Cybersecurity, Explainability, and Ethical AI

The architectural layer that seeks to ensure that the SCT functions within its stated sustainability goals and legal, ethical, and social requirements of the deploying organisation is governance. It is not a post-deployment compliance capability but an inherent design component: the systems, processes, and structures that are used to ensure accountability, transparency, and trust are built into the twin during the initial stages (Rasheed et al., 2020; Iliuta et al., 2024).

Cybersecurity governance concerns the integrity of the physical-digital nexus that enables the CDT to operate. A compromised twin - one with sensor data that is been manipulated, control outputs that have been compromised or knowledge models that have been poisoned - is not only an operational risk but a sustainability risk, since fabricated or falsified emissions data, compromised energy management, or falsified supply chain records can nullify the sustainability reporting of the organisation and leave the organisation vulnerable to regulatory and reputational liability. The SCT model has been forced to embrace defence-in-depth architecture of security and zero-trust network designs and systems that detect anomalies in both IT and OT data feeds.

Explainability governance is an issue of how cognitions of the twins can be interpreted. Structurally, black-box AI systems lack the sustainability decisions as the stakeholders require knowing the logic of the decisions, regulatory review, and ESG audit requirements. The SCT paradigm suggests that any decision solution should be complemented by explainability reports: ready records about the most important data sources, the line of inference, the made decision and the uncertainty degree which envelops the estimated performance. Such need is fulfilled with the help of explainable AI (XAI) techniques, including SHAP value attribution, attention mechanisms, and counterfactual explanation generation, which have a technical basis (Rathore et al., 2024).

Explainability is not equal to ethical AI control, which is more substantive to ensure that the objectives of the twin are in line with the social responsibility and human values. One particular ethical problem of sustainability-oriented CDTs is that bringing together and analysing operational, environmental, and supply chain data can reveal commercially sensitive information or enable invasive surveillance of its workers or destroy supply chain allies with lower digital maturity. SCT model would demand formal ethical impact evaluations on each of the key stages of deployment and constant algorithmic audits of bias and uncertainty.

6. SUSTAINABILITY GOVERNANCE AND DESIGN LOGIC

Section 5 outlined the architectural design of Sustainable Cognitive Twin (SCT). An underlying question would be based on the very design philosophy: why does sustainability have to be conceived as a design principle and not as a design aspect which is peripheral? Design aspect is normally added after the major system architecture is in place.

A design principle on the other hand guides architectural choices since the beginning and the development of all layers of the systems thus establishing sustainability as the core in the SCT framework.

By integrating sustainability into design rationales, it is ensured that sensor infrastructure, data scheme, optimisation algorithms, and governance will be filled with environmental, social, and governance (ESG) concerns. This means that the operational and sustainability goals are built together as opposed to being exclusive to each other once deployed. This part elaborates on the conceptual framework of sustainability integrated design and presents four fundamental principles on how sustainable cognitive digital twins should be developed.

6.1 Why Sustainability Cannot Be Retrofitted

Many digital twin deployments follow a sequential approach: systems are first implemented to improve operational efficiency, and sustainability features are added later. While this approach appears practical, it introduces structural limitations that are difficult to resolve.

Three key limitations can be identified:

- **Data Infrastructure Constraints:**

Sensor networks designed for operational monitoring often lack the capability to capture environmental variables at the required resolution. As a result, emissions, water usage, and material flow data may be incomplete or unreliable, limiting compliance with frameworks such as CSRD, ISSB, and TCFD.

- Algorithmic Inflexibility:

Early-stage optimisation models are typically trained to prioritise cost and efficiency. Introducing sustainability constraints later requires redesigning optimisation logic and retraining models, which is technically complex and may destabilise system performance.

- Governance Misalignment:

When sustainability is introduced after deployment, it becomes subject to organisational negotiation rather than being enforced as a technical constraint. This often leads to weakened environmental objectives.

Empirical evidence supports early integration. Organisations that embed sustainability within digital transformation strategies from the outset achieve stronger environmental outcomes than those attempting retrofitting approaches.

These limitations demonstrate that sustainability must be treated as a foundational design element rather than an add-on capability.

6.2 Sustainability as the Governing Objective of Cognition

Within the SCT framework, sustainability is treated as a governing constraint that defines the boundaries of optimisation. This does not imply that environmental goals override operational or economic objectives. Instead, sustainability establishes the limits within which optimisation can occur.

In practical terms:

- Environmental thresholds (e.g., emissions limits)
- Resource constraints (e.g., energy or water usage)
- Circularity targets

are embedded directly into the optimisation architecture. Any solution that violates these constraints is excluded before optimisation begins.

This approach reflects the transition from Industry 4.0 to Industry 5.0. While Industry 4.0 focuses on efficiency within existing systems, Industry 5.0 integrates environmental and societal goals into the structure of decision-making itself.

The SCT can therefore be understood as an Industry 5.0 system in which sustainability actively guides cognitive reasoning rather than being evaluated retrospectively.

6.3 Four Core Sustainability Design Principles

To operationalise sustainability-embedded design, the SCT framework introduces four cross-cutting principles:

Principle 1: Environmental Objective Specification

Sustainability objectives must be defined at the outset of system design. Organisations should translate commitments into measurable targets aligned with recognised standards such as the GHG Protocol or ISO frameworks.

These targets function as architectural constraints, ensuring that optimisation operates within sustainability boundaries from the beginning.

Principle 2: Whole-Lifecycle Scope

Sustainability analysis must extend beyond operational performance to include the full lifecycle of assets and materials, including:

- Extraction and manufacturing
- Operational use
- Refurbishment and reuse
- End-of-life recovery

This lifecycle perspective ensures that sustainability is embedded across all stages, from resource extraction to end-of-life management.

Lifecycle-integrated systems provide more accurate environmental insights and enable effective circular economy decision-making.

Principle 3: Radical Transparency and Auditability

Sustainability decisions must be transparent and auditable. Each system-generated decision should include:

- Data sources used
- Inference logic applied
- Alternative scenarios evaluated
- Associated uncertainty levels

This requirement directly influences model selection, favouring explainable approaches over opaque "black-box" systems in regulated environments.

Principle 4: Infrastructure Self-Accountability

Digital systems themselves consume energy and resources. The SCT framework therefore requires organisations to account for:

- Data centre energy consumption
- Carbon intensity of computation
- Hardware lifecycle impacts

Ignoring these factors can lead to misleading sustainability claims. As a result, green computing practices—such as energy-efficient infrastructure and carbon-aware scheduling—must be integrated into system design.

Illustrative Example: Smart Manufacturing Context

In a manufacturing environment, a sustainability-driven cognitive digital twin can continuously monitor energy consumption, emissions, and production efficiency across multiple production lines.

When predefined sustainability thresholds are exceeded, the system can recommend:

- Rescheduling high-energy operations
- Switching to energy-efficient machine configurations
- Adjusting resource allocation

Such capabilities enable organisations to optimise operational performance while maintaining compliance with environmental targets.

6.4 Organisational Preconditions for Sustainable CDT Deployment

Technical design alone is insufficient to ensure sustainability-driven implementation. Organisational conditions must also support sustainability integration.

Three key preconditions are required:

- Leadership Commitment:

Sustainability must be treated as a binding strategic objective rather than an aspirational goal. Executive support ensures that sustainability priorities are maintained during trade-offs.

- Cross-Functional Governance:

Digital twin development requires coordination across engineering, IT, sustainability, and operations. A joint governance structure ensures alignment throughout the design process.

- Aligned Incentives:

Performance metrics should incorporate sustainability outcomes. When incentives reflect sustainability goals, organisational behaviour aligns with system design.

Together, these conditions enable effective implementation of sustainability-embedded digital twin architectures in complex organisational environments.

7. ENTERPRISE TRANSFORMATION THROUGH A SUSTAINABILITY LENS

The implementation of digital twins cannot be seen as some technological modernisation, but an overhaul of an organisation. With sustainability incorporated into the design of the twin, this change flows to various levels of the business. The SCT framework identifies three main transformation levels, namely, operational, tactical, and strategic. All levels extend the capacity of its predecessor and increase the scope of sustainability governance.

7.1 Operational Transformation and Emissions Reduction

Sustainability on the operational level turns into a controllable variable instead of considering it a report measure on a retrospective level. Real-time sensing infrastructure makes it easy to monitor energy use, emissions, and flows of materials continuously. Digital twin systems are therefore able to identify the inefficiencies when they appear and propose corrective measures before the environmental levels are violated. This capacity has the potential to bring tangible improvements in the energy usage and resource consumption (Warke et al., 2021) in manufacturing settings. Similarly, reports of smart infrastructure demonstrate that AI-powered digital twins can significantly reduce the number of carbon emissions in cities due to better use of resources (Li, 2025). The major difference is in the architecture of the decision: the conventional monitoring systems spot sustainability issues after they occur, but sustainability-integrated twins base ESG constraints on the decision on their operations, which makes environmental performance an operational control variable of the system.

7.2 Tactical Coordination, Resilience, and Resource Efficiency

Digital twin capabilities are available at the tactical level, linking organisational units and facilities and allowing decisions to be made through the supply chains, manufacturing networks, and logistics systems together. These types of coordination are essential to an approach to challenges in sustainability that cannot be managed only at the asset level of supply-chain emissions, circular economy logistics, and energy optimisation across multi-facilities. Through empirical research, it is shown that the AI-based digital twins will be able to promote sustainability governance in the supply networks by incorporating data on operations of multiple partners (Sharma et al., 2022). Equally, Ali et al. (2025) note the significance of cross-organisational data visibility in providing circular recovery material process. The SCT framework provides these forms of network-level sustainability management through the use of semantic interoperability and shared data infrastructures.

7.3 Strategic Capital, Green Innovation, and ESG Planning

On the strategic level, digital twins with a sustainability focus affect investment selection, strategy of innovation, and corporate governance. The need to report in granular, verifiable data on the environmental aspects keeps increasing among organisational requirements, regulatory mandates and investor expectations. Such frameworks as the Corporate Sustainability Reporting Directive (CSRD), International Sustainability Standards Board (ISSB), and Task Force on Climate-Related Financial Disclosures (TCFD) require full disclosures related to emissions, use of resources, and exposure to the climate risk. Digital twins with sustainability inside of them provide the data structure required to support such demands. Constant monitoring of assets will enable organisations to produce auditable environmental performance indicators and determine the effect of strategic decisions before implementation. Studies have shown that those companies that combine digital infrastructure and sustainability governance tend to have better ESG outcomes and gains better access to the green capital market (Ben Youssef, 2025).

7.4 Alignment with Industry 5.0 and the Regenerative Enterprise

The paradigm shift of the SCT framework will be in line with the newly developed Industry 5.0 concept that focuses on sustainability, human-centered design, and resilient production systems. According to Barata and Kayser (2024), sustainability integration, human control, and transparent decision support are the three main features of Industry 5.0 digital infrastructures. These attributes are echoed in the SCT architecture which incorporates sustainability constraints into decision logic, human-in-the-loop governance structures and puts a premium on explainable analytical models. In this sense, run on sustainability digital twins assist in creating regenerative enterprises- organisations that are not only committed to reducing the impact of environmental degradation but also creating beneficial ecological and social value.

8. SUSTAINABLE COGNITIVE TWIN MATURITY MODEL

A maturity model provides organisations with an organized format of evaluating their existing digital twin capacities and possible development strategies in the future. Throughout the Sustainable Cognitive Twin Maturity Model (SCT -MM), unlike the generic digital maturity models that prioritize technology adoption, cognitive ability to stay on the technology path has altered and became sustained with the utilization of intelligence in a comprehension method. The two dimensions are co-evolved in the given framework as the sustainability-focused decision systems require sophisticated analytics as well as developed environmental data frameworks that have zero-tolerance to weak performance in this area. The SCT-MM, in its turn, thinks of the digital twin development as a gradual process of change, as opposed to one technological advancement. In overcoming the maturity levels of organisations, digital representations of assets may evolve into ecosystem-level cognitive systems able to serve the sustainability governance and regenerative enterprise approach. The model includes six levels of maturity with each of them having specific digital twin architecture, artificial intelligence potential and a range of sustainability performance indicators. Such levels help organisations to assess their status quo and define the level of technological and organisational competence needed to progress in the direction of sustainability-oriented digital transformation. Table 5 presents all the six stages of the Sustainable Cognitive Twin Maturity Model and showcases the type of digital twin, analytical capacity, and sustainability key performance indicator of each level.

Table 5. Sustainable Cognitive Twin Maturity Model — levels, capability, and sustainability KPIs

Level	Name	DT Type	AI Capability	Sustainability KPIs
1	Foundational Digitisation	Digital Model	None / rules-based	Asset inventory; energy baseline metering
2	Descriptive Analytics	Digital Shadow	Basic ML (classification, clustering)	Real-time energy dashboarding; emissions monitoring
3	Predictive Intelligence	Digital Twin	Predictive ML; anomaly detection	Predictive maintenance; waste forecasting; Scope 1 tracking
4	Prescriptive Autonomy	Cognitive DT (operational)	Reinforcement learning; multi-objective optimisation	Autonomous energy optimisation; Scope 1 & 2 reduction targets
5	Ecosystem-Level Sustainability Intelligence	Cognitive DT (strategic)	Knowledge graphs; causal inference; NLP	Scope 3 tracking; circular KPIs; supplier ESG scoring
6	Regenerative Enterprise	Ecosystem CDT	Generative AI; multi-agent systems; digital twin of organisations	Net positive impact metrics; SDG alignment; social value reporting

The maturity model that is presented in Table 5 shows that the sustainability capabilities are developed along with the development of digital twin intelligence. On the first tiers (Levels 12), the organisations focus on developing digital infrastructure and simple monitoring systems. The stages include above all; developing digital images of tangible assets, applying base energy measurements and introducing real-time monitoring panels on operational and environment indicators. These activities form the basis of data needed in more advanced sustainability management although the current level of analytical operations is minimal.

Level 3 and 4 (intermediate maturity levels) add predictive and prescriptive analytical factors. Level 3 has predictive machine learning models that allow organisations to forecast equipment malfunctions, resource usage trends, and waste generation to support proactive sustainability management instead of a reactionary one. Level 4 goes on to add prescriptive optimisation systems where reinforcement learning algorithms provide operational advice in accordance with sustainability goals, like achieving a lower intensity of emissions or greater energy efficiency.

The top levels of maturity (Levels 5-6) extends the sustainability intelligence outside the organisational boundaries. At Level 5, digital twins incorporate supply chain data, lifecycle information, and cross-enterprise knowledge networks to support ecosystem-level sustainability governance. This capability is particularly important for managing Scope 3 emissions, which often represent the largest share of an organisation's environmental footprint.

Further analytical approaches, such as knowledge graphs and causation inference models, would help organisations to analyse the effect of sustainability on complex value networks. Level 6 refers to the advanced stage of maturity, where the regenerative enterprise models are supported by the digital twin ecosystems. At this point of time, multi-agent digital infrastructures and generative AI systems can enable organisations to optimise sustainability choices in a continuum of stakeholders, encompassing suppliers, infrastructure operators, and regulators. Therefore, sustainability performance is not only a compliance imperative but a holistic part of the strategic value-generation dimension, which reduces to global sustainability objectives, in particular, the United Nations Sustainable Development Goals (SDGs). All these add up together to show how progressive development in capabilities is required as a result of sustainability-based digital transformation. Improved data infrastructure, artificial intelligence, governance models, and cross-organisation should be improved to realise organisations stepping beyond primitive monitoring systems towards complete visibility of integrated, sustainability-oriented cognitive digital twin ecosystems.

8.1 Foundational Digitisation

Level 1 The organisation creates online versions of its physical resources with no automatic data interconnection. Activities include at this phase include the construction of an assets catalogue, base-line energy and emission metering, data format standardisation and the definition of data-collection protocols. This base layer, cognitively limited as it is, is essential; without the accurate asset inventories and justifiable baselines, later layers lack the markers upon which a measure of improvement can be made, or an assertion of improvement can be made. Companies at Level 1 should invest simultaneously in sensor infrastructure and data-governance systems because they are the requirements of the further development.

8.2 Descriptive Analytics

Level 2 offers real time data flows between the physical assets and the virtual assets of a monitoring environment which leads to deployment of dashboards to monitor operations and sustainability parameters. Non-practical patterns and anomalies may be detected using basic machine-learning algorithms, namely classification and clustering of continuous data streams. At this level, the digital shadow is not yet used to facilitate independent intervention, although it shifts organisational awareness significantly, allowing sustainability managers to view, in both real-time and as asset-based visualisations, the energy use, emissions, and resource flow in operations and thus define what to focus on to reduce the impact evidence-based.

8.3 Predictive Intelligence

Level 3 adds a fully bidirectional digital twin, thus allowing predictive machine-learning models to predict the future conditions. Predictive maintenance is possible; equipment wear and tear can be predicted which will eliminate the temporal and carbon-consuming needs of emergency repairs supply chains. The waste and yield anticipatory models facilitate the scheduling of production processes, preventing material overruns. At this level, the cognitive input is a prospective one: the twin is aware not only of current phenomena but also of a likely futuristic event, which allows take a proactive position instead of a reactive one in the management of sustainability. Organisations are therefore able to set data validated, consequential short term goals that are geared towards emission reduction.

8.4 Prescriptive Autonomy

Level 4 is the embodiment of the real thinking ability. Recommendations that are created by the reinforcement-learning algorithms conditioned on sustainability goals include energy consumption reduction, equipment usage maximisation, emissions-per-unit-output reduction; these recommendations may be approved by human operators or automatically followed within established safety margins. The digital twin constantly improves the control policies with new operating conditions, thus maintaining performance and sustainability throughout the seasonal changes, demand variability, and technological changes. The Scope 1 and Scope 2 reduction goals are initially operationalised, because the digital cognitive twin (CDT) will be able to actively enforce and not just report it.

8.5 Ecosystem-Level Sustainability Intelligence

Level 5 expands cognitive potential outside of the organisation and incorporates the supply chains, product life cycles, and terminal material flows. The extended value network knowledge graphs in the organisation would have the power to make causal assumptions about the effects of sustainability to contextualise the implications of sustainability beyond the production site in space and time. The emissions that represent over three-quarters of the average amounts of carbon footprint by a manufacturer, known as scope 3 emissions, are manageable and are redefined. The objective measure used in the CDT can include the rates of material recovery, the indicator of product life extension, and the structure of waste-resource conversion, which are all indicators of a circular economy (Kristoffersen et al., 2020; Zhang et al., 2025). This degree requires a well-developed supply-chain data-sharing contract and semantic interoperability with partner systems, which require technical investment and innovation in contracts.

8.6 Maturity Assessment Matrix with Sustainability KPIs

To measure the maturity, organisations wishing to use this should assess capabilities based on five assessment dimensions, namely: (i) data infrastructure maturity, (ii) artificial-intelligence and analytics maturity, (iii) maturity of measuring sustainability, (iv) governance and auditability maturity, and (v) extent of cross-boundary integration. Each dimension corresponds to the 1–6 rating that is consistent with each of the above maturity levels to generate a radar profile illustrating the current position and the most urgent unsatisfied development requirements. The evaluation should be repeated after a year and checked by a collaboration of leadership on the basis of the functions of digital transformation, operations, and sustainability to provide balanced progress. Table 6 lists the illustrative KPIs of sustainability on each level within each dimension.

Table 6. Sustainability KPI examples by assessment dimension and maturity band

Assessment Dimension	Level 1–2 KPIs	Level 3–4 KPIs	Level 5–6 KPIs
Data Infrastructure	Asset inventory completeness; sensor coverage %	Data latency; data quality score	Cross-org data-sharing agreements; semantic interoperability score
AI & Analytics	Dashboard uptime; reporting frequency	Prediction accuracy (RMSE); anomaly recall rate	Causal model validity; multi-agent coordination efficiency
Sustainability Measurement	Energy baseline accuracy; Scope 1 completeness	Waste forecast error; Scope 2 reduction %	Scope 3 coverage %; circular material flow rate
Governance & Auditability	Data governance policy existence	Audit trail completeness; model explainability score	Third-party ESG verification; AI ethics compliance score
Cross-Boundary Integration	Internal system integration %	Supplier data connectivity; digital thread coverage	Ecosystem CDT nodes; SDG contribution score

As shown in Table 6, the maturity assessment matrix translates the conceptual maturity levels into measurable organisational indicators. The table organises evaluation across five key dimensions: data infrastructure, AI and analytics capability, sustainability measurement, governance and auditability, and cross-boundary integration. At lower maturity bands (Levels 1–2), organisations primarily focus on establishing reliable data infrastructure and baseline sustainability metrics. Intermediate levels (Levels 3–4) emphasise predictive analytics performance and measurable environmental improvements such as emissions reduction and waste forecasting accuracy. At the highest maturity band (Levels 5–6), assessment extends to ecosystem-level capabilities including supply chain data integration, circular resource flows, and third-party ESG verification. This structured KPI framework enables organisations to systematically evaluate progress and prioritise capability development toward sustainability-driven cognitive digital twin ecosystems.

9. IMPLEMENTATION ROADMAP FOR SUSTAINABLE COGNITIVE DIGITAL TWINS

The maturity model and conceptual framework explored in the foregoing sections cannot be applicable without having viable channels of organisational execution.

The creation of a Sustainability-Driven Cognitive Digital Twin (SCDT) should thus be considered a long-term capacity-building programme rather than a short-term technology implementation. The standard maturity phase of digital and analytical maturity organic development in organisations runs through incremental stages, each demonstrating potential value, sustainability, and institutional support for the next advancement of the occupation.

The maturity model, in this case, serves as a roadmap for implementation, enabling organisations to progressively enhance digital twin capabilities as they align them with sustainability goals. The roadmap focuses on four areas of strategic implementation: phased deployment and semantic knowledge base development; enterprise and edge-cloud integration; green data infrastructure; and modular service-oriented architecture design.

9.1 Phased Deployment and Semantic Knowledge Base

The implementation process usually starts by providing basic digital infrastructure.

Phase 1 (Months 1-12) aims to install sensors on high-impact assets, define the underlying data architecture, establish sustainability baselines, and develop a pilot digital twin system. Among the most vital and very often neglected consequences of this step, the creation of a semantic knowledge base must be counted. This body of knowledge is an ontologically defined definition of organisational assets, operational processes, and sustainability indicators. It provides a common lexicon that enables the mental system to make sense of sensor information, engage in analytical cognition, and report findings via human-machine interfaces. According to Ali et al. (2022), semantic interoperability is among the most challenging issues in enterprise digital twin systems. Companies with a slow pace in the development of semantic frameworks are likely to experience significant obstacles to integration in the later stages of the implementation process.

Phase 2 (Months 13-30) is when digital twin coverage is extended to other assets, and predictive analytics begin to emerge. The patterns of operations, anomalies, and trends in resource consumption are introduced as the machine learning models. These capabilities enable organisations to go beyond stagnant monitoring into early forms of sustainability-focused operational optimisation.

Phase 3 (Months 31-54) involves further development of the system, allowing it to incorporate prescriptive optimisation mechanisms and include supply chain data. Here, the digital twins begin to suggest decisions that balance operational efficiency and environmental goals. The coupling with supplier and logistics information also provides better visibility into Scope 3 emissions, which, in most cases, are the largest contributors to organisations' carbon footprints.

Lastly, Phase 4 (Month 55 and beyond) targets the development of ecosystem-level digital twin capabilities. Digital twin networks are expanded out of the organisation to lifecycle data, partner networks, and downstream product flows. Tao et al. (2019) highlight that this form of lifecycle integration presupposes that the data exchange data architecture can be developed in both directions at the initial stage of the implementation process. Companies that seek to incorporate supply chain data at late stages often adopt costly redesign and system integration.

9.2 Enterprise and Edge–Cloud Integration

SCDT requires effective integration with the existing enterprise information systems, which, in turn, is critical to its successful implementation. These are enterprise resource planning (ERP) solutions, manufacturing execution (MES), supply-chain management software, and ESG software.

Enterprise architecture management is crucial for aligning such integrations. Aier et al. (2021) emphasize that digital transformation projects often face barriers, including data model reconciliation, master data governance, and inconsistent application programming interfaces (APIs). Resolving them at the initial stages of the implementation process is a major boost towards the scalability of digital twin infrastructures.

Besides the integration of enterprise software, SCDT architectures entail an equal balance between edge and cloud. Edge computing machines process latency-sensitive operational data at the same location as physical assets, enabling rapid control responses. Cloud providers, on the other hand, facilitate large-scale analytics, machine learning model training, and strategy optimisation. As Leng et al. (2021) demonstrate, AI-powered digital twins in the context of smart manufacturing depend on this multi-layer interaction between operational technology (OT) and information technology (IT).

From a sustainability perspective, the computational task separation between edge and cloud systems also affects the environmental footprint of digital infrastructure. Computational efficiency is a relevant design factor of data centres, which use up huge quantities of electric power. On this basis, Bieser and Hilty (2018) argue that the energy consumption of digital infrastructure per se must be included in the sustainability evaluation of digital transformation projects.

9.3 Green Data Infrastructure

Green data infrastructure represents a fundamental requirement for implementing sustainability-driven cognitive digital twins. Rather than treating infrastructure emissions as a secondary concern, the SCDT framework emphasises the importance of designing digital systems that minimise their own environmental impact.

Several strategies support this objective. Organisations can prioritise cloud providers that operate data centres powered by renewable energy sources. Carbon-aware workload scheduling can also be implemented to shift non-urgent computational tasks to periods when renewable energy availability is higher. Additionally, energy-proportional computing architectures help reduce power consumption by aligning system activity with actual computational demand.

Hardware lifecycle management is another important component of sustainable digital infrastructure. Extending the lifespan of computing devices reduces the embodied carbon associated with hardware manufacturing and electronic waste generation. Bieser and Hilty (2018) emphasise that the indirect environmental impacts of digital systems—including energy consumption, hardware production, and rebound effects—should be explicitly incorporated into sustainability assessments.

Empirical research also demonstrates that green digital infrastructures can deliver both environmental and economic benefits. For example, Bibri and Huang (2025) show that AI-enabled digital infrastructures in smart building environments can simultaneously reduce their own energy consumption while optimising building energy management systems. These findings highlight that sustainable digital infrastructures are not only environmentally beneficial but also operationally efficient.

Ultimately, an SCDT that reduces operational emissions while generating excessive infrastructure-related emissions cannot be considered genuinely sustainable. Environmental performance must therefore be evaluated across both operational and digital infrastructure dimensions.

9.4 Modular and Service-Oriented Architectures

The SCDT framework strongly supports the adoption of **modular, service-oriented architectures (SOA)** for system implementation. Modular architectures allow organisations to deploy digital twin capabilities incrementally rather than attempting large-scale monolithic system implementations.

This approach offers several advantages. Individual system components can be developed, tested, and deployed independently, allowing organisations to generate measurable sustainability value even during partial system deployment. Such incremental development also reduces financial risk and allows organisations to adapt implementation strategies based on operational learning.

Rossmann and Hertweck (2022) highlight that modular digital twin architectures are particularly important for large enterprises where multiple asset domains, business processes, and organisational units must interact within a coherent digital ecosystem. Modular system design therefore improves both scalability and organisational flexibility.

API-first design principles further enhance system interoperability. By standardising data exchange mechanisms, organisations can ensure that digital twin systems remain compatible with future technologies, evolving regulatory requirements, and emerging sustainability measurement frameworks. Fuller et al. (2020) emphasise that openness and interoperability are essential characteristics of scalable digital twin ecosystems.

Recent research also suggests that sustainable digital transformation initiatives are most successful when system architectures remain **composable and adaptable**. Zakaria et al. (2025) argue that modular digital infrastructures support long-term ESG governance by enabling continuous technological upgrades without requiring complete system redesign.

Together, these implementation principles provide organisations with a practical pathway for developing sustainability-driven cognitive digital twins while maintaining architectural flexibility and long-term strategic value.

10. CHALLENGES AND BARRIERS

The introduction of cognitive digital twins as the endeavour to transform an enterprise towards sustainability faces a complicated environment of obstacles of technical, organisational, governance-related, and sustainability-measurement-specific kinds. This is because, before implementation programmes are made to navigate these barriers, such barriers must be understood first. This section provides the analysis of six types of barriers with a special focus on the ones that are specific to the sustainability aspect of the CDT deployment.

10.1 Technical Barriers

Technical obstacles to the implementation of sustainable CDT can be divided into three categories. The barriers to data quality and completeness are due to the heterogeneous, incomplete, and inconsistently formatted sensor and operational information, which characterizes most enterprise environments, especially in a legacy manufacturing and infrastructure environment. A challenge that generates model validity barriers is the difficulty in keeping correct digital twin models as a physical object grows older, is altered, or works outside its design conditions, and because it is necessary not only to keep calibrated sustainability impact models but also to keep models of operational state. The computational complexity challenge is due to a multi-objective nature of optimisation problem with multi-fold operational and sustainability goals in extended supply chains and product life that requires algorithmic solution to balance quality with computational affordability (Rathore et al., 2024; Rasheed et al., 2020).

10.2 Organisational Resistance

The resistance to CDT at organisational levels comes in several forms but has to be handled at purposeful change management than as an assumption that the resistance will disintegrate as technology becomes more mature. Independent CDT technologies can be viewed as a threat to professional knowledge and job safety by the operational staff. Middle managers can oppose the process of transparency that CDT-driven monitoring will create, especially in situations in which it points to inefficiencies or breaches of sustainability within their jurisdictions. Topmost managers might not like the extended implementation times, unpredictable payback curves, and inter-organizational governance demands of ardent CDT programmes. All of these resistance types demand specific sources of engagement: workforce communication and the development of capability in operational personnel; transparency-based governance frameworks, which assure managers against the punitive use of CDT information; and strict programme business cases, value realisation reporting, which are tied to milestones, and sustainability sponsorship of executive alignment in boards of directors (Rossmann and Hertweck, 2022).

10.3 Data Governance Issues

The problems of data governance are especially acute in CDT due to the fact that the twin is the aggregator of data of various sources the ownership of which might be conflicting, their sensitivity may be different, and their regulatory status may be different. The export control laws may apply to the industrial sensor data; the personal data are collected by the smart building occupancy monitoring can be treated by the data protection laws; the supply chain performance data disclosed by partners can be subjected to privacy rules that will not allow its use in the externally reported sustainability disclosure. Strong data management policies, such as stating how data is owned, allowed access, retention, consent, and audit protocols, are the precondition of the legal and ethical functioning of enterprise CDTs, and should be defined in Phase 1 before more ambitious features are added atop them.

10.4 Interoperability and Legacy Constraints

The interoperability with legacy systems is another issue that is always present and remains underestimated as an obstacle to the deployment of CDT. The majority of industrial enterprises involve physical and digital infrastructure that cuts across several technological generations, with systems that communicate using disparate protocols, data formats, and semantic models that are developed separately (in isolation) of each other. The expenditure invested in retrofitting old systems to the connectivity and the data quality of CDT architectures is disabling in the short run. Such pragmatic solutions as protocol bridges, data normalisation middleware, and selective asset modernisation in accordance with the importance of sustainability impact are found in the literature (Alcacer and Cruz-Machado, 2019; Warke et al., 2021) though they must be evaluated criteria in the case of the organisation and its technological and capital resources.

10.5 Sustainability Measurement Standardisation

Probably the most unique and significant obstacle within the framework of SCT is that there are no standardised systems of sustainability measures that are specific to CDT activities. Current sustainability reporting frameworks, such as those of GRI, SASB, and TCFD, along with those of CSRD, ISSB, were built to follow periodical, aggregate, and mainly financial-account based reporting, lack the nature of granular, real-time, computationally derived, and dynamically updated sustainability-related information that CDTs generate. The consequence is that there is a standards gap: current reporting structures are ill equipped to accept much richer sustainability data than can be produced by CDTs, and the structures themselves lack a clear view on how it should be measured, what should be included within the boundary, and how such generated information should be verified.

The concept of time-use-based methods of measuring the indirect environmental impacts of digitalisation that can be applied to CDT sustainability impact measurement as proposed by Bieser and Hilty (2018a, 2018b) should be seen as yet important methodological groundwork to this challenge. Nonetheless, before operation consensus can be reached, much additional development, by standards making bodies, regulators, and the research community, is needed. Organisations that undertake CDTs towards sustainability must pre-emptively become involved in emerging standards development processes in order to make sure that their experience of operation will influence the standards that will eventually govern their reporting requirements.

10.6 Human Trust and Adoption Barriers

Three factors that are inter-related in regard to human trust in CDT outputs are perceived competence (does the twin make reliable recommendations?), perceived transparency (can I understand why it made this recommendation?), and perceived alignment (are the objectives of the twin identical to my own values and responsibilities?). All aspects will be to be managed.

The transparency aspect is addressed by explainable AI or XAI research that makes the arguments of ML models comprehensible to non-expert users. The alignment dimension is met by value alignment techniques, which encode the human preferences and ethical constraints directly in the optimisation goals of the cognitive engine. The competence is determined by a high level of validation of the model with historical data, the openness of model performance monitoring during the live operation, and the conveying the model uncertainty (Intizar Ali et al., 2021; Iliuta et al., 2024). These three dimensions need to be invested in over the lifetime of the CDT deployment and not only at the early roll out stages. Organisations that fail to build trust will have the poorest result; a highly technical CDT that its outputs are systematically disregarded or overruled by the humans whose judgment it was meant to enlighten.

11. POLICY AND MANAGERIAL IMPLICATIONS

The concerns that pertain to the introduction of cognitive digital twins in the quest to bring about enterprise sustainability change have their implications that span far beyond the organisations that introduce them to encompass the societal policies, industry control, financial market, and artificial intelligence regulation. This section defines the most important implications that would befall five domains of policy and management.

Within the scope of Green Standards and Green Rules, green buildings incorporate green technologies, thus facilitating sustainable construction.

11.1 Green Standards and Regulation

The policy frames on the deployment of CDT to achieve sustainability need to cover three dimensions that are interrelated. First, the imperative of sustainability disclosure, as embodied in EU Corporate Sustainability Reporting Directive (CSRD), the ISSB standard-setting of IFRS S1 and S2, present a demand in the high-quality and granular and auditable sustainability data that can be delivered by the CDTs. This connection can and should be supported by policy, which should make technical guidance to use data generated by CDT in disclosures required by law and provide conditions such as data quality assurance and data modelling and independent checking. Second, sustainability standards of the operations of CDT themselves, dealing with the energy used, embodied carbon, and lifecycle of the computational infrastructure will have to be made to ensure that greenwashing via the digital can be avoided. Third, interoperability criteria to enable the circulation of CDT sustainability data across the organisational borders in supply chains are crucial to prompting the Scope 3 and circular economy value ecosystem-level CDT competence can convey.

11.2 Workforce Upskilling for Sustainable Digital Twins

To make sustainability through the successful implementation of CDTs possible, the workforce must possess the required capabilities in critically short supply: the capacity to bridge operational technology (OT) and information technology (IT), the familiarity with sustainability accounting, ESG reporting standards, and carbon markets, the skills to operate AI and data analytics at a level to implement and configure, as well as interpolate, the results of CDTs as of Section 10 in the organisation; and the organisational change management capabilities needed to negotiate the human aspects in the process of adopting the usage of CDTs as of The development of upskilling programmes which combine technical training with sustainability literacy and ability to manage change must be structured and resourced as part of projects in the deployment of CTD and not as an adjunct. The support of public policies, such as apprenticeship systems, incentives towards continuing professional development, and curriculum development in universities, is required to establish the workforce pipeline that will be necessary during the wide-scale sustainable deployment of CDT.

11.3 Green Infrastructure Incentives and Carbon Markets

The sustainable CDT use can be enhanced by governmental policy in a speedy manner by the use of specific financial instruments. The tax credit on investment or capital allowances on green data infrastructure in the form of renewable-powered edge computing, energy-efficient data centres, carbon-conscious scheduling systems would decrease the incremental cost of the sustainability-focused CDT architecture compared with the conventional implementations. The scale disadvantage on the part of SMEs in CDT investment would be mitigated by grants and risk-sharing facilities of pilot CDT programmes in SME supply routes. To the future, as soon as such CDT-generated, independently verified emissions data is introduced into voluntary and compliance carbon markets, where it could be used as the basis of high-quality carbon credits, this would establish a positive financial incentive to CDT-enabled sustainability performance that feeds on itself: revenue on carbon credits are used to develop further CDT, which effects further reduction of emissions.

11.4 Responsible AI and Sustainability Governance

The intellectual aspect of CDTs brings AI governance issues that have direct sustainability consequences. Biased training information may also yield CDTs, which systematically underestimate the impact of sustainability on marginalised communities, underrepresented geographies or future generations, a set of interests poorly represented in historical operational data. Indisputable AI-based decision-making may invalidate the validity of the sustainability claims based on the outputs of the CDT, especially when the claims are applicable in regulated disclosures or in transactions of carbon markets. Unintended sustainability outcomes could associate with autonomous CDT actions, which would have been avoided by human supervision. Fairness, accountability, transparency, safety, and environmental sustainability principles must be clearly articulated into the governance arrangements of all CDTs purporting to bring sustainability value, and implementation must also be governed by an independent audit. The CDT implementation will be regulated by the regulatory frameworks including the EU AI Act of which some of the applications of AI in industry are considered being risky will more and more come to govern the programme intended to be implemented.

11.5 Sustainability Reporting Mandates and Digital Twin Readiness

With the increasing presence of sustainability reporting regulations all over the world and organisations becoming more and more under pressure to deliver high quality auditable sustainability information on granularities and frequencies that are not supported by traditional reporting systems, the pressure on organisations is mounting. The CTDs are perfectly positioned to address this challenge: they provide the granular data that complex sustainability reporting requires because of their asset-level continuous monitoring, whilst the audit trail capabilities can be used to ensure the independent verification that is necessitated by mandatory disclosure. Nevertheless, the CDT sustainability reporting preparation of any type: data quality assurance, model documentation, boundary definition, and alignment with certain reporting standards needs to be actively constructed and sustained instead of being assumed. Organisations are advised to address preparedness to undertake CDT sustainability reporting as a programme governance criterion, and assign definite ownership to the mapping of CDT data outputs with the mandatory disclosure requirement as well as creating regular assurance reviews as a continuous governance cycle of the programme.

12. CONCLUSION

12.1 Summary of Contributions

This chapter develops the Sustainable Cognitive Twin (SCT) framework, which is a conceptual creation that places cognitive digital twins as an empowering infrastructure in a sustainable way of transforming enterprise. The framework reimagines digital twins as proposed not only as a monitoring structure but also as a decision architecture that can support complex sustainability goals. The framework illustrates how organisations can go beyond efficacy-based digitalisation and adopt a sustainability-based change by applying a sense of sustainability to the design logic of cognitive digital twins. There are a number of major contributions to this work. To start with, sustainability is positioned as a regulatory goal in the context of the digital twin design as opposed to the reporting layer. Secondly, the chapter presents a multi-level evolutionary approach that defines a connection between cognitive digital twin capacities and operational, tactical, strategic, and regenerative sustainability results. Third, Sustainable Cognitive Twin Maturity Model (SCT -MM) outlines six phases by which organisations may advance in their cognitive and sustainability intelligence development. Lastly, the chapter postulates an implementation roadmap, which highlights semantic data infrastructures, green computing architectures and modular system design as part of the requirements of implementing sustainability-oriented cognitive digital twins. Collectively, these aspects offer more systematic conceptual premise on how cognitive digital twin technologies can help transform sustainability at the enterprise level.

12.2 From Industry 4.0 Efficiency to Industry 5.0 Regeneration

The study has added to the literature about digital twins, artificial intelligence, and sustainable digital transformation. The current literature on digital twins has mostly been based on operational monitoring, predictive maintenance, and performance optimisation. In comparison, the SCT framework goes a step further to the same to theorise digital twins as cognitive sustainability infrastructures that can merge environmental, operational, and strategic decision-making. Two other aspects that the framework adds to are the current debates on Industry 5.0 that focuses on human-centred, sustainable and resilient industries (Xu et al., 2021; Barata and Kayser, 2024). In this regard, cognitive digital twins can also be used as coordination platforms that favor multi-objective decision-making in complex socio-technical systems. Another area where the maturity model can be used is to develop a theoretical perspective of how organisations can progressively make the transition between data-driven monitoring systems and environment-level cognitive infrastructures. This chapter can be regarded as providing a conceptual framework on interdisciplinary studies of sustainable digital innovation because it connects digital twin architecture, sustainability governance, and enterprise transformation.

12.3 Toward Regenerative Cognitive Enterprises

The research results in this chapter also provide a number of practical lessons to organisations, technology developers as well as policymakers. To organisations, the suggested framework is an organised avenue through which organisations can incorporate the aim of sustainability into digital transformation projects. The maturity model allows companies to evaluate what they have already done regarding digital twins and what needs the priority in terms of technological and organisational enhancement. The implementation roadmap identifies important design considerations to technology architects and those responsible to lead digital transformation, such as semantic knowledge frameworks, enterprise system integration, edge-cloud architecture, and green data infrastructures. By these means, the digital twin systems will be able to help sustainability goals without creating unintended environmental effects of digital infrastructure itself. The framework can also be of help to policymakers and regulators. With the continued government push to use sustainability reporting and ESG accountability as a way to ensure accountability, cognitive digital twins infrastructures may offer effective digital platforms to gauge environmental performance in complicated industrial ecosystems.

12.4 Future Research Directions

The Sustainable Cognitive Twin model requires continuous development of future studies that can be used to validate the framework via empirical experimental studies in various sectors like manufacturing sector, smart cities, and energy systems. Quantitative analyses would be able to assess how cognitive digital twins increase sustainability performance metrics, in terms of emissions reduction, resource efficiency and circularity outcomes. The next potential direction to pursue is to research governance systems of AI-enabled digital twins, especially as it concerns transparency, accountability, and ethical decision-making. Secondly, it is possible that future studies could help to explore how new technologies, including generative AI, traceability generated by blockchain, and federated learning, can be used to improve sustainability intelligence in digital twins ecosystems. Lastly, transorganisational digital twin cities are a little studied field. Further research might investigate the ways in which collaborative digital twins can facilitate sustainability transition at industry scales and optimisation at the ecosystem scale.

REFERENCES

Alcácer, V., & Cruz-Machado, V. (2019). Scanning the Industry 4.0: A literature review on technologies for manufacturing systems. *Engineering Science and Technology, an International Journal, 22*(3), 899–919. DOI: 10.1016/j.jestch.2019.01.006

Ali, M. I., Patel, P., Breslin, J. G., Harik, R., & Seth, A. (2021). Cognitive Digital Twins for Smart Manufacturing. *IEEE Intelligent Systems, 96–100*. Advance online publication. DOI: 10.1109/MIS.2021.3062437

Ali, Z. A., Zain, M., Hasan, R., Pathan, M. S., AlSalman, H., & Almisned, F. A. (2025). Digital twins: Cornerstone to circular economy and sustainability goals. *Environment, Development and Sustainability*. Advance online publication. DOI: 10.1007/s10668-025-06221-4

Barata, J., & Kayser, I. (2024). How will the digital twin shape the future of industry 5.0? *Technovation, 134*, 103025. Advance online publication. DOI: 10.1016/j.technovation.2024.103025

Bibri, S. E., & Huang, J. (2025). AI and AI-powered digital twins for smart, green, and zero-energy buildings: A systematic review of leading-edge solutions for advancing environmental sustainability goals. *Environmental Science and Ecotechnology, 28*, 100628. Advance online publication. DOI: 10.1016/j.ese.2025.100628 PMID: 41244699

Botín-Sanabria, D., Mihaita, A.-S., Peimbert-García, R. E., Ramírez-Moreno, M. A., Ramírez-Mendoza, R. A., & Lozoya-Santos, J. (2022). Digital twin technology challenges and applications: A comprehensive review. *Remote Sensing (Basel), 14*(6), 1335. DOI: 10.3390/rs14061335

Fuller, A., Fan, Z., Day, C., & Barlow, C. (2020). *Digital Twin: Enabling Technologies, Challenges and Open Research*. IEEE., DOI: 10.1109/ACCESS.2020.2998358

Grieves, M., & Vickers, J. (2017). Digital Twin: Mitigating Unpredictable, Undesirable Emergent Behavior in Complex Systems. In *Transdisciplinary Perspectives on Complex Systems*. Springer International Publishing., DOI: 10.1007/978-3-319-38756-7_4

Iliu ă, M. E., Moisescu, M. A., Pop, E., Ionita, A. D., Caramihai, S. I., & Mitulescu, T. C. (2024). Digital Twin—A Review of the Evolution from Concept to Technology and Its Analytical Perspectives on Applications in Various Fields. *Applied Sciences (Basel, Switzerland), 12*(8). Advance online publication. DOI: 10.3390/app14135454

Jones, D., Snider, C., Nassehi, A., Yon, J., & Hicks, B. (2020). Characterising the digital twin: A systematic literature review. *CIRP Journal of Manufacturing Science and Technology*, *29*, 36–52. DOI: 10.1016/j.cirpj.2020.02.002

Kristoffersen, E., Blomsma, F., Mikalef, P., & Li, J. (2020). The smart circular economy: A digital-enabled circular strategies framework for manufacturing companies. *Journal of Business Research*, *120*, 241–261. DOI: 10.1016/j.jbusres.2020.07.044

Kritzinger, W., Karner, M., Traar, G., Henjes, J., & Sihn, W. (2018). Digital twin in manufacturing: A categorical literature review and classification. [Elsevier.]. *IFAC-PapersOnLine*, *51*(11), 1016–1022. DOI: 10.1016/j.ifacol.2018.08.474

Leng, J., Wang, D., Shen, W., Li, X., Liu, Q., & Chen, X. (2021). Digital twins-based smart manufacturing system design in Industry 4.0: A review. *Journal of Manufacturing Systems*, *60*, 119–137. DOI: 10.1016/j.jmsy.2021.05.011

Li, Y. (2025). AI-Enhanced Digital Twins for Energy Efficiency and Carbon Footprint Reduction in Smart City Infrastructure. *Proceedings of the 3rd International Conference on Software Engineering and Machine Learning*, (p. 42). DOI: 10.54254/2755-2721/2025.20569

Pater, J., & Stadnicka, D. (2021). Towards Digital Twins Development and Implementation to Support Sustainability Systematic Literature Review. *Management and Production Engineering Review*, 63-73.

Qiu, S., Zaheer, Q., Ali, F., Wajid, S., Chen, H., Ai, C., & Wang, J. (2025). Exploring the impact of digital twin technology in infrastructure management: A comprehensive review. *Journal of Civil Engineering and Management*, *31*(4), 395–417. DOI: 10.3846/jcem.2025.23718

Rasheed, A., San, O., & Kvamsdal, T. (2020). Digital Twin: Values, Challenges and Enablers From a Modeling Perspective. *IEEE Access : Practical Innovations, Open Solutions*, *8*, 21980–22012. Advance online publication. DOI: 10.1109/ACCESS.2020.2970143

Rathore, M. M., Shah, S. A., Shukla, D., Bentafat, E., & Bakiras, S. (2021). The Role of AI, Machine Learning, and Big Data in Digital Twinning: A Systematic Literature Review, Challenges, and Opportunities. *IEEE Access : Practical Innovations, Open Solutions*, *9*, 32030–32052. Advance online publication. DOI: 10.1109/ACCESS.2021.3060863

Sacoto-Cabrera, E., Perez-Torres, A., Tello-Oquendo, L., & Cerrada, M. (2025). IoT, AI, and digital twins in smart cities: A systematic review for a thematic mapping and research agenda. *Smart Cities*, *8*(5), 175. Advance online publication. DOI: 10.3390/smartcities8050175

Sharma, A., Kosasih, E., Zhang, J., Brintrup, A., & Calinescu, A. (2022). Digital Twins: State of the art theory and practice, challenges, and open research questions. *Journal of Industrial Information Integration*, *30*, 100383. Advance online publication. DOI: 10.1016/j.jii.2022.100383

Tao, F., Zhang, H., Liu, A., & Nee, A. Y. (2019). Digital Twin in Industry: State-of-the-Art. *IEEE Transactions on Industrial Informatics*, *15*(4), 2405–2415. DOI: 10.1109/TII.2018.2873186

Teece, D. J. (2018). Business models and dynamic capabilities. *Long Range Planning*, *51*(1), 40–49. DOI: 10.1016/j.lrp.2017.06.007

Warke, V., Kumar, S., Bongale, A., & Kotecha, K. (2021). Sustainable Development of Smart Manufacturing Driven by the Digital Twin Framework: A Statistical Analysis. *Sustainability (Basel)*, *13*(18), 10139. Advance online publication. DOI: 10.3390/su131810139

Xu, X., Lu, Y., Voget-Heuser, B., & Wang, L. (2021). Industry 4.0 and Industry 5.0—Inception, conception and perception. *Journal of Manufacturing Systems*, *61*, 530–535. DOI: 10.1016/j.jmsy.2021.10.006

Youssef, A. B. (2025). Twin Transition: Digital Transformation Pathways for Sustainable Innovation. *Sustainability (Basel)*, *17*(21), 9491. Advance online publication. DOI: 10.3390/su17219491

Zhang, Z., Wei, Z., Court, S., Yang, L., Wang, S., Thirunavukarasu, A., & Zhao, Y. (2024). A Review of Digital Twin Technologies for Enhanced Sustainability in the Construction Industry. *Sustainability*, *16*. Advance online publication. DOI: 10.3390/su16041514

Compilation of References

Abdallah-Ou-Moussa, S., Wynn, M., & Rouaine, Z. (2026). Leveraging Blockchain and Digital Twins for Low-Carbon, Circular Supply Chains: Evidence from the Moroccan Manufacturing Sector. *Sustainability (Basel)*, *18*(2), 991. DOI: 10.3390/su18020991

Abdi, H., & Nozari, H. (2026). Energy challenges in transformative technologies-based super-smart city implementation. In *Energy-Efficient Transformative Technologies for Data-Driven Smart Cities* (pp. 71–90). Elsevier. DOI: 10.1016/B978-0-443-27618-7.00002-9

Abdulameer, L., Al-Khafaji, M. S., Al-Awadi, A. T., Al Maimuri, N. M., Al-Shammari, M., Al-Dujaili, A. N., & DhiyaAl-Jumeily, . (2025). Artificial intelligence in climate-resilient water management: A systematic review of applications, challenges, and future directions. *Water Conservation Science and Engineering*, *10*(1), 44. DOI: 10.1007/s41101-025-00371-2

Acuña-Duran, E., Oyanedel, J. C., & Pradenas-Wilson, D. (2022). Survey data of entrepreneurial intention and perceived social support from academics-scientists at Chilean universities. *Data in Brief*, *44*, 108537. DOI: 10.1016/j.dib.2022.108537 PMID: 36045643

Adcroft, A., Willis, R., & Dhaliwal, S. (2004). Missing the point? Management education and entrepreneurship. *Management Decision*, *42*(3), 521–531. DOI: 10.1108/00251740410518958

Adibi, S., Rajabifard, A., Shojaei, D., & Wickramasinghe, N. (2024). Enhancing healthcare through sensor-enabled digital twins in smart environments: A comprehensive analysis. *Sensors (Basel)*, *24*(9), 2793. DOI: 10.3390/s24092793 PMID: 38732899

Afolabi, A., Ogunrinde, O., & Zabihollah, A. (2025). Digital Twin and AI Models for Infrastructure Resilience: A Systematic Knowledge Mapping. *Applied Sciences (Basel, Switzerland), 15*(24), 13135. DOI: 10.3390/app152413135

Afrin, S., Roksana, S., & Akram, R. (2024). AI-Enhanced Robotic Process Automation: A Review of Intelligent Automation Innovations. *IEEE Access : Practical Innovations, Open Solutions*. Advance online publication. DOI: 10.1109/ACCESS.2024.3513279

Ahsen, R., Di Bitonto, P., Novielli, P., Magarelli, M., Romano, D., Diacono, D., Monaco, A., Amoroso, N., Bellotti, R., & Tangaro, S. (2025). Harnessing Digital Twins for Sustainable Agricultural Water Management: A Systematic Review. *Applied Sciences (Basel, Switzerland), 15*(8), 4228. DOI: 10.3390/app15084228

Ajunwa, I. (2020). An auditing imperative for automated hiring systems. *Harv. JL & Tech., 34*, 621. https://ssrn.com/abstract=3437631

Al-Ali, A. R., Gupta, R., Zaman Batool, T., Landolsi, T., Aloul, F., & Al Nabulsi, A. (2020). Digital twin conceptual model within the context of internet of things. *Future Internet, 12*(10), 163. DOI: 10.3390/fi12100163

Alam, M. D., Kabir, G., & Mirmohammadsadeghi, S. (2023). A digital twin framework development for apparel manufacturing industry. *Decis. Anal. J., 7*, 100252. Advance online publication. DOI: 10.1016/j.dajour.2023.100252

Al-Asady, H. A. J. (n.d.) A Cognitive Digital Twin Framework for Resilient Water Infrastructure under Climate Change. In *Smart Technologies for Sustainable Development Goals* (pp. 321–350). CRC Press. DOI: 10.1201/9781003714774-18

Alcácer, V., & Cruz-Machado, V. (2019). Scanning the Industry 4.0: A literature review on technologies for manufacturing systems. *Engineering Science and Technology, an International Journal, 22*(3), 899–919. DOI: 10.1016/j.jestch.2019.01.006

Alexander, R., Janet, A. M. V., Sri, R. V., Sasirekha, S., Kathyani, A., & Priya, V. S. (2025, September). A Multilingual LLM-Driven Digital Twin Framework for Climate Resilient Precision Agriculture. In 2025 6th International Conference on Electronics and Sustainable Communication Systems (ICESC) (pp. 1794-1805). IEEE.

Ali, M. I., Patel, P., Breslin, J. G., Harik, R., & Seth, A. (2021). Cognitive Digital Twins for Smart Manufacturing. *IEEE Intelligent Systems, 96–100*. Advance online publication. DOI: 10.1109/MIS.2021.3062437

Alipour, H., & Shojaei, S. (2025). Modeling Soil Futures: Integrating Classic and Emerging Approaches to Water Erosion. *Journal of Environmental Science and Engineering Technology, 13*, 18–27. DOI: 10.12974/2311-8741.2025.13.03

Ali, R., Hussain, A., Nazir, S., Khan, S., & Khan, H. U. (2023). Intelligent Decision Support Systems—An Analysis of Machine Learning and Multicriteria Decision-Making Methods. *Applied Sciences (Basel, Switzerland)*, *13*(22), 12426. DOI: 10.3390/app132212426

Ali, Z. A., Zain, M., Hasan, R., Pathan, M. S., AlSalman, H., & Almisned, F. A. (2025). Digital twins: Cornerstone to circular economy and sustainability goals. *Environment, Development and Sustainability*. Advance online publication. DOI: 10.1007/s10668-025-06221-4

Al-Jamal, M. Q., Alsarhan, A., Aljamal, Q., AlJamal, M., Khassawneh, B. S., Al Nuaim, A., & Al Nuaim, A. (2026). AI-Enabled System-of-Systems Decision Support: BIM-Integrated AI-LCA for Resilient and Sustainable Fiber-Reinforced Façade Design. *Information (Basel)*, *17*(2), 126. DOI: 10.3390/info17020126

Allen, K. E., & Mau, T. A. (2025). Breaking down traditional silos: Leading organizations from an interdependent mindset. *Humanistic Management Journal*, *10*(1), 125–139. DOI: 10.1007/s41463-024-00198-4

Alsaedi, F., & Masoud, S. (2025). Condition-based maintenance for degradation-aware control systems in continuous manufacturing. *Machines (Basel)*, *13*(2), 141. DOI: 10.3390/machines13020141

Alves, R. G., Maia, R. F., & Lima, F. (2023a). Development of a Digital Twin for smart farming: Irrigation management system for water saving. *Journal of Cleaner Production*, *388*, 135920. DOI: 10.1016/j.jclepro.2023.135920

Andreasyan, A., & Balyakin, A. (2022). Transformation of education through Big Data: Digital twins case study. *Journal of Physics: Conference Series*, *2210*(1), 012003. DOI: 10.1088/1742-6596/2210/1/012003

Arya, V., Colace, F., Lombardi, M., Lorusso, A., Santaniello, D., & Valentino, C. (2025). DTSafe-O: A Semantic Digital Twin Ontology for Construction Safety, International Journal on Semantic Web and Information Systems, 21, 1 36. .DOI: 10.4018/IJSWIS.395999

Asad, U., Khan, M., Khalid, A., & Lughmani, W. A. (2023). Human-centric digital twins in industry: A comprehensive review of enabling technologies and implementation strategies. *Sensors (Basel)*, *23*(8), 3938. DOI: 10.3390/s23083938 PMID: 37112279

Asthana, R. (2025). Driving User Adoption and Agility in PLM: A Data-Driven Approach to Change Management. https://www.theseus.fi/bitstream/handle/10024/887389/Asthana_Richa.pdf?sequence=2&isAllowed=y

Attah, R. U., Ogunsola, O. Y., & Garba, B. M. P. (2023). Leadership in the digital age: Emerging trends in business strategy, innovation, and technology integration. *Iconic Research and Engineering Journals*, *6*(9), 389–411.

Attaran, M. (2023). Digital twin: Benefits, use cases, challenges, and opportunities. *Decision Support Systems*, *162*, 113786. DOI: 10.1016/j.dss.2022.113786

Autio, E., Nambisan, S., Thomas, L. D. W., & Wright, M. (2018). Digital affordances, spatial affordances, and the genesis of entrepreneurial ecosystems. *Strategic Entrepreneurship Journal*, *12*(1), 72–95. DOI: 10.1002/sej.1266

Awais, M., Wang, X., Hussain, S., Aziz, F., & Mahmood, M. Q. (2025). Advancing precision agriculture through digital twins and smart farming technologies: A review. *AgriEngineering*, *7*(5), 137. DOI: 10.3390/agriengineering7050137

Awouda, A., Traini, E., Bruno, G., & Chiabert, P. (2024). IoT-Based Framework for Digital Twins in the industry 5.0 era. *Sensors (Basel)*, *24*(2), 594. DOI: 10.3390/s24020594 PMID: 38257686

Bacigalupo, M., Kampylis, P., Punie, Y., & Van Den Brande, L. (2016). EntreComp: The Entrepreneurship Competence Framework. EUR 27939 EN. Luxembourg (Luxembourg): Publications Office of the European Union; 2016. JRC101581, https://publications.jrc.ec.europa.eu/repository/handle/JRC101581 https://doi.org/ DOI: 10.2791/593884

Badakhshan, E., & Ball, P. (2023). Applying digital twins for inventory and cash management in supply chains under physical and financial disruptions. *International Journal of Production Research*, *61*(15), 5094–5116. DOI: 10.1080/00207543.2022.2093682

Bakó, K., Rácz, C., Dövényi-Nagy, T., Molnár, K., & Dobos, A. (2025). Advancements in Leaf Area Index Estimation for Maize Using Modeling and Remote Sensing Techniques: A Review. *Agronomy (Basel)*, *15*(3), 519. DOI: 10.3390/agronomy15030519

Balaji, K. "Exploring the Drivers and Effects on Supply Chain Resilience and Performance in an Emerging Market Using Artificial Intelligence," *2023 3rd International Conference on Smart Generation Computing, Communication and Networking (SMART GENCON)*, Bangalore, India, 2023, pp. 1-6, DOI: 10.1109/SMARTGENCON60755.2023.10442071

Balaji, K., & Rao, P. S. "Artificial Intelligence and Chatbots: Transforming User Experience in E-tailing," *2024 9th International Conference on Communication and Electronics Systems (ICCES)*, Coimbatore, India, 2024, pp. 1375-1379, DOI: 10.1109/ICCES63552.2024.10859474

Balaji, K. (2025). AI as a Game Changer: How Digital Entrepreneurs Are Redefining Business Practices. In Tunio, M. (Ed.), *Improving Entrepreneurial Processes Through Advanced AI* (pp. 21–48). IGI Global Scientific Publishing., DOI: 10.4018/979-8-3693-1495-1.ch002

Balaji, K. (2025). Harnessing AI and Blockchain in Sustainability Assurance: Trends Shaping the Future of Verification. In Khatib, S., & Abbas, A. (Eds.), *Navigating Trust in Sustainability Reporting and Assurance* (pp. 215–242). IGI Global Scientific Publishing., DOI: 10.4018/979-8-3373-0117-4.ch007

Balaji, K. (2025). Human-Centric Innovation: Building Sustainable Organizations for Industry 5.0. In Erturk, A., Özşahin, M., & Simovic, V. (Eds.), *Human-Centric, Sustainable, and Resilient Organizations in the Digital Age* (pp. 101–124). IGI Global Scientific Publishing., DOI: 10.4018/979-8-3693-8181-6.ch005

Balaji, K. (2025). IoT in Financial Services Innovations in Banking and Insurance. In Ho, R., Song, B., & Tee, P. (Eds.), *Managing Customer-Centric Strategies in the Digital Landscape* (pp. 75–104). IGI Global Scientific Publishing., DOI: 10.4018/979-8-3693-5668-5.ch004

Balaji, K. (2025). Neuromarketing and Sustainable Finance: Real-World Case Studies of Green Investment Promotions. In Taneja, S., Chahal, B., Johri, A., Ozen, E., & Kumar, P. (Eds.), *Neuromarketing's Role in Sustainable Finance* (pp. 169–204). IGI Global Scientific Publishing., DOI: 10.4018/979-8-3693-9117-4.ch010

Balaji, K. (2025). Revolutionizing High-Frequency Trading: The Impacts of Financial Technology and Data Science Innovations. In Chen, H. (Ed.), *Machine Learning and Modeling Techniques in Financial Data Science* (pp. 103–124). IGI Global Scientific Publishing., DOI: 10.4018/979-8-3693-8186-1.ch004

Banciu, C., & Florea, A. (2026). AIoT at the Frontline of Climate Change Management: Enabling Resilient, Adaptive, and Sustainable Smart Cities. *Climate (Basel), 14*(1), 19. DOI: 10.3390/cli14010019

Bandi, A., Kongari, B., Naguru, R., Pasnoor, S., & Vilipala, S. V. (2025). The rise of agentic ai: A review of definitions, frameworks, architectures, applications, evaluation metrics, and challenges. *Future Internet, 17*(9), 404. DOI: 10.3390/fi17090404

Banerjee, K., Dutta, S., Das, S., & Sadhukhan, R. (2025). Crop simulation models as decision tools to enhance agricultural system productivity and sustainability – a critical review. *Technology in Agronomy, 5*(1), 0–0. DOI: 10.48130/tia-0024-0032

Barata, J., & Kayser, I. (2024). How will the digital twin shape the future of industry 5.0? *Technovation, 134*, 103025. DOI: 10.1016/j.technovation.2024.103025

Barrier, K. M., Porche, D. J., Campbell, K. M., Byrd, T. O., Morris, M., Blalack, K. L., Ziglor, C., Tuazon, S., Mouton, C. P., & Southerland, J. H. (2025). Models of Diversity, Equity, Inclusion, and Belonging: Part II—Exploring Models of Inclusion from Other Health Professions for Dentistry. *Dental Clinics of North America, 69*(1), 55–68. DOI: 10.1016/j.cden.2024.08.010 PMID: 39603769

Barzigar, A., Hosseinalipour, S. M., & Mujumdar, A. S. (2025). AI-driven optimization of hybrid solar-wind dryers for climate-resilient agriculture. *Drying Technology, 43*(14), 2018–2019. DOI: 10.1080/07373937.2025.2540731

Barzigar, A., Hosseinalipour, S. M., & Mujumdar, A. S. (2025). Toward sustainable post-harvest practices: A critical review of solar and wind-assisted drying of agricultural produce with integrated thermal storage systems. *Drying Technology, 43*(10), 1463–1494. DOI: 10.1080/07373937.2025.2542440

Bashkin, V. N. (2026). Nitrogen management in the era of Industry 4.0: Progress, opportunities, and challenges for agricultural farming. In *Nonpoint Source Nitrogen Pollution* (pp. 493–510). Elsevier., DOI: 10.1016/B978-0-443-29070-1.00021-5

Bayasgalan, A., Park, Y. S., Koh, S. B., & Son, S.-Y. (2024). Comprehensive Review of Building Energy Management Models: Grid-Interactive Efficient Building Perspective. *Energies, 17*(19), 4794. DOI: 10.3390/en17194794

Becker, G. S. (1964). *Human capital: A theoretical and empirical analysis, with special reference to education*. University of Chicago Press.

Becker, G. S. (1964). *Human capital: A theoretical and empirical analysis*. University of Chicago Press.

Bedi, J., & Wani, N. A. (2026). Artificial intelligence for home digital twins. In *Home Digital Twins* (pp. 227–242). Elsevier.

Belfadel, A., Lacheroy, T., Creff, S., Hamida, A. B., & Didier, J. Y. (2026). XRTwin-4Industry: Empowering Industrial Digital Twins with XR and Instruction System-A Framework for Standardization and Interoperability. *Future Generation Computer Systems, 182*, 108476. DOI: 10.1016/j.future.2026.108476

Bessen, J., & Meurer, M. J. (2008). *Patent Failure: How Judges, Bureaucrats, and Lawyers Put Innovators at Risk*. Princeton University Press.

Bibri, S. E. (2021). A novel model for data-driven smart sustainable cities of the future: The institutional transformations required for balancing and advancing the three goals of sustainability. *Energy Informatics, 4*(1), 4. DOI: 10.1186/s42162-021-00138-8

Bibri, S. E., Alexandre, A., Sharifi, A., & Krogstie, J. (2023). Environmentally sustainable smart cities and their converging AI, IoT, and big data technologies and solutions: An integrated approach to an extensive literature review. *Energy Informatics, 6*(1), 9. DOI: 10.1186/s42162-023-00259-2 PMID: 37032812

Bibri, S. E., & Huang, J. (2025). AI and AI-powered digital twins for smart, green, and zero-energy buildings: A systematic review of leading-edge solutions for advancing environmental sustainability goals. *Environmental Science and Ecotechnology, 28*, 100628. Advance online publication. DOI: 10.1016/j.ese.2025.100628 PMID: 41244699

Black, P., & Wiliam, D. (1998). Assessment and classroom learning. *Assessment in Education: Principles, Policy & Practice, 5*(1), 7–74. DOI: 10.1080/0969595980050102

Blind, K., Edler, J., Frietsch, R., & Schmoch, U. (2006). Motives to patent: Empirical evidence from Germany. *Research Policy, 35*(5), 655–672. DOI: 10.1016/j.respol.2006.03.002

Böck, F., Ochs, M., Henrich, A., Landes, D., Leidner, J. L., & Sedelmaier, Y. (2025). Learner models: Design, components, structure, and modelling. *User Modeling and User-Adapted Interaction, 35*(4), 15. DOI: 10.1007/s11257-025-09434-4

Botha, M., & Taljaard, A. (2021). Exploring the Entrepreneurial Intention-Competency Model for Nascent Entrepreneurs: Insights From a Developing Country Context. *Frontiers in Psychology, 12*, 516120. DOI: 10.3389/fpsyg.2021.516120 PMID: 34393866

Botín-Sanabria, D., Mihaita, A.-S., Peimbert-García, R. E., Ramírez-Moreno, M. A., Ramírez-Mendoza, R. A., & Lozoya-Santos, J. (2022). Digital twin technology challenges and applications: A comprehensive review. *Remote Sensing (Basel), 14*(6), 1335. DOI: 10.3390/rs14061335

Breque, M., De Nul, L., & Petridis, A. (2021). Industry 5.0: Towards a sustainable, human-centric and resilient European industry. European Commission. DOI: 10.2777/308407

Brocca, L., Barbetta, S., Camici, S., Ciabatta, L., Dari, J., Filippucci, P., Massari, C., Modanesi, S., Tarpanelli, A., Bonaccorsi, B., Mosaffa, H., Wagner, W., Vreugdenhil, M., Quast, R., Alfieri, L., Gabellani, S., Avanzi, F., Rains, D., Miralles, D. G., & Fernandez, D. (2024a). A Digital Twin of the terrestrial water cycle: A glimpse into the future through high-resolution Earth observations. *Frontiers in Science, 1*, 1190191. DOI: 10.3389/fsci.2023.1190191

Brynjolfsson, E., & McAfee, A. (2014). *The second machine age: Work, progress, and prosperity in a time of brilliant technologies*. W. W. Norton & Company.

Bucci, I., Fani, V., & Bandinelli, R. (2024). Towards human-centric manufacturing: Exploring the role of human digital twins in industry 5.0. *Sustainability (Basel), 17*(1), 129. DOI: 10.3390/su17010129

Cao, Y., & Ming, X. (2025). Digital Twin-Driven Low-Carbon Service Design and Modularization in Central Air Conditioning Ecosystems: A Multi-Criteria and Co-Intelligence Approach. *Sustainability (Basel), 17*(21), 9877. DOI: 10.3390/su17219877

Celik, S. E., Chen, Z., Yildizbasi, A., & Sarkis, J. (2026). Generative artificial intelligence for adopting circular economy in supply chains: Opportunities, challenges, and future pathways. *International Journal of Generative Artificial Intelligence in Business, 1*(1-2), 8–35. DOI: 10.1504/IJGAIB.2026.151818

Cesco, S., Sambo, P., Borin, M., Basso, B., Orzes, G., & Mazzetto, F. (2023a). Smart agriculture and digital twins: Applications and challenges in a vision of sustainability. *European Journal of Agronomy, 146*, 126809. DOI: 10.1016/j.eja.2023.126809

Chadha, P., Gera, R., Khera, G. S., & Sharma, M. (2023). Challenges of Artificial Intelligence Adoption for Financial Inclusion. In *Artificial Intelligence, Fintech, and Financial Inclusion* (pp. 135-160). CRC Press. https://doi.org/DOI: 10.1201/9781003125204-9

Challoumis, C. (2024). the Future of Business-integrating AI Into the Financial Cycle. In *XIV International Scientific Conference* (pp. 44-78).

Chamola, V.. (2025). *Metaverse for education: Developments, challenges, and future direction. Computer Applications in Engineering Education Journal*. IEEE., DOI: 10.1002/cae.70018

Chataut, R., Nankya, M., & Akl, R. (2024). 6G networks and the AI revolution—Exploring technologies, applications, and emerging challenges. *Sensors (Basel), 24*(6), 1888. DOI: 10.3390/s24061888 PMID: 38544151

Chaudhary, P., Kandacharam, S., & Krishnan, B. S. (2025). Climate Change and Agriculture: Emerging Challenges and Far-Reaching Impacts. In Smart Agriculture for Sustainable Practices (pp. 67-90). Chapman and Hall/CRC.

Chehlarova, N., Tsochev, G., Kotseva, M., & Miltchev, R. (2021). Digital Competencies of Public Administration Employees Related to Cybersecurity. *12th National Conference with International Participation (ELECTRONICA)*, Sofia, Bulgaria, 2021, pp. 1-4, DOI: 10.1109/ELECTRONICA52725.2021.9513705

Chen, Q., Wang, X., & Liu, L. (2026). Next-Generation AI-Driven Digital Twin Framework for Market Volatility Warning and Risk Detection in Global Logistics Supply Chains. *IEEE Communications Standards Magazine*, 1–10. DOI: 10.1109/MCOMSTD.2025.3641867

Chen, Y., & Xin, J. (2026). Human-Digital Symbiosis and Green Intelligent Manufacturing: Organizational Digital Acceptance and Proficiency. *Journal of Global Information Management*, *34*(1), 1–43. DOI: 10.4018/JGIM.401375

Chernov, D., Sornette, D., Chernov, D., & Sornette, D. (2020). Specific features of risk management in the industrial and agricultural sectors. *Critical Risks of Different Economic Sectors: Based on the Analysis of More Than 500 Incidents, Accidents and Disasters*, 13-145. https://doi.org/DOI: 10.1007/978-3-030-25034-8_2

Chichi, O., & Mamad, M. (2025). Exploring synergies between Industry 4.0 and Industry 5.0 for digital, resilient and sustainable supply chains: a systems thinking perspective. *Continuity & Resilience Review*, 1-35. https://doi.org/10.1108/CRR-05-2025-0014

Chigbu, B. I., & Makapela, S. L. (2025). AI in education, sustainability, and the future of work: An integrative review of Industry 5.0, Education 5.0, and Work 5.0. *Journal of Open Innovation*, *11*(4), 100645. DOI: 10.1016/j.joitmc.2025.100645

Chung, D., Yun, J., Lee, J., & Jeon, Y. (2023). Predictive model of employee attrition based on stacking ensemble learning. *Expert Systems with Applications*, *215*, 119364. DOI: 10.1016/j.eswa.2022.119364

Cohen-Manrique, C. S., Villa-Ramírez, J. L., Camacho-León, S., Solano-Correa, Y. T., Alvarez-Month, A. A., & Coronado-Hernández, O. E. (2025). Simulation and optimisation using a digital twin for resilience-based management of confined aquifers. *Water (Basel)*, *17*(13), 1973. DOI: 10.3390/w17131973

Cordeiro, M., & Ferreira, J. C. (2025a). Beyond Traceability: Decentralised Identity and Digital Twins for Verifiable Product Identity in Agri-Food Supply Chains. *Applied Sciences (Basel, Switzerland)*, *15*(11), 6062. DOI: 10.3390/app15116062

Cravero, A., Sepúlveda, S., Gutiérrez, F., & Muñoz, L. (2026). From Precision Agriculture to Intelligent Agricultural Ecosystems: A Systematic Review of Machine Learning and Big Data Applications. *Agronomy (Basel)*, *16*(5), 516. DOI: 10.3390/agronomy16050516

Das, D., & Singh, A. (2023). Optimizing the delivery schedule of equipment in a project environment considering resource leveling of storage space and its constraint: A case study. *Journal of Advances in Management Research*, *20*(5), 781–800. DOI: 10.1108/JAMR-02-2023-0045

David, J., Lobov, A., & Lanz, M. (2018a), Leveraging Digital Twins for Assisted Learning of Flexible Manufacturing Systems, Proceedings of the 16TH IEEE International Conference on Industrial Informatics, INDIN, 2018. DOI: 10.1109/INDIN.2018.8472083

David, J., Lobov, A., Lanz, M. (2018b). Learning Experiences Involving Digital Twins. 3681-3686. DOI: 10.1109/IECON.2018.8591460

Davila-Gonzalez, S., & Martin, S. (2024). Human digital twin in industry 5.0: A holistic approach to worker safety and well-being through advanced AI and emotional analytics. *Sensors (Basel)*, *24*(2), 655. DOI: 10.3390/s24020655 PMID: 38276347

Davis, F. D. (1989). Perceived usefulness, perceived ease of use, and user acceptance of information technology. *Management Information Systems Quarterly*, *13*(3), 319–340. DOI: 10.2307/249008

Dawn, N., Ghosh, S., Saha, A., Chatterjee, S., Ghosh, T., Guha, S., Sarkar, S., Mukherjee, P., & Sanyal, T. (2023). A Review on Digital Twins Technology: A New Frontier in Agriculture. *Artificial Intelligence and Applications (Commerce, Calif.)*, *2*(4), 250–262. DOI: 10.47852/bonviewAIA3202919

De Jong, J. P. J., Parker, S. K., Wennekers, S., & Wu, C.-H. (2015). Entrepreneurial behavior in organizations: Does job design matter? *Entrepreneurship Theory and Practice*, *39*(4), 981–995. DOI: 10.1111/etap.12084

Dehghanimohammadabadi, M., Belsare, S., & Thiesing, R. (2021). Simulation–optimization of digital twin. *2021 Winter Simulation Conference (WSC)*, 1–10. https://ieeexplore.ieee.org/abstract/document/9715412/

Delfani, P., Thuraga, V., Banerjee, B., & Chawade, A. (2024). Integrative approaches in modern agriculture: IoT, ML and AI for disease forecasting amidst climate change. *Precision Agriculture*, *25*(5), 2589–2613. DOI: 10.1007/s11119-024-10164-7

Dhinakaran, D., Edwin Raja, S., Velselvi, R., & Purushotham, N. (2025). Intelligent IoT-Driven Advanced Predictive Maintenance System for Industrial Applications. *SN Computer Science*, *6*(2), 151. DOI: 10.1007/s42979-025-03695-x

Dihan, M. S., Akash, A. I., Tasneem, Z., Das, P., Das, S. K., Islam, M. R., Islam, M. M., Badal, F. R., Ali, M. F., Ahamed, M. H., Abhi, S. H., Sarker, S. K., & Hasan, M. M. (2024). Digital twin: Data exploration, architecture, implementation and future. *Heliyon*, *10*(5), e26503. Advance online publication. DOI: 10.1016/j.heliyon.2024.e26503 PMID: 38444502

Divya, A., Ramesh, S., Chakrasali, S. V., Kumar, M. K., Taj, M., Ganiya, R. K., Nageswara Rao, J., K, B., & Qamar, S. (2026). Federated Digital-Twin–Empowered Blockchain Architecture for Autonomous and Resilient Smart Water Management in Sustainable Cities. *Sustainable Computing : Informatics and Systems*, *50*, 101303. DOI: 10.1016/j.suscom.2026.101303

Djeundje, V. B., Crook, J., Calabrese, R., & Hamid, M. (2021). Enhancing credit scoring with alternative data. *Expert Systems with Applications*, *163*, 113766. DOI: 10.1016/j.eswa.2020.113766

Dolgui, A., Ivanov, D., & Sokolov, B. (2020). Digital supply chain twins: Managing the ripple effect, resilience, and disruption risks by data-driven optimization, simulation, and visibility. *International Journal of Production Research*, *58*(7), 1–15. DOI: 10.1080/0951192X.2020.1836677

Dos Santos, C. H., Montevechi, J. A. B., De Queiroz, J. A., Miranda, R. C., & Leal, F. (2022). Decision support in productive processes through DES and ABS in the digital twin era: A systematic literature review. *International Journal of Production Research*, *60*(8), 2662–2681. DOI: 10.1080/00207543.2021.1898691

Eirinakis, P., Lounis, S., Plitsos, S., Arampatzis, G., Kalaboukas, K., Kenda, K., Lu, J., Rožanec, J. M., & Stojanovic, N. (2022). Cognitive digital twins for resilience in production: A conceptual framework. *Information (Basel)*, *13*(1), 33. DOI: 10.3390/info13010033

Ekundayo, F. (2024). Leveraging AI-Driven Decision Intelligence for Complex Systems Engineering. *Int J Res Publ Rev*, *5*(11), 1–10. DOI: 10.55248/gengpi.5.1124.3343

El-Hajj, M. (2026). Semantic foundations for digital twins: The contribution of ontological analysis. *Frontiers of Computer Science*, *8*, 1757450. Advance online publication. DOI: 10.3389/fcomp.2026.1757450

Escamilla-García, P. E., Rivera, A. E., Rivera-González, G., & Ángeles Tovar, L. C. (2026). Digitalization-based nutrient management to achieve both food security and environmental sustainability. In *Point Source Nitrogen Pollution* (pp. 155–170). Elsevier., DOI: 10.1016/B978-0-443-27352-0.00019-7

Escribà-Gelonch, M., Liang, S., Van Schalkwyk, P., Fisk, I., Long, N. V. D., & Hessel, V. (2024). Digital Twins in Agriculture: Orchestration and Applications. *Journal of Agricultural and Food Chemistry*, *72*(19), 10737–10752. DOI: 10.1021/acs.jafc.4c01934

European Commission. (2021). *Industry 5.0: Towards a sustainable, human-centric and resilient European industry*. Publications Office of the European Union., DOI: 10.2777/308407

Eyring, A., Hoyt, N., Tenny, J., Domike, R., & Hovanski, Y. (2022). Analysis of a closed-loop digital twin using discrete event simulation. *International Journal of Advanced Manufacturing Technology*, *123*(1–2), 245–258. DOI: 10.1007/s00170-022-10176-5

Fang, B., Yu, J., Chen, Z., Osman, A. I., Farghali, M., Ihara, I., Hamza, E. H., Rooney, D. W., & Yap, P. S. (2023). Artificial intelligence for waste management in smart cities: A review. *Environmental Chemistry Letters*, *21*(4), 1959–1989. DOI: 10.1007/s10311-023-01604-3 PMID: 37362015

Farid, H. A., Rasti, M., Pongracz, E., & Anvari-Moghaddam, A. (2026). Digital Twin Assisted Real-Time Energy Management System For Smart Homes. *Energy and Building*, •••, 116978.

Feddoul, Y., Ragot, N., Duval, F., Havard, V., Baudry, D., & Assila, A. (2023). Exploring human-machine collaboration in industry: A systematic literature review of digital twin and robotics interfaced with extended reality technologies. *International Journal of Advanced Manufacturing Technology*, *129*(5), 1917–1932. DOI: 10.1007/s00170-023-12291-3

Fernández López, M., Gómez-Pérez, A., & Juristo, N. (1997). METHONTOLOGY: From Ontological Art Towards Ontological Engineering. Proceeding of the AAAI-97 Spring Symposium on ontological engineering. https://aaai.org/papers/0005-ss97-06-005-methontology-from-ontological-art-towards-ontological-engineering/

Figueroa, A. J., Poler, R., & Andres, B. (2024). Adaptive production rescheduling system for managing unforeseen disruptions. *Mathematics*, *12*(22), 3478. DOI: 10.3390/math12223478

Fischer, T., & Ringler, P. (2007). What patents are used for? Empirical evidence from the patent portfolios of Swiss firms. *Journal of Evolutionary Economics*, *17*(5), 553572.

Flaga, A., & Pacholczak, P. (2022). Demonstrator of a digital twin for educational and training purposes as a web application. Advances in Science and Technology – Research Journal16(5):110-119. DOI: https://doi.org/DOI: 10.12913/22998624/152927

Fogaça, D. R., Grijalvo, M., & Sacomano Neto, M. (2025). What are Industry 4.0 and Industry 5.0 all about? An integrative institutional model for the new industrial paradigms. *Administrative Sciences*, *15*(4), 118. DOI: 10.3390/admsci15040118

Fotouo Makouate, H., & Zude-Sasse, M. (2025). Advances in Growing Degree Days Models for Flowering to Harvest: Optimizing Crop Management with Methods of Precision Horticulture—A Review. *Horticulturae*, *11*(12), 1415. DOI: 10.3390/horticulturae11121415

Frantzén, M., Bandaru, S., & Ng, A. H. C. (2022). Digital-twin-based decision support of dynamic maintenance task prioritization using simulation-based optimization and genetic programming. *Decis. Anal. J.*, *3*, 100039. Advance online publication. DOI: 10.1016/j.dajour.2022.100039

Fuller, A., Fan, Z., Day, C., & Barlow, C. (2020). *Digital Twin: Enabling Technologies, Challenges and Open Research*. IEEE., DOI: 10.1109/ACCESS.2020.2998358

Gaffley, G. R., & Pelser, T. G. (2025). A digital transformation strategy model for leadership in manufacturing: Considering the technological innovations to advance industry 5.0 in smart manufacturing. *South African Journal of Business Management*, *56*(1), 1–16. DOI: 10.4102/SAJBM.v56i1.5449

Garro, A., & Sorrenti, A. (2026). Designing an ISO 23247-compliant Hybrid Digital Twin Architecture for industry. *Information and Software Technology*, *193*, 108039. DOI: 10.1016/j.infsof.2026.108039

Gassar, A. A. A., & Jafar, R. (2025). Artificial Intelligence-Enabled Heating, Ventilation, and Air Conditioning Systems Toward Zero-Emission Buildings: A Systematic Review of Applications, Challenges, and Future Directions. *Applied Sciences (Basel, Switzerland)*, *15*(19), 10497. DOI: 10.3390/app151910497

Ghansah, F. A., & Lu, W. (2025). Cyber-physical systems and digital twins for "cognitive building" in the construction industry. *Construction Innovation*, *25*(3), 787–818. DOI: 10.1108/CI-07-2022-0164

Girginova, K. (2026, February 13). When XR Meets Digital Twins: Redrawing the Conceptual Map of Extended Reality. *Frontiers in Virtual Reality*, *7*, 1762601. DOI: 10.3389/frvir.2026.1762601

Goodwin, T., Xu, J., Celik, N., & Chen, C.-H. (2024). Real-time digital twin-based optimization with predictive simulation learning. *Journal of Simulation, 18*(1), 47–64. DOI: 10.1080/17477778.2022.2046520

Gosztonyi, M., & Csákné Filep, J. (2022). Profiling (non-)nascent entrepreneurs in Hungary based on machine learning approaches. *Sustainability (Basel), 14*(6), 3571. DOI: 10.3390/su14063571

Goto, M. (2021). Collective professional role identity in the age of artificial intelligence. *Journal of Professions and Organization, 8*(1), 86–107. DOI: 10.1093/jpo/joab003

Goundar, S. (Ed.). (2023). *Edge Computing: Technology, Management and Integration*. BoD–Books on Demand., DOI: 10.5772/intechopen.105637

Grieves, M. (2014). Digital twin: Manufacturing excellence through virtual factory replication. *White Paper, 1*, 1–7.

Grieves, M. (2014). Digital twin: Manufacturing excellence through virtual factory replication. White Paper, Florida Institute of Technology.

Grieves, M. (2014). *Digital twin: Manufacturing excellence through virtual factory replication*. Florida Institute of Technology.

Grieves, M., & Vickers, J. (2017). Digital twin: Mitigating unpredictable, undesirable emergent behavior in complex systems. In Kahlen, F.-J., Flumerfelt, S., & Alves, A. (Eds.), *Transdisciplinary Perspectives on Complex Systems* (pp. 85–113). Springer International Publishing., DOI: 10.1007/978-3-319-38756-7_4

Gupta, A., Singh, B., & Saini, H. K. (2025). Enhancing Crop Resilience to Climate Change with AI. In *AI and Data Analytics in Precision Agriculture for Sustainable Development* (pp. 133–151). Springer Nature Switzerland. DOI: 10.1007/978-3-031-93087-4_8

Hair, J. F., Black, W. C., Babin, B. J., & Anderson, R. E. (2010). *Multivariate Data Analysis* (7th ed.). Pearson.

Hall, B. H., Helmers, C., Rogers, M., & Sena, V. (2014). The choice between formal and informal intellectual property: A review. *Journal of Economic Literature, 52*(2), 375–423. DOI: 10.1257/jel.52.2.375

Hatami-Marbini, A., Varzgani, N., Sajadi, S. M., & Kamali, A. (2022). An emergency medical services system design using mathematical modeling and simulation-based optimization approaches. *Decis. Anal. J., 3*, 100059. Advance online publication. DOI: 10.1016/j.dajour.2022.100059

He, J. (2026). The Remote Sensing Geostatistical Paradigm: A Review of Key Technologies and Applications. *Remote Sensing (Basel), 18*(4), 600. DOI: 10.3390/rs18040600

Henriksen, H. J., Schneider, R., Koch, J., Ondracek, M., Troldborg, L., Seidenfaden, I. K., Kragh, S., Bøgh, E., & Stisen, S. (2022). A new digital twin for climate change adaptation, water management, and disaster risk reduction (HIP digital twin). *Water (Basel), 15*(1), 25. DOI: 10.3390/w15010025

He, Q., Li, L., Li, D., Peng, T., Zhang, X., Cai, Y., Zhang, X., & Tang, R. (2024). From digital human modeling to human digital twin: Framework and perspectives in human factors. *Chinese Journal of Mechanical Engineering, 37*(1), 9. DOI: 10.1186/s10033-024-00998-7

He, Q., Wang, P., Peng, T., Wang, B., & Tang, R. (2025). Human digital twin in human-centric smart manufacturing. In *Human-Centric Smart Manufacturing Towards Industry 5.0* (pp. 59–76). Springer Nature Switzerland. DOI: 10.1007/978-3-031-82170-7_3

Hevner, A. R., March, S. T., Park, J., & Ram, S. (2004). Design science in information systems research. *Management Information Systems Quarterly, 28*(1), 75–105. DOI: 10.2307/25148625

He, Y. (2026). Rewiring industrial futures: The role of AI-enabled digital twins in energy productivity transitions. *International Review of Economics & Finance, 105*, 104898. DOI: 10.1016/j.iref.2026.104898

Hilal, W., Gadsden, S. A., & Yawney, J. (2022). Financial fraud: A review of anomaly detection techniques and recent advances. *Expert Systems with Applications, 193*, 116429. DOI: 10.1016/j.eswa.2021.116429

Ho, M.-H. R., Uy, M. A., Kang, B. N. Y., & Chan, K.-Y. (2018). Impact of entrepreneurship training on entrepreneurial efficacy and alertness among adolescent youth. *Frontiers in Education, 3*, 13. DOI: 10.3389/feduc.2018.00013

Hosamo, H. H., Imran, A., Cardenas-Cartagena, J., Svennevig, P. R., Svidt, K., & Nielsen, H. K. (2022). A review of the digital twin technology in the AEC-FM industry. *Advances in Civil Engineering, 2185170*(1), 1–17. DOI: 10.1155/2022/2185170

Hossain, M. I., Hossan, M. R., Shaon, Z. H., & Ferdous, M. N. (2026). Linking digital twin paradigm for urban heat monitoring and policy integration to building smart city climate resilience. *Discover Cities, 3*(1), 1. DOI: 10.1007/s44327-025-00179-8

Howari, F. (2025). *From Ground Truth to Digital Twins: AI-Driven Remote Sensing for Scalable Precision Agrochemical Management*. Biology and Life Sciences. DOI: 10.20944/preprints202510.1798.v1

Hozdić, E., & Makovec, I. (2023). Evolution of the human role in manufacturing systems: On the route from digitalization and cybernation to cognitization. *Applied System Innovation, 6*(2), 49. DOI: 10.3390/asi6020049

Hrouga, M., & Sbihi, A. (2023). Logistics 4.0 for supply chain performance: Perspectives from a retailing case study. *Business Process Management Journal, 29*(6), 1892–1919. DOI: 10.1108/BPMJ-03-2023-0183

Huang, Z., Yan, H., Zhan, Q., Yang, S., Zhang, M., Zhang, C., Lei, Y., Liu, Z., Liu, Q., & Wang, Y. (2025). *A Survey on Remote Sensing Foundation Models: From Vision to Multimodality* (Version 1). arXiv. DOI: 10.48550/ARXIV.2503.22081

Huang, Y., Ren, F., & Wang, Y. (2025). Evaluation and pathways for achieving agricultural resilience under the framework of climate-smart agriculture. *Humanities & Social Sciences Communications, 13*(1), 105. DOI: 10.1057/s41599-025-06409-z

Hu, F., Qiu, X., Jing, G., Tang, J., & Zhu, Y. (2023). Digital twin-based decision making paradigm of raise boring method. *Journal of Intelligent Manufacturing, 34*(5), 2387–2405. DOI: 10.1007/s10845-022-01941-0

Hurmat, K. B., Rani, L., & Srivastav, A. L. (2026). Modelling the Future: Digital Twins in Sustainable and Climate-Aware Farming. In Digital Twin Technology for Sustainable Agriculture: Applications, Implementation and Future Trends (pp. 179-193). Singapore: Springer Nature Singapore.

Iliuţă, M.-E., Moisescu, M.-A., Pop, E., Ionita, A.-D., Caramihai, S.-I., & Mitulescu, T.-C. (2024). Digital Twin—A Review of the Evolution from Concept to Technology and Its Analytical Perspectives on Applications in Various Fields. *Applied Sciences (Basel, Switzerland), 14*(13), 5454. DOI: 10.3390/app14135454

Ionescu, D., Filipescu, A., Simion, G., Mincă, E., Cernega, D., olea, R., & Filipescu, A. (2022). Communication and control of an assembly, disassembly and repair flexible manufacturing technology on a mechatronics line assisted by an autonomous robotic system. *Inventions (Basel, Switzerland), 7*(2), 43. DOI: 10.3390/inventions7020043

Islam, M. T., Sepanloo, K., Woo, S., Woo, S. H., & Son, Y. J. (2025). A review of the industry 4.0 to 5.0 transition: Exploring the intersection, challenges, and opportunities of technology and human–machine collaboration. *Machines (Basel), 13*(4), 267. DOI: 10.3390/machines13040267

Ismail, S., Natarajan, S. K., & Voleti, R. (2024, November). Leveraging Predictive Analytics of Age Demographics for Engagement and Revenue Forecasting. In *2024 2nd International Conference on Advances in Computation, Communication and Information Technology (ICAICCIT)* (Vol. 1, pp. 749-754). IEEE. http://doi.org/ DOI: 10.1109/ICAICCIT64383.2024.10912167

Ivanov, D., & Dolgui, A. (2021). A digital supply chain twin for managing the disruption risks and resilience in the era of Industry 4.0. *Production Planning and Control*, *32*(9), 775–788. DOI: 10.1080/09537287.2020.1768450

Jagtap, N., Singh, T., & Sonawane, E. (2026). The pivotal role of artificial intelligence in digital twins: A case study. In Transforming Industries, Empowering Societies (pp. 187-198). Elsevier. DOI: 10.1016/B978-0-443-32878-7.00014-6

Jan, A., & Jan, A. U. (2025). Harnessing Digital Agriculture to Combat Climate Change and Feed the Future. In Climate Resilient and Sustainable Agriculture: Volume 2: Social and Transformative Strategies (pp. 313-332). Cham: Springer Nature Switzerland. DOI: 10.1007/978-3-032-04141-8_12

Jazi, A. E., Taheri Andani, M., Hasanzadeh, A., & Ameri, F. (2025). A Semantic-Driven Digital Twin Framework for Manufacturing. In Proceedings of the ASME 2025 International Design Engineering Technical Conferences (Vol. 2A, Paper No. DETC2025-169558). ASME. https://doi.org/DOI: 10.1115/DETC2025-169558

Jeong, H., Chen, K. B., Jeon, M., & Stanley, L. (2025). Human-Centered eXtended Reality for Occupational Applications in the Era of Industry 5.0: Introduction to the Special Issue. *IISE Transactions on Occupational Ergonomics and Human Factors*, *13*(2), 75–78. DOI: 10.1080/24725838.2025.2512681 PMID: 40464284

John, B. I. (2023). Data-driven resource optimization approaches enhancing capacity planning, labor utilization, material efficiency and continuous improvement across manufacturing project lifecycles. *GSC Advanced Research and Reviews*, *17*(3), 220–236. DOI: 10.30574/gscarr.2023.17.3.0467

Jones, D., Snider, C., Nassehi, A., Yon, J., & Hicks, B. (2020). Characterising the digital twin: A systematic literature review. *CIRP Journal of Manufacturing Science and Technology*, *29*, 36–52. DOI: 10.1016/j.cirpj.2020.02.002

Kabashkin, I. (2025). AI-Based Digital Twins of Students: A New Paradigm for Competency-Oriented Learning Transformation. *Information (Basel)*, *16*(10), 846. DOI: 10.3390/info16100846

Kadyan, P., & Singh, R. (2025). Smart HRM: Unleashing the Power of Business Intelligence for Workforce Excellence. In *Harnessing Business Intelligence for Modern Talent Management* (pp. 275-308). IGI Global Scientific Publishing. http://doi.org/DOI: 10.4018/979-8-3373-1942-1.ch011

Kalla, K. R., Giri Teja, B., Vijaya Kumar, D., Ganga, A., & Tarra, J. R. (2026). Digital Twin-Driven Crop Monitoring and Management: A Machine Learning Approach for Crop Yield Optimization. In Srivastav, A. L., Kumar, N., & Zinicovscaia, I. (Eds.), *Digital Twin Technology for Sustainable Agriculture* (pp. 309–331). Springer Nature Singapore., DOI: 10.1007/978-981-95-5915-2_17

Kalubanga, M., & Gudergan, S. (2022). The impact of dynamic capabilities in disrupted supply chains—The role of turbulence and dependence. *Industrial Marketing Management, 103*, 154–169. DOI: 10.1016/j.indmarman.2022.03.005

Kalusivalingam, A. K., Sharma, A., Patel, N., & Singh, V. (2020). Optimizing Workforce Planning with AI: Leveraging Machine Learning Algorithms and Predictive Analytics for Enhanced Decision-Making. *International Journal of Artificial Intelligence and Machine Learning, 1*(3). https://cognitivecomputingjournal.com/index.php/IJAIML-V1/article/view/53

Kamau, E., Myllynen, T., Damilola Mustapha, S., Opeyemi Babatunde, G., & Adeleye Alabi, A. (2024). A Conceptual Model for Real-Time Data Synchronization in Multi-Cloud Environments. *International Journal of Multidisciplinary Research and Growth Evaluation, 5*(1), 1139–1150. DOI: 10.54660/.IJMRGE.2024.5.1.1139-1150

Kane, G. C., Palmer, D., Phillips, A. N., Kiron, D., & Buckley, N. (2015). Strategy, not technology, drives digital transformation. *MIT Sloan Management Review, 14*(1), 1–25.

Kanike, U. K. (2023). Factors disrupting supply chain management in manufacturing industries. *Journal of Supply Chain Management Science, 4*(1–2), 1–24. DOI: 10.18757/jscms.2023.6986

Kantaros, A., Ganetsos, T., Pallis, E., & Papoutsidakis, M. (2025). From Mathematical Modeling and Simulation to Digital Twins: Bridging Theory and Digital Realities in Industry and Emerging Technologies. *Applied Sciences (Basel, Switzerland), 15*(16), 9213. DOI: 10.3390/app15169213

Karabulut, E., Pileggi, S. F., Groth, P., & Degeler, V. (2024). Ontologies in digital twins: A systematic literature review. *Future Generation Computer Systems, 153*, 442–456. DOI: 10.1016/j.future.2023.12.013

Karanam, S. A. K., & Hartman, N. W. (2025). A systematic review of Digital Twin (DT) and virtual learning environments (VLE) for smart manufacturing education. *Manufacturing Letters*, *44*(Supplement), 1597–1608. DOI: 10.1016/j.mfglet.2025.06.179

Karkaria, V., Tsai, Y.-K., Chen, Y.-P., & Chen, W. (2025). An optimization-centric review on integrating artificial intelligence and digital twin technologies in manufacturing. *Engineering Optimization*, *57*(1), 161–207. DOI: 10.1080/0305215X.2024.2434201

Karunaratne, T., Ajiero, I. R., Joseph, R., Farr, E., & Piroozfar, P. (2025). Evaluating the Economic Impact of Digital Twinning in the AEC Industry: A Systematic Review. *Buildings (Basel, Switzerland)*, *15*(14), 2583. DOI: 10.3390/buildings15142583

Khdhir, R., Magdich, A., Elloumi, M., Asklany, S., Hamdaoui, R., Abbas, G., & Elrashidi, A. (2026). Building resilient national critical infrastructure: A digital twin-based framework for comprehensive insider and external threat detection. *Journal of King Saud University. Computer and Information Sciences*, *38*(3), 118. DOI: 10.1007/s44443-026-00464-5

Khuong, M. N., & Hoang, Q. T. (2024). Entrepreneurial competencies – A systematic literature review. *Journal of International Entrepreneurship*, *23*(1), 54–130. DOI: 10.1007/s10843-024-00356-7

Kiasari, M., & Aly, H. (2026). Agentic Artificial Intelligence for Smart Grids: A Comprehensive Review of Autonomous, Safe, and Explainable Control Frameworks. *Energies*, *19*(3), 617. DOI: 10.3390/en19030617

Kinsner, W., & Saracco, R. (2019, July). Towards evolving symbiotic cognitive education based on digital twins. In 2019 IEEE 18th International Conference on Cognitive Informatics & Cognitive Computing (ICCI* CC) (pp. 13-21). IEEE. DOI: 10.1109/ICCICC46617.2019.9146095

Klisaroski, M. Naumova - Mihajlovska, K., Josimovska, T. (2025) Digital Transformation and HR: Challenges and Opportunities for Companies in North Macedonia, 10th International Scientific-Business Conference LIMEN 2024 – Selected Papers, pp.39-48, DOI: https://doi.org/DOI: 10.31410/LIMEN.SP2024.39

Kong, M., Liu, X., Chen, D., Xu, Y., Zhang, J., Chen, X., Liang, J., & Xu, H. (2025). Biochar Synergy with Smart Agriculture and Environment: From Soil Amendment to Precision Regulation Systems. In Ferreira Mendes, K., & Nogueira De Sousa, R. (Eds.), *Biochar—Applications in Agriculture and Environment*. IntechOpen., DOI: 10.5772/intechopen.1011847

Kowalczuk, Z., & Czubenko, M. (2023). Cognitive motivations and foundations for building intelligent decision-making systems. *Artificial Intelligence Review*, *56*(4), 3445–3472. DOI: 10.1007/s10462-022-10255-9

Kristoffersen, E., Blomsma, F., Mikalef, P., & Li, J. (2020). The smart circular economy: A digital-enabled circular strategies framework for manufacturing companies. *Journal of Business Research*, *120*, 241–261. DOI: 10.1016/j.jbusres.2020.07.044

Kritzinger, W., Karner, M., Traar, G., Henjes, J., & Sihn, W. (2018). Digital twin in manufacturing: A categorical literature review and classification. *IFAC-PapersOnLine*, *51*(11), 1016–1022. DOI: 10.1016/j.ifacol.2018.08.474

Krupas, M., Kajati, E., Liu, C., & Zolotova, I. (2024). Towards a human-centric digital twin for human–machine collaboration: A review on enabling technologies and methods. *Sensors (Basel)*, *24*(7), 2232. DOI: 10.3390/s24072232 PMID: 38610442

Kumar, K., Hoque, A., Padhiary, M., & Saikia, P. (2025). Emerging Trends in Nanotechnology, Autonomous Farms, and Collaborative Agri-Industry Marketing: The Future of AI in Food Production. In Whig, P., & Elngar, A. (Eds.), (pp. 239–278). Advances in Computational Intelligence and Robotics. IGI Global Scientific Publishing., DOI: 10.4018/979-8-3373-0842-5.ch010

Kumar, M., Saranya, G., Radder, U., Sundaramurthy, V., Divya, A., & Masih, R. K. (2026). Reinforcement-led digital twin optimization for secure operation in cyber-physical grids. *Electric Power Systems Research*, *253*, 112423. DOI: 10.1016/j.epsr.2025.112423

Kummari, D. N. (2025). *Advanced Practices in Auditing, Regulatory Compliance, and Smart Manufacturing Systems*. Deep Science Publishing., DOI: 10.70593/978-93-7185-253-1

Lanjouw, J. O., & Schankerman, M. (2001). Characteristics of patent litigation: A window on competition. *The RAND Journal of Economics*, *32*(1), 129–151. DOI: 10.2307/2696401

Lattanzi, L., Raffaeli, R., Peruzzini, M., & Pellicciari, M. (2021). Digital twin for smart manufacturing: A review of concepts towards a practical industrial implementation. *International Journal of Computer Integrated Manufacturing*, *34*(6), 567–597. DOI: 10.1080/0951192X.2021.1911003

Laveglia, S., Altieri, G., Genovese, F., Matera, A., & Di Renzo, G. C. (2024). Advances in Sustainable Crop Management: Integrating Precision Agriculture and Proximal Sensing. *AgriEngineering*, *6*(3), 3084–3120. DOI: 10.3390/agriengineering6030177

Lazaroiu, G., Androniceanu, A., Grecu, I., Grecu, G., & Neguri ă, O. (2022). Artificial intelligence-based decision-making algorithms, Internet of Things sensing networks, and sustainable cyber-physical management systems in big data-driven cognitive manufacturing. *Oeconomia Copernicana, 13*(4), 1047–1080. DOI: 10.24136/oc.2022.030

Lee, J., Ni, J., Singh, J., Jiang, B., Azamfar, M., & Feng, J. (2020). Intelligent maintenance systems and predictive manufacturing. *Journal of Manufacturing Science and Engineering, 142*(11), 110805. DOI: 10.1115/1.4047856

Lehmann, J., Lober, A., Häußermann, T., Rache, A., Ollinger, L., Baumgärtel, H., & Reichwald, J. (2023). The Anatomy of the Internet of Digital Twins: A Symbiosis of Agent and Digital Twin Paradigms Enhancing Resilience (Not Only) in Manufacturing Environments. *Machines (Basel), 11*(5), 504. DOI: 10.3390/machines11050504

Leirmo, T. L. (2024). Digital twins for Industry 5.0: Unlocking the human potential. *Procedia CIRP, 130*, 761–766. DOI: 10.1016/j.procir.2024.10.161

Leng, J., Wang, D., Shen, W., Li, X., Liu, Q., & Chen, X. (2021). Digital twins-based smart manufacturing system design in Industry 4.0: A review. *Journal of Manufacturing Systems, 60*, 119–137. DOI: 10.1016/j.jmsy.2021.05.011

Levander, A., & Raccuia, I. (2001). *Entrepreneurial profiling: A cognitive approach to entrepreneurship*. Stockholm School of Economics.

Li, Y., Zhang, C., Guo, Z., Yang, N., Yuan, D., & Ding, W. (2026). SafeTwinNet: Secure and Adaptive Learning Ecosystems for Digital Twins. IEEE Communications Standards Magazine.

Liang, S., Long, N. V. D., Parvin, M. I., Escribà-Gelonch, M., Lantin, S., & Hessel, V. (2025). Digital twin for lettuce growth in six climate zones ('Climate 2050'). *Computers and Electronics in Agriculture, 238*, 110880. DOI: 10.1016/j.compag.2025.110880

Liang, X., Yu, S., Ju, Y., Wang, Y., & Yin, D. (2025). Multi-scale remote-sensing phenomics integrated with multi-omics: Advances in crop drought–heat stress tolerance mechanisms and perspectives for climate-smart agriculture. *Plants, 14*(18), 2829. DOI: 10.3390/plants14182829 PMID: 41011981

Li, M., Yang, C. M., Lo, W., & Kao, Y. W. (2026). A Digital-Twin-Enabled AI-Driven Adaptive Planning Platform for Sustainable and Reliable Manufacturing. *Machines (Basel), 14*(2), 197. DOI: 10.3390/machines14020197

Liu, Y., Zhu, L., & Chen, C. (2025). *The mechanisms and optimization strategies of tillage practices affecting soil pore structure* (No. 0). Resources Economics Research Board. DOI: 10.50908/grb.4.0_507

Liu, H., Wen, W., Gou, W., Lu, X., Ma, H., Zhu, L., Zhang, M., Wu, S., & Guo, X. (2025a). Research on Intelligent Control Technology for a Rail-Based High-Throughput Crop Phenotypic Platform Based on Digital Twins. *Agriculture, 15*(11), 1217. DOI: 10.3390/agriculture15111217

Liu, J.-J., Wu, H., & Riaz, I. (2025). Advanced Technologies for Smart Fertilizer Management in Agriculture: A Review. *IEEE Access : Practical Innovations, Open Solutions, 13*, 139766–139790. DOI: 10.1109/ACCESS.2025.3594361

Liu, Z., Meyendorf, N., Blasch, E., Tsukada, K., Liao, M., & Mrad, N. (2025). The role of data fusion in predictive maintenance using digital twins. In Meyendorf, N., Ida, N., Singh, R., & Vrana, J. (Eds.), *Handbook of Nondestructive Evaluation 4.0* (pp. 1–23). Springer Nature., DOI: 10.1007/978-3-030-48200-8_65-1

Li, Y. (2025). AI-Enhanced Digital Twins for Energy Efficiency and Carbon Footprint Reduction in Smart City Infrastructure. *Proceedings of the 3rd International Conference on Software Engineering and Machine Learning*, (p. 42). DOI: 10.54254/2755-2721/2025.20569

López, E. J., Ruiz, J. E. P., Chávez, O. L., Muñoz, F., Velásquez, L. A. G., & Lugo, J. G. C. (2026). The Evolution of Mechatronics Engineering and Its Relationship with Industry 3.0, 4.0, and 5.0. *Technologies, 14*(2), 81. DOI: 10.3390/technologies14020081

Louati, A., Louati, H., Kariri, E., Neifar, W., Hassan, M. K., Khairi, M. H., Farahat, M. A., & El-Hoseny, H. M. (2024). Sustainable smart cities through multi-agent reinforcement learning-based cooperative autonomous vehicles. *Sustainability (Basel), 16*(5), 1779. DOI: 10.3390/su16051779

Lv, Z. (2023). Digital twins in industry 5.0. *Research, 6*, 0071. DOI: 10.34133/research.007

Maldonado-Romo, J., Bradley, R., Anthony, B., Juárez, L., Yoon, J., Amador, A., Guerrero, E., Bergmann, N., Montesinos, L., & Ponce, P. (2026). A collaborative and sustainable engineering design framework enabled by agentic AI. *International Journal of Sustainable Engineering, 19*(1), 2624166. DOI: 10.1080/19397038.2026.2624166

Mălina , C. D., Matei, F., Pop, I. D., Sălăgean, T., & Mălina , A. (2025). Individual and Synergistic Contributions of GIS, Remote Sensing, and AI in Advancing Climate-Resilient Agriculture. *AgriEngineering*, *7*(7), 230. DOI: 10.3390/agriengineering7070230

Mammone, V., Binetti, M. S., & Massarelli, C. (2026). Digital Twin in Territorial Planning: Comparative Analysis for the Development of Adaptive Cities. *Urban Science (Basel, Switzerland)*, *10*(2), 80. DOI: 10.3390/urbansci10020080

Manivasagam, V. S. (2025). From Bytes to Farm: Transferability of Industrial Digital Twins in Agricultural Systems. *Journal of Biosystems Engineering*, *50*(1), 130–144. DOI: 10.1007/s42853-025-00256-1

Manoli, E., Mattera, G., Caggiano, A., Marzano, A., & Nele, L. (2026). From Digital Twins to Immersive Manufacturing: XR and Gesture-Based Control for Enhancing Human–Robot Collaboration. *IET Collaborative Intelligent Manufacturing*, *8*(1), e70055. DOI: 10.1049/cim2.70055

Marah, H., & Challenger, M. (2024). MADTwin: A framework for multi-agent digital twin development: smart warehouse case study. *Annals of Mathematics and Artificial Intelligence*, *92*(4), 975–1005. DOI: 10.1007/s10472-023-09872-z

Marini, R., Mikhaylov, K., Pasolini, G., & Buratti, C. (2022a). Low-Power Wide-Area Networks: Comparison of LoRaWAN and NB-IoT Performance. *IEEE Internet of Things Journal*, *9*(21), 21051–21063. DOI: 10.1109/JIOT.2022.3176394

Mattamala, M., Chebrolu, N., Frey, J., Freißmuth, L., Oh, H., Casseau, B., Hutter, M., & Fallon, M. (2025). Building Forest Inventories With Autonomous Legged Robots—System, Lessons, and Challenges Ahead. *IEEE Transactions on Field Robotics*, *2*, 418–436. DOI: 10.1109/TFR.2025.3583972

Maurya, S. K., Kalhapure, A., Verma, V. K., Tiwari, A., Chaubey, C., Maurya, D. K., & Kumar, M. (2024). Irrigation Scheduling and Cultivar Management for Increasing Water Productivity under Dryland Condition: A Review. *International Journal of Environment and Climate Change*, *14*(1), 461–470. DOI: 10.9734/ijecc/2024/v14i13856

Mayer, A., Greif, L., Häußermann, T. M., Otto, S., Kastner, K., El Bobbou, S., Chardonnet, J.-R., Reichwald, J., Fleischer, J., & Ovtcharova, J. (2025). Digital twins, extended reality, and artificial intelligence in manufacturing reconfiguration: A systematic literature review. *Sustainability (Basel)*, *17*(5), 2318. DOI: 10.3390/su17052318

Ma, Z. G., Billanes, J. D., & Jørgensen, B. N. (2025). Climate Resilience and Energy Flexibility in Industrial Systems: A Scoping Review of Concepts, Technologies, Applications, and Policy Links. *Energies, 18*(18), 4985. DOI: 10.3390/en18184985

McConnachie, G. (1997). The Management of Intellectual Assets: Delivering Value to the Business. *Journal of Knowledge Management, 1*(1), 56–62. DOI: 10.1108/EUM0000000004580

Mehta, A. R., Kumar, P., Tyagi, S., Garg, K., & Aggarwal, S. (2026). Digital Twins for Sustainable Agriculture: Advancing Productivity, Resilience, and Equity. In *Digital Twin Technology for Sustainable Agriculture: Applications, Implementation and Future Trends* (pp. 393–415). Springer Nature Singapore. DOI: 10.1007/978-981-95-5915-2_21

Melesse, T. Y. (2025). Digital twin-based applications in crop monitoring. *Heliyon, 11*(2), e42137. DOI: 10.1016/j.heliyon.2025.e42137

Melnyk, L., Kubatko, O., Matsenko, O., Balatskyi, Y., & Serdyukov, K. (2021). Transformation of the human capital reproduction in line with Industries 4.0 and 5.0. *Problems and Perspectives in Management, 19*(2), 480–494. DOI: 10.21511/ppm.19(2).2021.38

Michailidis, P., Michailidis, I., & Kosmatopoulos, E. (2025). Reinforcement Learning for Electric Vehicle Charging Management: Theory and Applications. *Energies, 18*(19), 5225. DOI: 10.3390/en18195225

Mincer, J. (1974). *Schooling, experience, and earnings*. Columbia University Press.

Mishra, P., & Singh, G. (2023). Energy management systems in sustainable smart cities based on the internet of energy: A technical review. *Energies, 16*(19), 6903. DOI: 10.3390/en16196903

Mittal, V., Mehta, P., Relan, D., & Gabrani, G. (2020). Methodology for resume parsing and job domain prediction. *Journal of Statistics and Management Systems, 23*(7), 1265–1274. DOI: 10.1080/09720510.2020.1799583

Modoni, G. E., & Sacco, M. (2023). A human digital-twin-based framework driving human centricity towards industry 5.0. *Sensors (Basel), 23*(13), 6054. DOI: 10.3390/s23136054 PMID: 37447903

Mohanty, S., Hazra, S., & Paul, S. (2020). Intelligent prediction of engine failure through computational image analysis of wear particle. *Engineering Failure Analysis, 116*, 104731. DOI: 10.1016/j.engfailanal.2020.104731

Mourtzis, D., Panopoulos, N., & Angelopoulos, J. D. (2022). A hybrid teaching factory model towards personalized education 4.0. *International Journal of Computer Integrated Manufacturing*, *36*(12), 1739–1759. DOI: 10.1080/0951192X.2022.2145025

Mücke, N. T. (2024a). *Deep Learning for Real-Time Inverse Problems and Data Assimilation with Uncertainty Quantification for Digital Twins*. [Dr., Utrecht University], DOI: 10.33540/2380

Mumford, E. (2006). The story of socio-technical design: Reflections on its successes, failures and potential. *Information Systems Journal*, *16*(4), 317–342. DOI: 10.1111/j.1365-2575.2006.00221.x

Mun, H.-S., Lagua, E. B., Hong, S.-K., Ryu, S.-B., Sharifuzzaman, M., Hasan, M. K., Kim, Y.-H., & Yang, C.-J. (2025). Energy-Efficient Technologies and Strategies for Feasible and Sustainable Plant Factory Systems. *Sustainability (Basel)*, *17*(7), 3259. DOI: 10.3390/su17073259

Nadaf, H., Jabade, M. V., Jamadar, K., Jogdeo, B., & Jamdade, V. (2025). Improving Emergency Response: A Comparative Analysis of Traditional vs Artificial Intelligence-assisted Triage Systems in Health Care and Their Impact. *Indian Journal of Critical Care Medicine : Peer-Reviewed, Official Publication of Indian Society of Critical Care Medicine*, *29*(11), 925–929. DOI: 10.5005/jp-journals-10071-25083 PMID: 41368447

Nair, A., Pillai, S. V., & Senthil Kumar, S. A. (2024). Towards emerging Industry 5.0–a review-based framework. *Journal of Strategy and Management*, •••, 1–19. https://doi.org/10.1108/JSMA-04-2023-0067

Nambisan, S. (2017). Digital entrepreneurship: Toward a digital technology perspective of entrepreneurship. *Entrepreneurship Theory and Practice*, *41*(6), 1029–1055. DOI: 10.1111/etap.12254

Narzari, R., Choudhury, B. U., Singhal, G., & Choudhary, K. K. (2025). A critical review of how UAVs can transform precision agriculture in the realm of Agroecology. *Discover Soil*, *2*(1), 28. DOI: 10.1007/s44378-025-00055-2

Nasirahmadi, A., & Hensel, O. (2022). Toward the Next Generation of Digitalization in Agriculture Based on Digital Twin Paradigm. *Sensors (Basel)*, *22*(2), 498. DOI: 10.3390/s22020498

Nechesov, A., Dorokhov, I., & Ruponen, J. (2025). Virtual cities: From digital twins to autonomous AI societies. *IEEE Access : Practical Innovations, Open Solutions*, *13*, 13866–13903. DOI: 10.1109/ACCESS.2025.3531222

Neffati, O. S., Sengan, S., Thangavelu, K. D., Kumar, S. D., Setiawan, R., Elangovan, M., & Velayutham, P. (2021). Migrating from traditional grid to smart grid in smart cities promoted in developing country. *Sustainable Energy Technologies and Assessments*, *45*, 101125. DOI: 10.1016/j.seta.2021.101125

Negri, E., Fumagalli, L., & Macchi, M. (2021). A review of the roles of digital twin in CPS-based production systems. *Procedia Manufacturing*, *11*, 939–948. DOI: 10.1016/j.promfg.2017.07.198

Nguyen, T. D., & Adhikari, S. (2026). Advancing Digital Twins for Building Lifecycle Management in Construction: A Systematic Literature Review. *Buildings (Basel, Switzerland)*, *16*(6), 1151. DOI: 10.3390/buildings16061151

Nicoletti, B., & Appolloni, A. (2025). A digital twin framework for enhancing human–agentic AI–machine collaboration. *Journal of Intelligent Manufacturing*, •••, 1–17.

Noy, N. F., & McGuinness, D. L. (2001). Ontology development 101: A guide to creating your first ontology (Stanford University Knowledge Systems Laboratory Technical Report KSL-01-05). Stanford University.

Nozari, H. (2025). Integrating Neuromorphic Intelligence with Fuzzy Multi-Objective Models in Resilient Green Supply Chain Design. Process Integration and Optimization for Sustainability, 1-14.

Nozari, H. Integrating Neuromorphic Intelligence with Fuzzy Multi-Objective Models in Resilient Green Supply Chain Design. Process Integr Optim Sustain (2025). https://doi.org/DOI: 10.1007/s41660-025-00598-5

Nozari, H., & Yordanova, Z. (2025). A fuzzy–digital twin optimization framework for simultaneous management of waste and energy consumption in sustainable manufacturing. An International Journal of Optimization and Control: Theories & Applications, 25360152.

Nozari, H., & Yordanova, Z. (2025). A fuzzy–digital twin optimization framework for simultaneous management of waste and energy consumption in sustainable manufacturing. *An International Journal of Optimization and Control: Theories & Applications*, 25360152. https://doi.org/10.36922/IJOCTA025360152

Nozari, H., & Yordanova, Z. (2025). An intelligent digital twin approach for optimizing multi-channel supply chains in uncertain environments. Green Technologies and Sustainability, 100309.

Nozari, H., & Yordanova, Z. (2025). An intelligent digital twin approach for optimizing multi-channel supply chains in uncertain environments. *Green Technologies and Sustainability*, 100309. https://doi.org/10.1016/j.grets.2025.100309

Nozari, H., & Yordanova, Z. (2026). Digital twin–driven decision support for human capital management under uncertainty. Green Technologies and Sustainability, 100353.

Nozari, H., & Yordanova, Z. (2026). Digital twin–driven decision support for human capital management under uncertainty. *Green Technologies and Sustainability*, 100353. https://doi.org/10.1016/j.grets.2026.100353

Nozari, H. (2024). Green supply chain management based on artificial intelligence of everything. *Journal of Economics and Management*, *46*(1), 171–188. DOI: 10.22367/jem.2024.46.07

Nozari, H. (2026). Digital twin-based predictive maintenance in cold chain logistics. *Operations Research*, *60*(1), 481–499. DOI: 10.1051/ro/2025167

Nozari, H., Abdi, H., Szmelter-Jarosz, A., & Rahmaty, M. (2025). A digital twin-based neutrosophic multi-objective optimization model for smart supply chains in the food industry. *Journal of Intelligent Manufacturing*, 1–15. DOI: 10.1007/s10845-025-02689-z

Nozari, H., & Nassar, S. (2025). *Real-time data-driven multi-objective optimization in B2B marketing using digital twins*. OPSEARCH., DOI: 10.1007/s12597-025-01044-3

Nozari, H., Nassar, S., & Szmelter-Jarosz, A. (2025). Fuzzy Multi-Objective Optimization Model for Resilient Supply Chain Financing Based on Blockchain and IoT. *Digital*, *5*(3), 32. DOI: 10.3390/digital5030032

Nozari, H., Rahmaty, M., Foukolaei, P. Z., Movahed, H., & Bayanati, M. (2025). Optimizing cold chain logistics with artificial intelligence of things (AIoT): A model for reducing operational and transportation costs. *Future Transportation*, *5*(1), 1. DOI: 10.3390/futuretransp5010001

Nozari, H., & Szmelter-Jarosz, A. (2025). A Predictive Maintenance Approach for Composting Plants Based on ERP and Digital Twin Integration. *Machines (Basel)*, *13*(12), 1123. DOI: 10.3390/machines13121123

Nozari, H., Szmelter-Jarosz, A., & Weiland, D. (2025). A Fuzzy Multi-Objective Sustainable and Agile Supply Chain Model Based on Digital Twin and Internet of Things with Adaptive Learning Under Environmental Uncertainty. *Applied Sciences (Basel, Switzerland)*, *15*(19), 10399. DOI: 10.3390/app151910399

Nozari, H., & Yordanova, Z. (2025). Hybrid digital twin and quantum AI with fuzzy multiobjective modeling in supply chain management. *Edelweiss Applied Science and Technology*, *9*(10), 609–628. DOI: 10.55214/2576-8484.v9i10.10482

Nozari, H., & Yordanova, Z. (2026). A fuzzy–digital twin optimization framework for simultaneous management of waste and energy consumption in sustainable manufacturing. *An International Journal of Optimization and Control: Theories & Applications*, *16*(1), 160–175. DOI: 10.36922/IJOCTA025360152

Nozari, H., & Yordanova, Z. (2026). A self-deciding adaptive digital twin framework using agentic AI for fuzzy multi-objective optimization of food logistics. *Algorithms*, *19*(3), 218. DOI: 10.3390/a19030218

Nozari, H., & Yordanova, Z. (2026). An Adaptive Fuzzy Multi-Objective Digital Twin Framework for Multi-Depot Cold-Chain Vehicle Routing in Agri-Biotech Supply Networks. *Logistics (Basel)*, *10*(2), 27. DOI: 10.3390/logistics10020027

Nozari, H., & Yordanova, Z. (2026). Digital twin-driven decision support for human capital management under uncertainty. *Green Technologies and Sustainability*, *4*(2), 100353. DOI: 10.1016/j.grets.2026.100353

Nozari, H., & Yordanova, Z. (2026). Fuzzy multi-objective digital twin framework for dynamic pricing and assortment optimization in metaverse-based retail environments. *Journal of Revenue and Pricing Management*. Advance online publication. DOI: 10.1057/s41272-026-00586-5

Ntshangase, S. D., & Ezeuduji, I. O. (2020). Profiling entrepreneurial behaviour based on demographic variables: Tourism-related entrepreneurs in Mtubatuba Local Municipality, South Africa. *Geo Journal of Tourism and Geosites*, *31*(3), 944–950. DOI: 10.30892/gtg.31302-525

Nweye, K., Kaspar, K., Buscemi, G., Fonseca, T., Pinto, G., Ghose, D., Duddukuru, S., Pratapa, P., Li, H., Mohammadi, J., Lino Ferreira, L., Hong, T., Ouf, M., Capozzoli, A., & Nagy, Z. (2025). CityLearn v2: Energy-flexible, resilient, occupant-centric, and carbon-aware management of grid-interactive communities. *Journal of Building Performance Simulation*, *18*(1), 17–38. DOI: 10.1080/19401493.2024.2418813

O. Okimi, T., Adeoye, O., & Omoboye, O. (2026). Integrating digital twins and decision support systems for predictive risk management in urban renewal. Digital Twin, 2612464.

Okimi, T. O. (2025). Human centric digital twins for sustainable supply chain in construction. *Smart Constr. Sustain. Cities*, *3*(1), 31. DOI: 10.1007/s44268-025-00078-2

Omia, E., Bae, H., Park, E., Kim, M. S., Baek, I., Kabenge, I., & Cho, B.-K. (2023). Remote Sensing in Field Crop Monitoring: A Comprehensive Review of Sensor Systems, Data Analyses and Recent Advances. *Remote Sensing*, *15*(2), 354. https://doi.org/10.3390/rs15020354

Omrany, H., & Al-Obaidi, K. M. (2025). Application of digital twin technology for Urban Heat Island mitigation: Review and conceptual framework. *Smart and Sustainable Built Environment*, *14*(7), 2145–2175. DOI: 10.1108/SASBE-05-2024-0189

Orlikowski, W. J. (1992). The duality of technology: Rethinking the concept of technology in organizations. *Organization Science*, *3*(3), 398–427. DOI: 10.1287/orsc.3.3.398

Othman, A. A. E., & Mohamed Abdrabou, N. (2026). A digital twin framework for enhancing building maintainability during the design process in architectural design firms in Egypt. *International Journal of Construction Management*, 1–18. DOI: 10.1080/15623599.2026.2624416

Padhiary, M., Prasad, G., Hoque, A., Kumar, K., & Sahu, B. (2025). Advances in Vertical Farming: The Role of Artificial Intelligence and Automation in Sustainable Agriculture. *LatIA*, *3*, 131. DOI: 10.62486/latia2025131

Pajpach, M., Pribiš, R., Drahoš, P., Kučera, E., & Haffner, O. (2025). Methodology and pilot implementation of a modular educational-development platform for interoperable digital twin. *Scientific Reports*, *15*(1), 28020. Advance online publication. DOI: 10.1038/s41598-025-13579-y PMID: 40745468

Panda, D., Mohanty, S., Das, S., Senapaty, J., Sahoo, D. B., Mishra, B., Baig, M. J., & Behera, L. (2025). From spectrum to yield: Advances in crop photosynthesis with hyperspectral imaging. *Photosynthetica*, *63*(2), 196–233. DOI: 10.32615/ps.2025.012

Parewai, I., & Köppen, M. (2025). A Digital Twin Approach for Soil Moisture Measurement with Physically Based Rendering Simulations and Machine Learning. *Electronics (Basel)*, *14*(2), 395. DOI: 10.3390/electronics14020395

Park, J.-H., & Kim, S.-J. (2025). Entrepreneurial Competencies in the Era of Digital Transformation: A Systematic Literature Review. *Digital*, *2025*(5), 46. DOI: 10.3390/digital5040046

Parveen, N., & Basit, F. (2023). Securing Data in Motion and at Rest: AI and Machine Learning Applications in Cloud and Network Security. https://doi.org/DOI: 10.13140/RG.2.2.35114.84163

Pasmore, W. A., Francis, C. A., Haldeman, J., & Shani, A. B. (1982). Sociotechnical systems: A North American reflection on empirical studies of the seventies. *Human Relations; Studies Towards the Integration of the Social Sciences*, *35*(12), 1179–1204. DOI: 10.1177/001872678203501207

Patel, K. (2024). Ethical reflections on data-centric AI: balancing benefits and risks. *Available at SSRN* 4993089. http://dx.doi.org/DOI: 10.2139/ssrn.4993089

Patel, U. R., Ghaffarianhoseini, A., GhaffarianHoseini, A., & Burgess, A. (2026). Towards a well-being-oriented framework for urban digital twins. *Cities (London, England)*, *169*, 106579. DOI: 10.1016/j.cities.2025.106579

Pater, J., & Stadnicka, D. (2021). Towards Digital Twins Development and Implementation to Support Sustainability Systematic Literature Review. *Management and Production Engineering Review*, 63-73.

Pati, A. K. (2025). Agentic AI: A comprehensive survey of technologies, applications, and societal implications. *IEEE Access : Practical Innovations, Open Solutions*, *13*, 151824–151837. DOI: 10.1109/ACCESS.2025.3585609

Peladarinos, N., Piromalis, D., Cheimaras, V., Tserepas, E., Munteanu, R. A., & Papageorgas, P. (2023). Enhancing Smart Agriculture by Implementing Digital Twins: A Comprehensive Review. *Sensors (Basel)*, *23*(16), 7128. DOI: 10.3390/s23167128

Perno, M., Hvam, L., & Haug, A. (2023). A machine learning digital twin approach for critical process parameter prediction in a catalyst manufacturing line. *Computers in Industry*, *151*, 151. DOI: 10.1016/j.compind.2023.103987

Peruzzi, G., & Pozzebon, A. (2022). Combining LoRaWAN and NB-IoT for Edge-to-Cloud Low Power Connectivity Leveraging on Fog Computing. *Applied Sciences (Basel, Switzerland)*, *12*(3), 1497. DOI: 10.3390/app12031497

Peruzzini, M., Prati, E., & Pellicciari, M. (2024). A framework to design smart manufacturing systems for Industry 5.0 based on the human-automation symbiosis. *International Journal of Computer Integrated Manufacturing*, *37*(10-11), 1426–1443. DOI: 10.1080/0951192X.2023.2257634

Petkov, K. (2024). Logical framework method for conducting pre- and post-money valuations of equity crowdfunded companies leveraging AI. Industrial Growth Conference 2024. Az-buki Publ. pp: 207-219. https://doi.org/DOI: 10.53656/igc-2024.14

Pham, H. S. T., & Li, S. B. (2026). Towards Industry 5.0: A conceptual model for leading organisational change in digital age. *Journal of Organizational Change Management*, *39*(1), 125–161. DOI: 10.1108/JOCM-03-2025-0264

Pilipczuk, O. (2022). Building the Cognitive Enterprise in the Energy Sector. *Energies, 15*(24), 9479. DOI: 10.3390/en15249479

Piras, G., Agostinelli, S., & Muzi, F. (2024). Digital twin framework for built environment: A review of key enablers. *Energies, 17*(2), 436. DOI: 10.3390/en17020436

Podsakoff, P. M., MacKenzie, S. B., Lee, J. Y., & Podsakoff, N. P. (2003). Common method biases in behavioral research: A critical review of the literature and recommended remedies. *The Journal of Applied Psychology, 88*(5), 879–903. DOI: 10.1037/0021-9010.88.5.879 PMID: 14516251

Poór, P., Ženíšek, D., & Basl, J. (2019, July). Historical overview of maintenance management strategies: Development from breakdown maintenance to predictive maintenance in accordance with four industrial revolutions. In *Proceedings of the international conference on industrial engineering and operations management, Pilsen, Czech Republic* (pp. 23-26). https://ieomsociety.org/pilsen2019/papers/135.pdf?_gl=1*uinndd*_ga*MzIxMjkzMDIuMTc2MzM3ODE2OA

Poornima, G., & Nasurudeen Ahamed, N. (2025). Reimagining Autonomy: Agentic AI in the Age of Human-Centric Innovation. In *The Power of Agentic AI: Redefining Human Life and Decision-Making: In Industry 6.0* (pp. 291–307). Springer Nature Switzerland. DOI: 10.1007/978-3-031-89424-4_19

Pospieszny, P., & Brodowicz, D. P. (2025). Toward Agentic Environments: GenAI and the Convergence of AI, Sustainability, and Human-Centric Spaces. *Sustainable Development (Bradford)*, sd.70544. DOI: 10.1002/sd.70544

Prajapat, N., Turner, C., Tiwari, A., Tiwari, D., & Hutabarat, W. (2020). Real-time discrete event simulation: A framework for an intelligent expert system approach utilising decision trees. *International Journal of Advanced Manufacturing Technology, 110*(11), 2893–2911. DOI: 10.1007/s00170-020-06048-5

Purcell, W., Neubauer, T., & Mallinger, K. (2023a). Digital Twins in agriculture: Challenges and opportunities for environmental sustainability. *Current Opinion in Environmental Sustainability, 61*, 101252. DOI: 10.1016/j.cosust.2022.101252

Qin, M., Zhang, C., Ma, G., Ma, Y., Feng, X., Li, P., & Feng, K. (2025a). Digital twin-enabled intelligent irrigation-drainage system for precision water-salt management in saline agroecosystems. *Agricultural Water Management, 322*, 109957. DOI: 10.1016/j.agwat.2025.109957

Qiu, S., Zaheer, Q., Ali, F., Wajid, S., Chen, H., Ai, C., & Wang, J. (2025). Exploring the impact of digital twin technology in infrastructure management: A comprehensive review. *Journal of Civil Engineering and Management*, *31*(4), 395–417. DOI: 10.3846/jcem.2025.23718

Radanliev, P., De Roure, D., Nicolescu, R., Huth, M., & Santos, O. (2022). Digital twins: Artificial intelligence and the IoT cyber-physical systems in Industry 4.0. *International Journal of Intelligent Robotics and Applications*, *6*(1), 171–185. DOI: 10.1007/s41315-021-00180-5

Rane, N., Choudhary, S., & Rane, J. (2024). Artificial intelligence driven approaches to strengthening Environmental, Social, and Governance (ESG) criteria in sustainable business practices: a review. *Social, and Governance (ESG) criteria in sustainable business practices: a review*. DOI: 10.2139/ssrn.4843215

Rasheed, A., San, O., & Kvamsdal, T. (2020). Digital Twin: Values, Challenges and Enablers From a Modeling Perspective. *IEEE Access : Practical Innovations, Open Solutions*, *8*, 21980–22012. DOI: 10.1109/ACCESS.2020.2970143

Rastgou, M., Jin, X., Jiang, Q., Liu, S., Lou, R., Wang, J., Tang, R., & He, Y. (2025). Optimizing deep neural networks for estimating soil water retention curves: A comparison of metaheuristic and numerical algorithms. *Vadose Zone Journal*, *24*(5), e70035. DOI: 10.1002/vzj2.70035

Rathore, M. M., Shah, S. A., Shukla, D., Bentafat, E., & Bakiras, S. (2021). The Role of AI, Machine Learning, and Big Data in Digital Twinning: A Systematic Literature Review, Challenges, and Opportunities. *IEEE Access : Practical Innovations, Open Solutions*, *9*, 32030–32052. Advance online publication. DOI: 10.1109/ACCESS.2021.3060863

Ratković, T., Šlogar, H., & Šokčević, S. (2022). Entrepreneurial competencies of university students. *Economics & Sociology (Ternopil)*, *15*(4), 129–145. DOI: 10.14254/2071-789X.2022/15-4/6

Ravichandran, N., Inaganti, A. C., & Muppalaneni, R. (2023). AI-Driven sentiment analysis for employee engagement and retention. *Journal of Computing Innovations and Applications*, *1*(01), 1–9. https://ciajournal.com/index.php/jcia/article/view/6

Riaz, K., McAfee, M., & Gharbia, S. S. (2023). Management of Climate Resilience: Exploring the Potential of Digital Twin Technology, 3D City Modelling, and Early Warning Systems. *Sensors (Basel)*, *23*(5), 2659. DOI: 10.3390/s23052659

Ristvej, J., Halúsková, B., Nováková, K., & Chovanec, D. (2026). Implementation Maturity Levels of Digital Twin Technology and Data Content Design for Flood Digital Twin. *Smart Cities*, *9*(2), 28. DOI: 10.3390/smartcities9020028

Ritto, T. G., & Rochinha, F. A. (2021). Digital twin, physics-based model, and machine learning applied to damage detection in structures. *Mechanical Systems and Signal Processing*, *155*, 155. DOI: 10.1016/j.ymssp.2021.107614

Rizo-Maestre, C., Flores-Moreno, J. M., Nebot-Sanz, A., & Huesca-Tortosa, J. A. (2026). Multi-Agent Systems and Digital Twins as a Basis for Smart Buildings with Integrated Sustainable Efficient Ventilation. *Buildings (Basel, Switzerland)*, *16*(5), 1026. DOI: 10.3390/buildings16051026

Rizqi, Z. U., Chou, S.-Y., & Cahyo, W. N. (2024). A simulation-based digital twin for smart warehouse: Towards standardization. *Decis. Anal. J.*, *12*, 100509. Advance online publication. DOI: 10.1016/j.dajour.2024.100509

Rodrigues, H., Ceddia, M. B., Tassinari, W., Vasques, G. M., Brandão, Z. N., Morais, J. P. S., Oliveira, R. P., Neves, M. L., & Tavares, S. R. L. (2024). Remote and Proximal Sensors Data Fusion: Digital Twins in Irrigation Management Zoning. *Sensors (Basel)*, *24*(17), 5742. DOI: 10.3390/s24175742

Rojek, I., Mikołajewski, D., & Dostatni, E. (2025). Artificial intelligence in digital twin applications. *Applied Sciences (Basel, Switzerland)*, *15*(2), 1120.

Rojek, I., Mikołajewski, D., Piszcz, A., Małolepsza, O., & Kozielski, M. (2025). Role of Generative AI in AI-Based Digital Twins in Industry 5.0 and Evolution to Industry 6.0. *Applied Sciences (Basel, Switzerland)*, *15*(18), 10102. DOI: 10.3390/app151810102

Rosati, R., Romeo, L., Cecchini, G., Tonetto, F., Viti, P., Mancini, A., & Frontoni, E. (2023). From knowledge-based to big data analytic model: A novel IoT and machine learning based decision support system for predictive maintenance in industry 4.0. *Journal of Intelligent Manufacturing*, *34*(1), 107–121. DOI: 10.1007/s10845-022-01960-x

Sacavém, A., de Bem Machado, A., dos Santos, J. R., Palma-Moreira, A., Belchior-Rocha, H., & Au-Yong-Oliveira, M. (2025). Leading in the digital age: The role of leadership in organizational digital transformation. *Administrative Sciences*, *15*(2), 43. DOI: 10.3390/admsci15020043

Sacoto-Cabrera, E., Perez-Torres, A., Tello-Oquendo, L., & Cerrada, M. (2025). IoT, AI, and digital twins in smart cities: A systematic review for a thematic mapping and research agenda. *Smart Cities*, *8*(5), 175. Advance online publication. DOI: 10.3390/smartcities8050175

Saha, B. (2025). AI-driven predictive analytics for financial risk management. *International Journal of Advance Research and Innovation (IJARI, 2347-3258)*, *13*(01), 12-22. https://doi.org/DOI: 10.69996/ijari.2025002

Sahoh, B., & Choksuriwong, A. (2023). The role of explainable Artificial Intelligence in high-stakes decision-making systems: A systematic review. *Journal of Ambient Intelligence and Humanized Computing*, *14*(6), 7827–7843. DOI: 10.1007/s12652-023-04594-w PMID: 37228699

Saikia, P., Sahu, B., Prasad, G., Kumar, S., Suman, S., & Kumar, K. (2025). Smart Infrastructure Systems: A Review of IoT-Enabled Monitoring and Automation in Civil and Agricultural Engineering. *Asian Journal of Research in Computer Science*, *18*(4), 24–44. DOI: 10.9734/ajrcos/2025/v18i4606

Saivineeth, M., & Venkateswarlu, B. (2025). Linear Programming-Based Decision Support System of Solar Power Reserve Management in Pulsation Free Current Circulation (PFCC) for Agricultural Systems. *IEEE Access : Practical Innovations, Open Solutions*, *13*, 44510–44525. DOI: 10.1109/ACCESS.2025.3549464

Santander, O., Kuppuraj, V., Harrison, C. A., & Baldea, M. (2023). A machine learning framework for improving refinery production planning. *AIChE Journal. American Institute of Chemical Engineers*, *69*(8), e18109. Advance online publication. DOI: 10.1002/aic.18109

Sapes, G., Schroeder, L., Scott, A., Clark, I., Juzwik, J., Montgomery, R. A., Guzmán, Q. J. A., & Cavender-Bares, J. (2024). Mechanistic links between physiology and spectral reflectance enable previsual detection of oak wilt and drought stress. *Proceedings of the National Academy of Sciences of the United States of America*, *121*(7), e2316164121. DOI: 10.1073/pnas.2316164121

Scataglini, S., & Truijen, S. (2023, July). Critical appraisal of using digital human model, virtual human, human digital twin and digital twin. In *International conference on digital human modeling* (pp. 154-158). Cham: Springer Nature Switzerland. DOI: 10.1007/978-3-031-37848-5_17

Schmitt, U. (2025, October). Platform Powering Cognitive and Memetic Digital Twins: To Boldly Go Where Memes May Live Long and Prosper. In International Conference on Cooperative Design, Visualization and Engineering (pp. 112-122). Cham: Springer Nature Switzerland.

Schott, C., van Roekel, H., & Tummers, L. G. (2020). Teacher leadership: A systematic review, methodological quality assessment and conceptual framework. *Educational Research Review, 31*, 100352. DOI: 10.1016/j.edurev.2020.100352

Schultz, T. W. (1961). Investment in human capital. *The American Economic Review, 51*(1), 1–17.

Sedrakyan, G., Snoeck, M., & Poelmans, S. (2014). Assessing the effectiveness of feedback-enabled simulation in teaching conceptual modeling. *Computers & Education, 78*, 367–382. DOI: 10.1016/j.compedu.2014.06.014

Sendas, N., & Rajale, D. (2025). Future Trends in MLOps. In *The Definitive Guide to Machine Learning Operations in AWS: Machine Learning Scalability and Optimization with AWS* (pp. 371-423). Berkeley, CA: Apress. https://doi.org/DOI: 10.1007/979-8-8688-1076-3

Shamim, S., Yang, Y., Zia, N. U., Khan, Z., & Shariq, S. M. (2023). Mechanisms of cognitive trust development in artificial intelligence among front line employees: An empirical examination from a developing economy. *Journal of Business Research, 167*, 114168. DOI: 10.1016/j.jbusres.2023.114168

Shapiro, C. (2000). Navigating the Patent Thicket: Cross Licenses, Patent Pools, and Standard Setting. A. B. Jaffe, J. Lerner, and S. Stern (Eds.) 2000, Innovation policy and the economy (Vol. 1, pp. 119–150). MIT Press.

Sharma, A., Kosasih, E., Zhang, J., Brintrup, A., & Calinescu, A. (2022). Digital Twins: State of the art theory and practice, challenges, and open research questions. *Journal of Industrial Information Integration, 30*, 100383. Advance online publication. DOI: 10.1016/j.jii.2022.100383

Sharma, H., & Kaur, S. (2025). Deep learning for sustainable development across climate, energy, agriculture and urban systems. *Discover Sustainability, 6*(1), 1408. DOI: 10.1007/s43621-025-02186-6

Shi, Y., Han, L., Zhang, X., Sobeih, T., Gaiser, T., Thuy, N. H., Behrend, D., Srivastava, A. K., Halder, K., & Ewert, F. (2025). *Deep Learning Meets Process-Based Models: A Hybrid Approach to Agricultural Challenges* (Version 1). arXiv. DOI: 10.48550/ARXIV.2504.16141

Shoaib, M., Khan, S. U., AbdelHameed, H., & Qahmash, A. (2026). Plant stress detection using multimodal imaging and machine learning: From leaf spectra to smartphone applications. *Frontiers in Plant Science, 16*, 1670593. DOI: 10.3389/fpls.2025.1670593

Shobhana, N. (2024). AI-powered supply chains towards greater efficiency. In *Complex AI Dynamics and Interactions in Management* (pp. 229–249). IGI Global., DOI: 10.4018/979-8-3693-0712-0.ch011

Shukla, R. M., & Sengupta, S. (2022). A novel machine learning pipeline to detect malicious anomalies for the Internet of Things. *Internet of Things : Engineering Cyber Physical Human Systems*, *20*, 100603. DOI: 10.1016/j.iot.2022.100603

Silva, L., Conceição, L. A., Lidon, F. C., Patanita, M., D'Antonio, P., & Fiorentino, C. (2023). Digitization of Crop Nitrogen Modelling: A Review. *Agronomy (Basel)*, *13*(8), 1964. DOI: 10.3390/agronomy13081964

Silva, L., Rodríguez-Sedano, F., Baptista, P., & Coelho, J. P. (2023). The Digital Twin Paradigm Applied to Soil Quality Assessment: A Systematic Literature Review. *Sensors (Basel)*, *23*(2), 1007. DOI: 10.3390/s23021007

Şimşek, Ö. (2025). Harvesting sustainability: Innovations and practices in modern agriculture. *Green Technologies and Sustainability*, *3*(3), 100192. DOI: 10.1016/j.grets.2025.100192

Singh, B., & Kaunert, C. (2024). Dynamic landscape of artificial general intelligence (AGI) for advancing renewable energy in urban environments: synergies with SDG 11—Sustainable cities and communities lensing policy and governance. In *Artificial general intelligence (AGI) security: Smart applications and sustainable technologies* (pp. 247–270). Springer Nature Singapore.

Singh, M., Fuenmayor, E., Hinchy, E. P., Qiao, Y., Murray, N., & Devine, D. (2022). Digital Twin: Origin to Future. *Applied System Innovation*, *4*(2), 36. DOI: 10.3390/asi4020036

Singh, M., Srivastava, R., Fuenmayor, E., Kuts, V., Qiao, Y., Murray, N., & Devine, D. (2022a). Applications of Digital Twin across Industries: A Review. *Applied Sciences (Basel, Switzerland)*, *12*(11), 5727. DOI: 10.3390/app12115727

Sitokonstantinou, V., Porras, E. D. S., Bautista, J. C., Piles, M., Athanasiadis, I., Kerner, H., Martini, G., Sweet, L., Tsoumas, I., Zscheischler, J., & Camps-Valls, G. (2024). *Causal machine learning for sustainable agroecosystems* (Version 1). arXiv. DOI: 10.48550/ARXIV.2408.13155

Sofiullah, M., Gomes Vale, E., & Darr, D. (2023). Effectiveness of an interactive start-up simulation to foster entrepreneurial intentions among undergraduate university students: A quasi-experimental study. *Entrepreneurship Education*, *6*(4), 445–467. DOI: 10.1007/s41959-023-00108-z

Solmaz, S., Mahy, W., & Zogopoulos, V. (2026). Shop floor environment digital twin scan for extended reality applications. *Procedia CIRP, 138*, 1127–1132. DOI: 10.1016/j.procir.2026.01.194

Sompolgrunk, A., Golzad, H., Liu, H., Banihashemi, S., & Najafi, M. (2026). The evolution of digital twins within the big BIM data landscape. Smart and Sustainable Built Environment, 1-28.

Sörensen, A., Mitra, R., Hulthen, E., Hartmann, T., & Clausen, E. (2022). Bringing the entrepreneurial mindset into mining engineering education. *Mining, Metallurgy & Exploration, 39*(4), 2013–2027. DOI: 10.1007/s42461-022-00620-1

Sorochak, O., & Gvozd, M. (2025). Education 4.0: Innovative approaches to competence development in digital organizations. *Journal of Education and Learning, 14*(2), 1–15. DOI: 10.33271/nvngu/2025-5/188

Sousa, M. J., & Rocha, Á. (2019). Skills for disruptive digital business. *Journal of Business Research, 94*, 257–263. DOI: 10.1016/j.jbusres.2017.12.051

Spaniol, M. J., & Rowland, N. J. (2023). AI-assisted scenario generation for strategic planning. *Futures & Foresight Science, 5*(2), e148. DOI: 10.1002/ffo2.148

Srivastav, A. K., & Das, P. (2025). Digital Twins in Agriculture. In Dr. A. K. Srivastav & Dr. P. Das, *Biotechnology and IoT in Agriculture and Food Production* (pp. 241–248). Apress. DOI: 10.1007/979-8-8688-1469-3_24

Srivastava, M. K., & Gnyawali, D. R. (2011). When Do Relational Resources Matter? Leveraging Portfolio Technological Resources for Breakthrough Innovation. *Academy of Management Journal, 54*(4), 797–810. DOI: 10.5465/amj.2011.64870140

Stacchio, L., Angeli, A., & Marfia, G. (2022). Empowering digital twins with extended reality collaborations. *Virtual Reality & Intelligent Hardware, 4*(6), 487–505. DOI: 10.1016/j.vrih.2022.06.004

Stach, C. (2023). Data is the new oil–sort of: A view on why this comparison is misleading and its implications for modern data administration. *Future Internet, 15*(2), 71. DOI: 10.3390/fi15020071

Sterev, N., & Yordanov, D. (2024). Academic entrepreneurship: practical results and training, Strategies for Policy in Science & Education, vol. 32, no. 6 s, pp. 150 – 162. DOI: https://doi.org/DOI: 10.53656/str2024-6s-12-aca

Sterev, N., Kisimov, V., Stoyanova, T., & Andonov, V. (2019). Multidimensional framework for cross-corporate business social networks (BSN). In *Proceedings of the 11th IADIS International Conference Information Systems* (pp. 169–176). IADIS Press. ISBN 978-989-8533-74-6

Sthapit, I., & Olbina, S. (2026). Digital Twin (DT) and Extended Reality (XR) in the Construction Industry: A Systematic Literature Review. *Buildings (Basel, Switzerland), 16*(3), 517. DOI: 10.3390/buildings16030517

Su, C., Tang, X., Han, Y., Wang, T., & Jiang, D. (2024). Cognitive digital twin in manufacturing process: Integrating the knowledge graph for enhanced human-centric Industry 5.0. *International Journal of Production Research*, 1–22. DOI: 10.1080/00207543.2024.2435583

Suman, S., Ray, S., & Prusty, A. K. C, U., Rath, G. P., Patnaik, S., ... & M, L. (2025, July). Digital Twins in Agriculture: Revolutionizing Climate Resilience with AI and IoT. In International Conference on ICT for Sustainable Development (pp. 444-454). Cham: Springer Nature Switzerland.

Sunkar, G. S. (2025). Professional Adaptation and Success in the Artificial Intelligence Era: A Paradigm Shift in Workforce Evolution. *European Modern Studies Journal*, *9*(4), 291–298. DOI: 10.59573/emsj.9(4).2025.30

Suthar, O. P., Parihar, V., Kumar, M., Verma, N. K., & Kashyap, A. K. (2026). Digital Twins and AI: Synergizing Intelligence for Sustainable Farm Management. Digital Twin Technology for Sustainable Agriculture: Applications, Implementation and Future Trends, 347-372.

Tao, F., Xiao, B., Qi, Q., Cheng, J., & Ji, P. (2022). Digital twin modeling. *Journal of Manufacturing Systems*, *64*, 372–389. DOI: 10.1016/j.jmsy.2022.06.015

Tao, F., Zhang, H., Liu, A., & Nee, A. Y. C. (2019). Digital twin in industry: State-of-the-art. *IEEE Transactions on Industrial Informatics*, *15*(4), 2405–2415. DOI: 10.1109/TII.2018.2873186

Tao, F., Zhang, M., Liu, Y., & Nee, A. Y. C. (2019). Digital twin driven product design, manufacturing and service. *International Journal of Advanced Manufacturing Technology*, *94*(9-12), 3563–3576. DOI: 10.1007/s00170-017-0233-1

Tarafdar, M., Cooper, C. L., & Stich, J. F. (2019). The technostress trifecta—Techno eustress, techno distress and design: Theoretical directions and an agenda for research. *Information Systems Journal*, *29*(1), 6–42. DOI: 10.1111/isj.12169

Tariqul Islam, A. F. M., Saha, K. K., & Mahaboob, M. G. (2026). Computational Intelligence in Climate-Adaptive Agriculture: Pathways to Resilient Food Systems. In *IoT and AI in Agriculture: Strategies to Develop Smart Agricultural Space for In-field and In-house Crop Production* (pp. 67–88). Springer Nature Singapore. DOI: 10.1007/978-981-95-5218-4_5

Teece, D. J. (2018). Business models and dynamic capabilities. *Long Range Planning*, *51*(1), 40–49. DOI: 10.1016/j.lrp.2017.06.007

Teece, D. J., Pisano, G., & Shuen, A. (1997). Dynamic capabilities and strategic management. *Strategic Management Journal*, *18*(7), 509–533. DOI: 10.1002/(SICI)1097-0266(199708)18:7<509::AID-SMJ882>3.0.CO;2-Z

Timperi, M., Kokkonen, K., Hannola, L., & Elfvengren, K. (2023). Impacts of digital twins on new business creation: Insights from manufacturing industry. *Measuring Business Excellence*, *27*(3), 433–448. DOI: 10.1108/MBE-09-2022-0104

Tjahyadi, H., Prasetya, K., & Murwantara, I. M. (2023). Digital Twin Based Laboratory for Control Engineering Education. *International Journal of Information and Education Technology (IJIET)*, *13*(4), 704–711. DOI: 10.18178/ijiet.2023.13.4.1856

Tlili, F., Ayed, S., & Fourati, L. C. H. A. A. R. I. (2026). Integration paradigm of intelligent digital twin into UAVs systems. *Soft Computing*, *30*(4), 1–18. DOI: 10.1007/s00500-025-11065-1

Tooth, J., Watson, J. D. M., & Tuptuk, N. (2024). Transdisciplinary Perspectives on Human-Centric Digital Twins in Industry 4.0 & 5.0. *Transdisciplinary Perspectives on Human-Centric Digital Twins in Industry, 4*. http://dx.doi.org/10.2139/ssrn.5011257

Tortorella, G., Gloet, M., Samson, D., Kurnia, S., Fogliatto, F. S., & Anzanello, M. J. (2025). Food supply chain resilience through digital transformation: A mixed-method approach. *International Journal of Logistics Management*, *36*(2), 381–412. DOI: 10.1108/IJLM-01-2024-0030

Trajtenberg, M. (1990). A Penny for Your Quotes: Patent Citations and the Value of Innovations. *The RAND Journal of Economics*, *21*(1), 172–187. DOI: 10.2307/2555502

Tran, G. T., Le Dinh, T., & Le, T. D. (2026). Agentic AI Across Technologies, Applications, and Development Domains: A Systematic Literature Review. In *International Conference on Data Analytics & Management* (pp. 587-601). Springer, Cham. DOI: 10.1007/978-3-032-03769-5_44

Trevisan, A., Dukovska-Popovska, I., Acerbi, F., Terzi, S., & Sassanelli, C. (2026). Teaching methods for building human capital for the twin transition: a skill-based approach, Technology in Society, Volume 85, 2026, 103172, ISSN 0160-791X, https://doi.org/DOI: 10.1016/j.techsoc.2025.103172

Trist, E. L., & Bamforth, K. W. (1951). Some social and psychological consequences of the longwall method of coal-getting. *Human Relations; Studies Towards the Integration of the Social Sciences*, *4*(1), 3–38. DOI: 10.1177/001872675100400101

Trivedi, A., & Juyal, S. A. (2026). Smart Cities and Urban Circularity: Advancing Sustainable Transformation through Digital Twins, AI, and Capacity Building. In *Green Horizons: Advancing Climate Solutions Through Green Finance, Circular Economy, and ESG* (pp. 339–356). Springer Nature Switzerland. DOI: 10.1007/978-3-032-03484-7_18

Tsakiridis, N. L., Samarinas, N., Kalopesa, E., & Zalidis, G. C. (2023). Cognitive Soil Digital Twin for Monitoring the Soil Ecosystem: A Conceptual Framework. *Soil Systems*, *7*(4), 88. DOI: 10.3390/soilsystems7040088

Tseng, Y. H., Lin, C. J., & Lin, Y. I. (2007). Text mining techniques for patent analysis. *Information Processing & Management*, *43*(5), 1216–1230. DOI: 10.1016/j.ipm.2006.11.011

Tubis, A. A., & Werbińska-Wojciechowska, S. (2026). House of resilience for energy supply chains: A digitalization-based approach to enhancing supply chain robustness. *Environment Systems & Decisions*, *46*(1), 1. DOI: 10.1007/s10669-025-10054-x

Uddin, M. K. S. (2024). A Review of Utilizing Natural Language Processing and AI For Advanced Data Visualization in Real-Time Analytics. *Global Mainstream Journal*, *1*(4), 10–62304. DOI: 10.62304/ijmisds.v1i04.185

Uschold, M., & Gruninger, M. (1996). Ontologies: Principles, Methods and Applications. *The Knowledge Engineering Review*, *11*(2), 93–136. DOI: 10.1017/S0269888900007797

van der Valk, H., Strobel, G., Winkelmann, S., Hunker, J., & Tomczyk, M. (2022). Supply chains in the era of digital twins–a review. *Procedia Computer Science*, *204*, 156–163. DOI: 10.1016/j.procs.2022.08.019

Van Laar, E., Van Deursen, A. J. A. M., Van Dijk, J. A. G. M., & De Haan, J. (2017). The relation between 21st-century skills and digital skills: A systematic literature review. *Computers in Human Behavior*, *72*, 577–588. DOI: 10.1016/j.chb.2017.03.010

Van Lier, Q. D. J., De Melo, M. L. A., & Pinheiro, E. A. R. (2024). Stochastic analysis of plant available water estimates and soil water balance components simulated by a hydrological model. *Vadose Zone Journal*, *23*(3), e20306. DOI: 10.1002/vzj2.20306

Van Zeebroeck, N., & Graham, S. J. H. (2010). Comparing Patent Litigation Across Europe: A First Look. *Stanford Technology Law Review : STLR : An Online High-Technology Law Journal from Stanford Law School.*

Varma, A., Dawkins, C., & Chaudhuri, K. (2023). Artificial intelligence and people management: A critical assessment through the ethical lens. *Human Resource Management Review*, *33*(1), 100923. DOI: 10.1016/j.hrmr.2022.100923

Varma, A., Pereira, V., & Patel, P. (2024). Artificial intelligence and performance management. *Organizational Dynamics*, *53*(1), 101037. DOI: 10.1016/j.orgdyn.2024.101037

Verhoef, P. C., Broekhuizen, T., Bart, Y., Bhattacharya, A., Dong, J. Q., Fabian, N., & Haenlein, M. (2021). Digital transformation: A multidisciplinary reflection and research agenda. *Journal of Business Research*, *122*, 889–901. DOI: 10.1016/j.jbusres.2019.09.022

Vial, G. (2019). Understanding digital transformation: A review and research agenda. *The Journal of Strategic Information Systems*, *28*(2), 118–144. DOI: 10.1016/j.jsis.2019.01.003

Vikram, C., & Manojkumar, R. (2026). AI-Digital Twin Synergy for Smart and Autonomous Wind Energy Systems. In Advanced Wind Energy Systems: Grid Integration, Markets, and Sustainable Infrastructure (pp. 137-174). IGI Global Scientific Publishing.

Villegas, L. F., Romero, D., Macchi, M., & Skoogh, A. (2026). Towards a Digital Twin Lifecycle Management Framework. *Enterprise Information Systems*, •••, 2616681. DOI: 10.1080/17517575.2026.2616681

Vital-Soto, A., & Olivares-Aguila, J. (2023). Manufacturing systems for unexpected events: An exploratory review for operational and disruption risks. *IEEE Access*. https://ieeexplore.ieee.org/abstract/document/10237110/

Voipio, V., Elfvengren, K., Korpela, J., & Vilko, J. (2023). Driving competitiveness with RFID-enabled digital twin: Case study from a global manufacturing firm's supply chain. *Measuring Business Excellence*, *27*(1), 40–53. DOI: 10.1108/MBE-06-2021-0084

Wang, B., Zhou, H., Li, X., Yang, G., Zheng, P., Song, C., Yuan, Y., Wuest, T., Yang, H., & Wang, L. (2024). Human Digital Twin in the context of Industry 5.0. *Robotics and Computer-integrated Manufacturing*, *85*, 102626. DOI: 10.1016/j.rcim.2023.102626

Wang, L., Zhang, X., & He, W. (2026). The construction and optimization of resilient community living sphere driven by digital twins. *Scientific Reports*, *16*(1), 4268. DOI: 10.1038/s41598-025-34455-9 PMID: 41547970

Warke, V., Kumar, S., Bongale, A., & Kotecha, K. (2021). Sustainable Development of Smart Manufacturing Driven by the Digital Twin Framework: A Statistical Analysis. *Sustainability (Basel)*, *13*(18), 10139. Advance online publication. DOI: 10.3390/su131810139

Warner-Søderholm, G., & Kuoppamäki, M. (2025). Beyond Industry 5.0: Leadership 5.0—Driving Future-Ready Organizations. *Businesses*, *5*(4), 56. DOI: 10.3390/businesses5040056

Wei, Y. (2025). A digital twin model for grain enterprise financial shared service centers based on distributed deep learning and neural symbolic reasoning. *Scientific Reports*, *15*(1), 40558. DOI: 10.1038/s41598-025-24350-8 PMID: 41254019

Westerman, G., Bonnet, D., & McAfee, A. (2011). Digital transformation: A roadmap for billion-dollar organizations. MIT Center for Digital Business and Capgemini Consulting, 1–68.

Widanage, C., & Kim, K. P. (2026). Digital Twin for Sustainable Social Housing: Integrating BIM and MMC Towards Industry 5.0. *Encyclopedia*, *6*(2), 30. DOI: 10.3390/encyclopedia6020030

Wirtz, J. J., & George, R. Z. (2022). Assessing futures intelligence: Looking back on global trends 2025. *Political Science Quarterly*, *137*(3), 481–510. DOI: 10.1002/polq.13354

Wu, Q., Tan, W., Zhou, L., Deveci, M., Ding, W., Martínez, L., & Pamucar, D. (2025). An integrated consensus-oriented decision-making framework for exploring the barriers and applications of medical digital twins. *Expert Systems with Applications*, *284*, 127683. Advance online publication. DOI: 10.1016/j.eswa.2025.127683

Xu, M., David, J. M., & Kim, S. H. (2021). The fourth industrial revolution: Opportunities and challenges. *International Journal of Financial Research*, *9*(2), 90–95. DOI: 10.5430/ijfr.v9n2p90

Xu, X., Gao, F., Xiong, D., Fan, Z., Xiong, S., Dong, P., Qiao, H., & Ma, X. (2025a). Digital twin-based winter wheat growth simulation and optimization. *Field Crops Research, 329*, 109953. DOI: 10.1016/j.fcr.2025.109953

Xu, X., Lu, Y., Voget-Heuser, B., & Wang, L. (2021). Industry 4.0 and Industry 5.0—Inception, conception and perception. *Journal of Manufacturing Systems, 61*, 530–535. DOI: 10.1016/j.jmsy.2021.10.006

Yazdi, M. (2024). Reliability-centered design and system resilience. In *Advances in Computational Mathematics for Industrial System Reliability and Maintainability* (pp. 79–103). Springer Nature Switzerland., DOI: 10.1007/978-3-031-53514-7_5

Ye, H., Zhang, N., Chen, H., & Chen, J. (2022). Ontology-enhanced Prompt-tuning for Few-shot Learning. In *Proceedings of the ACM Web Conference 2022 (WWW '22)*. ACM. DOI: 10.1145/3485447.3511921

Yigitcanlar, T., Desouza, K. C., Butler, L., & Roozkhosh, F. (2020). Contributions and risks of artificial intelligence (AI) in building smarter cities: Insights from a systematic review of the literature. *Energies, 13*(6), 1473. DOI: 10.3390/en13061473

Yordanova, Z., & Nozari, H. (2025). Harnessing cyber-physical systems and digital twins for industrial innovation and economic resilience. In *Dynamic and Safe Economy in the Age of Smart Technologies* (pp. 33–48). IGI Global Scientific Publishing. DOI: 10.4018/979-8-3693-4369-2.ch003

Yordanova, Z., & Nozari, H. (2026). An Agentic Digital Twin Framework for Fuzzy Multi-Objective Optimization in Dynamic Humanitarian Logistics. *Algorithms, 19*(3), 198. DOI: 10.3390/a19030198

Yordanova, Z., & Nozari, H. (2026). Enhancing Innovation and Resilience in Entrepreneurial Ecosystems Using Digital Twins and Fuzzy Optimization. *Digital, 6*(1), 25. DOI: 10.3390/digital6010025

Yorucu, V., Bekun, F. V., & Yitmen, I. (2026). Interdisciplinary Studies on Urban Resilience: Urban Digital Twins and Construction 5.0 for Resilient, Sustainable and Smart Cities. In *Modern Approach of Resilient and Sustainable Smart Cities* (pp. 141–155). Springer Nature Switzerland. DOI: 10.1007/978-3-032-05715-0_9

Youssef, A. B. (2025). Twin Transition: Digital Transformation Pathways for Sustainable Innovation. *Sustainability (Basel), 17*(21), 9491. Advance online publication. DOI: 10.3390/su17219491

Youvan, D. C. (2026). From Ancient Mnemonics to Neural Symbiosis: The Historical and Futuristic Evolution of Cognitive Tools Shaping Human Thought.

Yuan, X., Niu, G., Yin, J., & Xie, Y. (2025). Data Assimilation in Hydrological Models: Methods, Challenges and Emerging Trends. *Hydrology, 12*(12), 323. DOI: 10.3390/hydrology12120323

Yu, J., Wu, Y., Xu, L., Peng, J., Chen, G., Shen, X., Lan, R., Zhao, C., & Zhangzhong, L. (2022a). Evaluating the Hydrus-1D Model Optimized by Remote Sensing Data for Soil Moisture Simulations in the Maize Root Zone. *Remote Sensing (Basel), 14*(23), 6079. DOI: 10.3390/rs14236079

Zahid, A., Ferraro, A., Petrillo, A., & De Felice, F. (2025). Exploring the role of digital twin and industrial metaverse technologies in enhancing occupational health and safety in manufacturing. *Applied Sciences (Basel, Switzerland), 15*(15), 8268. DOI: 10.3390/app15158268

Zhan, C. (2024). *Microbial Decomposition and Soil Health: Mechanisms and Ecological Implications.* Molecular Soil Biology., DOI: 10.5376/msb.2024.15.0007

Zhang, N., Bahsoon, R., & Theodoropoulos, G. (2020, October). Towards engineering cognitive digital twins with self-awareness. In 2020 IEEE international conference on systems, man, and cybernetics (SMC) (pp. 3891-3891). IEEE.

Zhang, D., Pei, Y., & Liu, Q. (2022). Vulnerability evaluation for a smartphone digital twin workshop under temporal and spatial disruptions. *Machines (Basel), 10*(9), 752. Advance online publication. DOI: 10.3390/machines10090752

Zhang, H., Lin, Y., & Nguyen, G. (2025). AI-driven Sustainable Urban Intelligence: Integrating Smart Technologies for Efficient, Resilient, and Inclusive City Management. Journal of Science. *Innovation & Social Impact, 1*(1), 201–208.

Zhang, R., Zhu, H., Chang, Q., & Mao, Q. (2025). A Comprehensive Review of Digital Twins Technology in Agriculture. *Agriculture, 15*(9), 903. DOI: 10.3390/agriculture15090903

Zhang, Y., Liu, X., & Chen, Z. (2025). Real-time digital twin systems and industrial applications. *IEEE Transactions on Industrial Informatics, 21*(4), 2345–2356.

Zhang, Z., Wei, Z., Court, S., Yang, L., Wang, S., Thirunavukarasu, A., & Zhao, Y. (2024). A Review of Digital Twin Technologies for Enhanced Sustainability in the Construction Industry. *Sustainability, 16.* Advance online publication. DOI: 10.3390/su16041514

Zhao, C., Liu, G., Zhang, R., Liu, Y., Wang, J., Kang, J., Niyato, D., Li, Z., Shen, X., Han, Z., Sun, S., Yuen, C., & Kim, D. I. (2026). Edge General Intelligence Through World Models, Large Language Models, and Agentic AI: Fundamentals, Solutions, and Challenges. *IEEE Transactions on Cognitive Communications and Networking*, *12*, 5649–5675. DOI: 10.1109/TCCN.2026.3658762

Zhao, Y., Jiang, H., Cai, Y., & Deng, Y. (2024). A generalized data assimilation architecture of digital twin for complex process industrial systems. *Measurement Science & Technology*, *35*(6), 066003. DOI: 10.1088/1361-6501/ad27c6

Zheng, X., Lu, J., & Kiritsis, D. (2022). The emergence of cognitive digital twin: Vision, challenges and opportunities. *International Journal of Production Research*, *60*(24), 7610–7632. DOI: 10.1080/00207543.2021.2014591

Zimmerman, B. J. (2000). Attaining self-regulation: A social cognitive perspective. In Boekaerts, M., Pintrich, P. R., & Zeidner, M. (Eds.), *Handbook of self-regulation* (pp. 13–39). Academic Press. DOI: 10.1016/B978-012109890-2/50031-7

About the Contributors

Zornitsa Yordanova is an academic, entrepreneur, and digital transformation expert specializing in AI, innovation management, digital twins, ICT, and organizational digitization. She currently serves as an Associate Professor at the University of National and World Economy (UNWE) in Sofia—the largest economic university in the Balkans—where she teaches Innovation & Technology Management and Digital Transformation / MIS. Her academic work includes more than 120 publications, with over 70 indexed in Scopus and Web of Science, and she holds an H-index of 12. She has successfully led multiple national research projects in education, innovation, and digital transformation. Beyond academia, Dr. Yordanova works at the intersection of research and industry. She has extensive experience as a project manager in international ICT initiatives, a trainer and consultant supporting organizations in their innovation and digital transformation journeys, and a founder of technology-driven ventures.

Salam Al-E'mari earned her Ph.D. in Cybersecurity from Universiti Sains Malaysia (USM) in 2022. She is the Chair of the Information Security Department and an Assistant Professor at the University of Petra (UoP). With expertise in Blockchain, Deep Learning, and Network Security, she has contributed significantly to cybersecurity research. She previously earned her Bachelor's and Master's in Computer Science from Yarmouk University, Jordan, shaping her strong foundation in security and technology.

K. Balaji is an Assistant Professor in Business and Management, Christ University, Bangalore. He did Ph.D. from KLU Business School, KL University,Vijayawada. He is having 12 years of teaching experience. He has to his credit 23 Research Articles published in reputed National and International Journals. He has participated and presented more than 12 papers in both National and International Seminars/Conferences and also attended more than 50 workshops/ FDPS organized by various reputed organizations. He has also participated in more than 50 seminars / webinars organized by various reputed organizations. He has also attended in 6 Short Term Training programs.He has successfully conducted 3 webinars, one 5 day FDP and one International Conference on Reimaging marketing in New Normal. He is a ratified as Assistant Professor by both JNTUH, JNTU Anantapur and Sri Venkateswara University,Tirupati. His specialized areas include, Accounting and Financial management, Marketing Management, Retailing Management, Consumer Behavior studies, Entrepreneurship and Business Laws.

Joytu Debnath is a researcher at the Department of Agricultural Engineering, Triguna Sen School of Technology, Assam University, Silchar, India. His research areas are artificial intelligence, machine learning, automation, and agriculture.

Simeon Kolyandov is a university lecturer and researcher in the field of strategic management, industrial business, and digital marketing, focusing on the transformation of organizations in the context of digitalization and the development of marketing 5.0/6.0. His scientific work is aimed at researching strategic and technological factors and developing contemporary theoretical and applied models for marketing and management development.

Kundan Kumar is a researcher at the Department of Agricultural Engineering, Triguna Sen School of Technology, Assam University, Silchar, India. His research areas are artificial intelligence, machine learning, automation, simulation, and agriculture.

Ahad Ahmed Laskar is a researcher at the Department of Agricultural Engineering, Triguna Sen School of Technology, Assam University, Silchar, India. His research areas are artificial intelligence, machine learning, automation, and agriculture.

Radoslav Mitchev is Professor at the Department of "Theory of Mechanisms and Machines" at the Technical University - Sofia for the specialties "Digital Industrial Technologies", "Computer Design and Technologies in Mechanical Engineering". He holds certificates: Person VUE Certified Test Administrator; Cisco Certified Network Associate Routing and Switching, certificate Verification No. 437223582874FNZL, Cisco ID No. CSCO13059868 2018 Cisco CyberOps, Certificate of completion; Intertek, CQI-IRCA Certified ISO 9001:2015 Auditor/ Lead Auditor Training course (Quality Management Systems); Linux Professional Institute, NDG Linux Essential Course, Certificate of Completion. Professionally deals with: Programming languages: VB.NET, Visual Basic 6.0, Visual C++ 6.0, Visual Basic for Application, DIGITAL Visual Fortran, Borland Delphi, Pascal, Java, Python; Cloud services: Microsoft Azure, Office 365; CAD applications:

AutoCAD, Inventor Professional, SolidWorks, Blender, SketchUp, VectorWorks; GIS applications: QGIS, GRASS, ArcGIS; Databases: MS SQL Server, MS Access, MySQL, Oracle; Web programming: ASP, PHP, JavaScript, VBScript; Scientific software: R, Matlab, ANSYS, QuickField, FEMM Has over 120 publications in refereed and indexed journals

Miglena Molhova-Vladova is a member of the academic staff of the Industrial Business Department at the Business Faculty of the University of National and World Economy in Sofia. She is an associate professor in the professional field 3.8. "Economics", with the scientific specialty "Economics and Management (software technologies, innovation, patents and licensing)". She holds a Master's degree in Economics of Intellectual Property, Master's degree in Software Engineering and a PhD in Economics. Currently she is also the Manager of Technology and Knowledge Transfer Office at Medical University Plovdiv, under Strategic Research and Innovation Program for the Development of MU-Plovdiv (SRIPD-MUP), with main responsibilities Innovation Development Support, State-of-the-Art Searches, Patent Drafting, Patent Protection, Technology Transfer of IP, IP Strategizing. Fields of expertise: intellectual property, innovation and technology management, licensing, state-of-the-art and novelty searches, freedom-to-operate analysis, competitive technology analysis, software patents. She has more than 60 publications.

Teodora Nikolova is an expert in software testing with 10 years of experience as QA engineer. In 2022, she received a Master's degree in Software Engineering. She is currently a PhD student in the Department of Computer Sciences at ULSIT.

Hamed Nozari is a distinguished researcher and innovator in industrial engineering, supply chain optimization, and artificial intelligence applications. Holding a Ph.D. in Industrial Engineering, he has made significant contributions to the fields of multi-objective decision-making, smart supply chains, and digital twin technologies. As the developer of the Greedy Man Optimization Algorithm (GMOA), Dr. Nozari has introduced a novel approach to problem-solving inspired by human behavior and parasitic resistance. His research spans various interdisciplinary areas, including predictive maintenance in green supply chains, AI-driven marketing optimization, cybersecurity in smart economies, and autonomous AI for sustainable last-mile delivery

Mrutyunjay Padhiary is an Assistant Professor in the Department of Agricultural Engineering at Assam University, Silchar, India, with a deep academic background. His previous positions included significant teaching and research duties at the Faculty of Technology, UBKV, in West Bengal, and Tribhuvan University in Nepal, as well as research experience at IIT Kharagpur. He has been awarded multiple awards, including an International NUS Research Fellowship and an excellent score on the 2001 GATE Exam, demonstrating his proficiency in precision agriculture. He has made substantial contributions to biofuel production, artificial intelligence vision, aquaculture, precision agriculture, agricultural automation, and artificial intelligence. He is currently working on several research projects, which range from autonomous surface vessels to precision agriculture all-terrain vehicles.

Priyanka Roy is a researcher at the Department of Agricultural Engineering, Triguna Sen School of Technology, Assam University, Silchar, India. His research areas are artificial intelligence, machine learning, automation, and agriculture.

Yousef Sanjalawe is an Assistant Professor specializing in Cybersecurity and Cloud Computing at the University of Jordan (JU), Jordan. He earned his PhD in Cloud Computing from Universiti Sains Malaysia (USM) in 2020, where his research focused on enhancing cloud service broker selection using optimization algorithms. Dr. Sanjalawe has extensive experience in academia, research, and professional training. He has supervised Ph.D. students in Cybersecurity, Cloud Computing, IoT, Fog Computing, Optimization, and Artificial Intelligence (AI). From 2023 to 2024, he served as the Chair of the Cybersecurity Department at the American University of Madaba (AUM), Jordan, and from 2022 to 2024, he was an Assistant Professor in Cybersecurity at AUM.

Nikolay Sterev is full professor at University of National and World Economy – Sofia. He is Head of Industrial Business Department and his main interests cover: industrial development, industrial growth, regenerative economy, Industry 4.0 and Industry 5.0 and others. He has PhD in marketing management and habilitation in marketing leadership and DSc. (econ) in Business economics within Industry 5.0. He is chairman of the (Bulgarian) Association of Professors in Economics and Management of Industry (APIUI) and chairman of Organization Committee of Industrial Growth Conferences. Professor Nikolay Sterev has publisher more than 100 publications and has participated in more than 25 international projects. He has been a keynote speaker on different international conferences and has been appointed as guest lectors in different universities worldwide.

Manuel Villasalero is an associate professor of Business Organization at the University of Castilla-La Mancha (UCLM), Spain, from which he received his Ph.D. He has experience as a director in several technology-based firms. Currently, he is the co-head of the MEISA Research Unit for Digitalization of Manufacturing Services, a business-university partnership on digital transformation at UCLM. His research interests include knowledge management, open innovation, and digital transformation. He has published articles in the European Journal of Innovation

Management, Journal of Intellectual Capital, Journal of Knowledge Management, Journal of Technology Transfer, Management Decision, Production Planning & Control, Journal of Business Research, and International Journal of Human Resource Management.

Daniel Yordanov is professor at University of National and World Economy – Sofia, Transport and Energy Department. His main interests cover: business competitiveness, academic entrepreneurship, financial models, management od SMEs. He has PhD in financial models of SMEs and habilitation in Business financial models in transport and energy. Associate professor Daniel Yordanov has publisher more than 30 publications and has participated in 10 national and international projects. He has been appointed as business consultant in business management in transport industry.

Index

A

C

D

E

F

G

H

U

W

www.ingramcontent.com/pod-product-compliance
Lightning Source LLC
LaVergne TN
LVHW062331300926
849161LV00007B/1650
* 9 7 9 8 2 6 0 0 1 6 6 9 5 *